Advances in Technologies for Producing Food-Relevant Polyphenols

Contemporary Food Engineering

Series Editor

Professor Da-Wen Sun, Director

Food Refrigeration & Computerized Food Technology
National University of Ireland, Dublin
(University College Dublin)
Dublin, Ireland
http://www.ucd.ie/sun/

Advances in Heat Transfer Unit Operations: Baking and Freezing in Bread Making, *edited by Georgina Calderon-Dominguez, Gustavo F. Gutierrez-Lopez, and Keshavan Niranjan* (2016)

Innovative Processing Technologies for Foods with Bioactive Compounds, *edited by Jorge J. Moreno* (2016)

Edible Food Packaging: Materials and Processing Technologies, *edited by Miquel Angelo Parente Ribeiro Cerqueira, Ricardo Nuno Correia Pereira, Oscar Leandro da Silva Ramos, Jose Antonio Couto Teixeira, and Antonio Augusto Vicente* (2016)

Handbook of Food Processing: Food Preservation, *edited by Theodoros Varzakas and Constantina Tzia* (2015)

Handbook of Food Processing: Food Safety, Quality, and Manufacturing Processes, *edited by Theodoros Varzakas and Constantina Tzia* (2015)

Edible Food Packaging: Materials and Processing Technologies, *edited by Miquel Angelo Parente Ribeiro Cerqueira, Ricardo Nuno Correia Pereira, Oscar Leandro da Silva Ramos, Jose Antonio Couto Teixeira, and Antonio Augusto Vicente* (2015)

Advances in Postharvest Fruit and Vegetable Technology, *edited by Ron B.H. Wills and John Golding* (2015)

Engineering Aspects of Food Emulsification and Homogenization, *edited by Marilyn Rayner and Petr Dejmek* (2015)

Handbook of Food Processing and Engineering, Volume II: Food Process Engineering, *edited by Theodoros Varzakas and Constantina Tzia* (2014)

Handbook of Food Processing and Engineering, Volume I: Food Engineering Fundamentals, *edited by Theodoros Varzakas and Constantina Tzia* (2014)

Juice Processing: Quality, Safety and Value-Added Opportunities, *edited by Víctor Falguera and Albert Ibarz* (2014)

Engineering Aspects of Food Biotechnology, *edited by José A. Teixeira and António A. Vicente* (2013)

Engineering Aspects of Cereal and Cereal-Based Products, *edited by Raquel de Pinho Ferreira Guiné and Paula Maria dos Reis Correia* (2013)

Fermentation Processes Engineering in the Food Industry, *edited by Carlos Ricardo Soccol, Ashok Pandey, and Christian Larroche* (2013)

Modified Atmosphere and Active Packaging Technologies, *edited by Ioannis Arvanitoyannis* (2012)

ADVANCES IN TECHNOLOGIES FOR PRODUCING FOOD-RELEVANT POLYPHENOLS

Edited by
José Cuevas-Valenzuela • José Rodrigo Vergara-Salinas
José Ricardo Pérez-Correa

CRC Press
Taylor & Francis Group
Boca Raton London New York

CRC Press is an imprint of the
Taylor & Francis Group, an **informa** business

CRC Press
Taylor & Francis Group
6000 Broken Sound Parkway NW, Suite 300
Boca Raton, FL 33487-2742

ISBN-13: 978-1-4987-1497-6 (hbk)

Library of Congress Cataloging-in-Publication Data

Names: Cuevas-Valenzuela, José. | Vergara-Salinas, José Rodrigo. | Pérez-Correa, José Ricardo.
Title: Advances in technologies for producing food-relevant polyphenols / José Cuevas-Valenzuela, José Rodrigo Vergara-Salinas, and José Ricardo Pérez-Correa.
Description: Boca Raton : Taylor & Francis, 2017. | Series: Contemporary food engineering ; 39 | "A CRC title."
Identifiers: LCCN 2016008771 | ISBN 9781498714976 (alk. paper)
Subjects: LCSH: Polyphenols. | Chemical inhibitors. | Functional foods.
Classification: LCC QK898.P764 C84 2017 | DDC 572/.2--dc23
LC record available at https://lccn.loc.gov/2016008771

Visit the Taylor & Francis Web site at
http://www.taylorandfrancis.com

and the CRC Press Web site at
http://www.crcpress.com

To our colleague, outstanding professor and researcher Dr. Hugo Segura. Rest in peace dear friend.

Contents

Foreword

Polyphenolic compounds are constituents of plant-based foods and diets, which during the past two decades have attracted great interest and attention by scientists and technologists. The scope of research on polyphenols (PP) has expanded significantly from traditional studies that focused on their contribution to the sensory properties of foods and drinks (taste, astringency, aroma, color) to the predominant current approach regarding their biological and health-related properties.

From the publication of the French paradox (Renaud and De Lorgeril, *Lancet*, 1992), questions and debates have emerged about the role of polyphenols and the cardiovascular effects attributed to red wine. Consequently, the number of articles published on polyphenols rose from 623 in 1994 to 7360 in 2014 (Web of Science), addressing basic aspects (biological properties, bioavailability, metabolism, health-related properties) and potential applications in nutrition and health, supported by clinical and observational studies.

On the other hand, the demand for natural antioxidants to replace synthetic antioxidants in foods has led to the extraordinary development of processes for obtaining polyphenolic extracts and individual polyphenols for use in the prevention of food lipid oxidation and as ingredients in functional foods. More than 4600 patents on polyphenols have been registered in the last 20 years.

Hundreds of phenolic substances with different molecular structures and molecular sizes have been reported in vegetable materials. Food technologists are focusing on the development of suitable methods to extract and isolate specific groups (flavonoids, phenolic acids, stilbenes, hydrolysable tannins, and proanthocyanidins) or individual phenols (catechins, ferulic acid, resveratrol, tyrosol, punicalagins, etc.).

In this context, this book addresses the latest procedures for polyphenol extracts and updated techniques for their analysis, purification, and fractionation.

This book will be of interest to food scientists, technologists, engineers, chemists, biochemists, nutritionists, and R&D professionals. The basic aspects including sources of PP and the available databases of content and antioxidant capacity of PP in food are discussed; metabolism and mechanisms of action are also addressed. A detailed description of the procedures for obtaining extracts of PP is presented, including low and high pressure, microwave and ultrasound assisted, supercritical fluids, and pressurized hot water—an emerging green technology. Practical information to develop industrial procedures (design, scale-up, equipment suppliers) is also provided. Key aspects for developing commercial products are covered: an efficient analysis and characterization of the extracts and fractionation and isolation of the target compounds—specific PP groups or individual PP—until a dry powder is obtained. Finally, the chapters address characteristics related to the inclusion of PP in food matrices to favor PP bioaccessibility.

In summary, this book presents a complete and practical approach for obtaining polyphenol extracts, including aspects not previously reported in other texts in this field.

Fulgencio Saura-Calixto
Institute of Food Science, Technology and Nutrition (*ICTAN-CSIC*)
Madrid, Spain

Preface

The growing concern for human well-being has generated an increase in the demand for polyphenols, secondary plant metabolites that exhibit different bioactive properties. This increasing demand is mainly due to the current applications in the food industry, where polyphenols are considered essential ingredients for generating natural products for human health and nutrition. Given the importance of these compounds in the food industry today, we decided to gather experts in the area and review the advances in polyphenol processing technologies.

The aim of this book is to provide researchers, scientists, engineers, and professionals involved in the food industry with the latest methodologies and equipment, which are useful in extracting, isolating, purifying, and analyzing polyphenols from various available sources, such as herbs, flora, vegetables, fruits, and agro-industrial waste. The technologies, which are currently used to add polyphenols to diverse food matrices are also included. Hence, this book is a reference for developing and optimizing processes for obtaining polyphenols and producing polyphenol-rich foods.

The first two chapters describe the various polyphenols and their main sources and effects on human health and nutrition. Particularly, in Chapter 1, the identification and characterization of several classes of polyphenols, their major health-related bioactivity, their principal plant sources, and related databases are provided. This includes commonly used experimental assays that measure the polyphenolic content and antioxidant capacity of plant and food materials. Insights on polyphenol bioactive properties (e.g., antioxidant, antibacterial, antiviral, and anticancer properties) and bioavailability are presented in Chapter 2.

Chapters 3 through 5 cover the main and emergent technologies for obtaining crude polyphenol extracts from different sources, such as solid–liquid extraction (SLE) at low pressure, pressurized liquid extraction (PLE), and supercritical fluid extraction (SCFE). Within the SLE technologies (Chapter 3), the use of organic solvents and ultrasound and microwave-assisted techniques are analyzed. In the case of PLE (Chapter 4), the laboratory and pilot-scale methods and equipment that take advantage of the enhanced solvating properties of solvents (mainly water and hydroalcoholic mixtures) at high temperatures and pressure are reviewed. Chapter 5 covers the SCFE of polyphenols, analyzing and describing the use of carbon dioxide and modifier mixtures at supercritical conditions. Special focus is given to existing industrial applications and methods to scale-up the extraction processes.

Once the crude polyphenol extracts are obtained through SLE, PLE, or SCFE technologies, they should be carefully analyzed and characterized to determine their composition, bioactive potential, and suitability for specific applications. Adequate technologies to accomplish these tasks, such as spectrophotometry, chromatography, and mass spectrometry techniques, are addressed in Chapter 6.

Chapters 7 and 8 cover the main technologies and processes for the purification and fractionation of polyphenols. Fundamentally, the use of adsorption (stirred vessels and packed beds) and membrane technologies (microfiltration, ultrafiltration, nanofiltration, and reverse osmosis) to isolate polyphenols from other compounds present in crude extracts are analyzed. In addition, the application of chromatographic techniques is reviewed, particularly countercurrent chromatography (CCC) to the fractionation of purified extracts into different classes of polyphenols. Particular attention is given to large-scale applications and scale-up strategies.

After purification and fractionation processes, polyphenols are usually encapsulated to improve their stability and shelf life. Microencapsulation by spray drying, which is thoroughly described in Chapter 9, is one of the most suitable and useful techniques for polyphenols. In addition, the chapter focuses on the opportunities and future trends for this topic.

Finally, Chapter 10 discusses the processes used for incorporating polyphenols in food matrices, such as dietary supplements and functional foods. The methods of baking, frying, and extrusion to fabricate formulated snacks, in particular, are discussed. The effects of different manufacturing conditions on the stability and bioavailability of polyphenols are also described.

The editors thank Anillo en Ciencia y Tecnologia ACT1105 and Fondecyt Postdoctorado 3150622 Projects of CONICYT (National Commission of Research in Science and Technology of Chile) for the financial support to our research activities related to polyphenol processing. We also thank our colleagues and collaborators worldwide who have supported the design and development of this book.

José Cuevas-Valenzuela, PhD

José Rodrigo Vergara-Salinas, PhD

José Ricardo Pérez-Correa, PhD

Contributors

Elena M. Balboa
Department of Chemical Engineering
Universidade de Vigo
Ourense, Spain

Pedro Bouchon
Department of Chemical and Bioprocess Engineering
Pontificia Universidad Católica de Chile
Santiago, Chile

Eduardo Caballero-Valdés
Regional Center for Studies in Food and Health
Valparaíso, Chile

María P. Casas
Department of Chemical Engineering
Universidade de Vigo
Ourense, Spain

Enma Conde
Department of Chemical Engineering
Universidade de Vigo
Ourense, Spain

José Cuevas-Valenzuela
Department of Chemical and Bioprocess Engineering
Pontificia Universidad Católica de Chile
Santiago, Chile

José M. del Valle
Department of Chemical and Bioprocess Engineering
Pontificia Universidad Católica de Chile
Santiago, Chile

Beatriz Díaz-Reinoso
Department of Chemical Engineering
Universidade de Vigo
Ourense, Spain

Herminia Domínguez
Department of Chemical Engineering
Universidade de Vigo
Ourense, Spain

Eva Dorta
Faculty of Chemical Sciences
Pontificia Universidad Católica de Chile
Santiago, Chile

Verónica Dueik
Department of Chemical and Bioprocess Engineering
Pontificia Universidad Católica de Chile
Santiago, Chile

Ulrich H. Engelhardt
Institute of Food Chemistry
Technische Universität Braunschweig
Braunschweig, Germany

Carolina Fredes
Department of Food Science and Chemical Technology
Universidad de Chile
Santiago, Chile

Jocelyn Fuentes
Nutrition and Food Technology Institute
Universidad de Chile
Santiago, Chile

Paula García
Department of Food Science and Chemical Technology
Universidad de Chile
Santiago, Chile

Camilo Lopez-Alarcón
Faculty of Chemical Sciences
Pontificia Universidad Católica de Chile
Santiago, Chile

Gilberto Mercado-Mercado
Research Laboratory in Food Science
National Technology of Mexico/Tepic Institute of Technology
Nayarit, Mexico

Andrés Moure
Department of Chemical Engineering
Universidade de Vigo
Ourense, Spain

Araceli Olivares-Miralles
Regional Center for Studies in Food and Health
Valparaíso, Chile

Juan Carlos Parajó
Department of Chemical Engineering
Universidade de Vigo
Ourense, Spain

José Ricardo Pérez-Correa
Department of Chemical and Bioprocess Engineering
Pontificia Universidad Católica de Chile
Santiago, Chile

Jara Pérez-Jiménez
Institute of Food Science, Technology and Nutrition
Spanish Research Council
Madrid, Spain

Sara Ramos-Romero
Institute for Advanced Chemistry of Catalonia
Consejo Superior de Investigaciones Científicas
Barcelona, Spain

Paz Robert
Department of Food Science and Chemical Technology
Universidad de Chile
Santiago, Chile

Sonia Guadalupe Sáyago-Ayerdi
Research Laboratory in Food Science
National Technology of Mexico/Tepic Institute of Technology
Nayarit, Mexico

Carmen Soto-Maldonado
Regional Center for Studies in Food and Health
Valparaíso Chile

María Luisa Soto
Department of Chemical Engineering
Universidade de Vigo
Ourense, Spain

Hernán Speisky
Nutrition and Food Technology Institute
Universidad de Chile
Santiago, Chile

Josep Lluís Torres
Institute for Advanced Chemistry of Catalonia
Consejo Superior de Investigaciones Científicas
Barcelona, Spain

Edgar Uquiche
Department of Chemical Engineering
Universidad de La Frontera
Temuco, Chile

Loreto Valenzuela
Department of Chemical and Bioprocess Engineering
Pontificia Universidad Católica de Chile
Santiago, Chile

José Rodrigo Vergara-Salinas
Escuela de Ingeniería Bioquímica
Pontificia Universidad Católica de Valparaíso
Valparaíso, Chile

Peter Winterhalter
Institute of Food Chemistry
Technische Universität Braunschweig
Braunschweig, Germany

María Elvira Zúñiga-Hansen
Escuela de Ingeniería Bioquímica
Pontificia Universidad Católica de Valparaíso
Valparaíso, Chile

1 Polyphenols
Sources and Main Characteristics

Hernán Speisky, Jocelyn Fuentes, Eva Dorta, and Camilo Lopez-Alarcón

CONTENTS

1.1 INTRODUCTION

Polyphenols, found abundantly in many edible plants, have long attracted the interest of biomedical and food chemistry researchers due to the potential of polyphenol-rich foods to significantly lower the relative risk of developing various cardiovascular and tumoral diseases. The biological basis to explain the health benefits of polyphenols relates to their well-established antioxidant properties. In recent years, however, several other remarkable bioactivities have been described for some of these compounds, broadening their application scope and thereby innovation potential. This chapter includes a basic introduction to the chemistry of polyphenols with a brief structural classification and description of the major forms of polyphenols, which occur within plant-derived foodstuffs. We also describe the major direct and indirect mechanisms by which polyphenols exert their antioxidant action, and briefly deal with bioactivities that are not necessarily linked to the antioxidant properties of these compounds. Emphasis is also placed on the main methodologies currently employed to assess "polyphenolic richness" and antioxidant capacity, as applicable to edible plants. Finally, a specific portion of this chapter has been devoted to describe some of the currently available major databases on polyphenol content and antioxidant capacity of plant foods.

1.2 DEFINITION AND STRUCTURAL CLASSIFICATION OF POLYPHENOLS

Polyphenols are molecules whose structure contains one or more benzene rings to which at least two hydroxyl groups are attached. However, the term also applies to some simple phenols in which only one free hydroxyl group is attached to a single benzene ring. Polyphenols are found in nature primarily as secondary plant metabolites with over 8000 structures having been described (Landete, 2012; Strack and Wray, 1992). Within plants, some of these compounds can play myriad functions, including protecting against UV radiation and from herbivores and microbial infections, and acting as attractants for pollinators and seed-dispersing animals (Mierziak et al., 2014; Strack, 1997; Wasson et al., 2006). Polyphenols constitute the largest group of nonenergetic substances present in foods of plant origin. Due to their ubiquitous distribution in the plant kingdom and the broad scope of health-related bioactivities attributed to polyphenols (aspect briefly referred to in the following text and dealt with in further detail in Chapter 2), these compounds have remained for decades among the most intensively investigated molecules by food chemistry and biochemistry researchers. If only the most regularly consumed types of plant foods are considered, over 500 different polyphenols can be described to be part of our diet (Kühnau, 1976; Liu, 2013; Perez-Jimenez et al., 2010b; Yao et al., 2004).

From a structural point of view, polyphenols can be classified into two major distinguishable classes. The first and largest one is the flavonoids, whose basic structure (a diphenylpropane skeleton or flavan nucleus, C6–C3–C6) includes two benzene rings (A and B) that are linked through three carbon atoms that frequently form an oxygenated pyran heterocyclic ring (C) (Figure 1.1). Flavonoids are the most abundant type of polyphenols in the diet, constituting approximately two-thirds of intake (Beecher, 2003; Corcoran et al., 2012; Scalbert and Williamson, 2000; Visioli et al., 2011). Most naturally occurring flavonoids possess at least 15 carbon atoms and often exhibit 3 or more hydroxyl groups, of which 2 are on ring A, at positions 5 and 7, and 1 or more on ring B, frequently in positions 3 or 4. Within the flavonoid class, six major subclasses can be distinguished as follows (examples of some are given in parentheses, and their corresponding Roman-numbered structures are shown in Figure 1.2 and Table 1.1): (1) flavonols [e.g., quercetin (I), kaempferol (II), myricetin (III), morin (IV)]; (2) flavones, which are structurally similar to flavonols, except that they lack oxygenation at C3 [e.g., luteolin (V), apigenin (VI), chrysin (VII), baicalein (VIII)]; (3) flavanones, which are characterized by the absence of a Δ2,3 double bond and the presence of a chiral center at C2 [e.g., naringenin (IX),

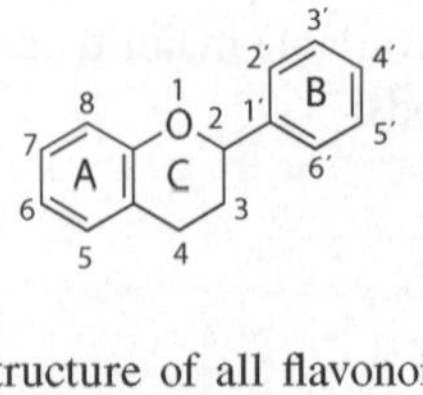

FIGURE 1.1 Flavan, the basic structure of all flavonoids, consists of two benzene rings (A and B) and an oxygenated pyran heterocyclic ring (C).

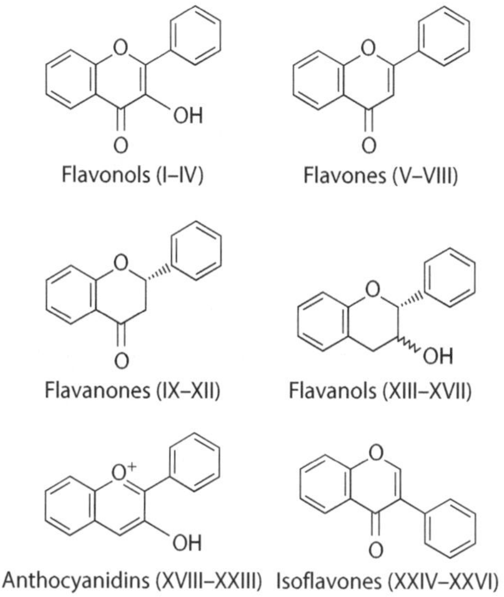

FIGURE 1.2 Basic structural formulae of the six major subclasses of flavonoids. For each subclass, Roman-numbered examples, as quoted in the text, are given in parentheses (I–XXVI). The substituents of each of these cases are presented in Table 1.1.

hesperetin (X), eriodictyol (XI), taxifolin (XII)]; (4) flavanols also known as flavan-3-ols [e.g., catechin (XIII), epicatechin (XIV), gallocatechin (XV), epigallocatechin (XVI), epigallocatechin-3-gallate (XVII)]; (5) anthocyanidins [e.g., pelargonidin (XVIII), cyanidine (XIX), delphinidin (XX), peonidin (XXI), petunidin (XXII), malvidin (XXIII)]; and (6) isoflavones, which have the B-ring attached at the C3 rather than the C2 position [e.g., genistein (XXIV), daidzein (XXV), formononetin (XXVI)]. Because of their structural similarity to estrogen, isoflavones are also known as phytoestrogens. Flavonoids, like flavanols or anthocyanidins, can exist both in monomeric and oligomeric forms (e.g., proanthocyanidins, theaflavins, and thearubigins). Many proanthocyanidins are oligomers of catechin and epicatechin and their gallic acid esters. More complex polyphenols, which have the same polymeric building block, form the tannins. Although some flavonoids are found in food plants mainly in free nonconjugated or aglycone form (i.e., flavonols and flavones), most often these compounds are present in glycone form as 3-*O*-glycosides or bound to other sugar conjugates. The most common glycosidic unit is glucose, though other sugar conjugates often include galactose, arabinose, glucorhamnose, and rhamnose. The glycone and aglycone forms can be chemically separated by acid hydrolysis and can also be hydrolyzed by alkali. Numerous enzymes are capable of forming and breaking glycosidic bonds. The most important cleavage enzymes are the glycoside hydrolases, and the most significant among synthetic enzymes are the glycosyltransferases. Given the many permutations in glycoside types and linkages, the number of unique flavonoids that have been identified so far amounts to more than 5000 compounds. In addition to adding hydrophilic sugars to their structure, some flavonoids can undergo prenylation, becoming compounds that are more hydrophobic

TABLE 1.1
Structural Substituents of the Examples of Flavonoids Described in Figure 1.2 as Follows: Flavonols (I–IV), Flavones (V–VIII), Flavanones (IX–XII), Flavanols (XIII–XVII), Anthocyanidins (XVIII–XXIII), and Isoflavones (XXIV–XXVI) Are Presented

Subclass	Examples	Substitutions	IUPAC Name
Flavonol	Quercetin (I)	3′,4′,5,7-Tetrahydroxy-	2-(3,4-Dihydroxyphenyl)-3,5,7-trihydroxy-4*H*-chromen-4-one
	Kaempferol (II)	4′,5,7-Trihydroxy-	3,5,7-Trihydroxy-2-(4-hydroxyphenyl)-4*H*-chromen-4-one
	Myricetin (III)	3′,4′,5′,5,7-Pentahydroxy-	3,5,7-Trihydroxy-2-(3,4,5-trihydroxyphenyl)-4-chromenone
	Morin (IV)	2′,4′,5,7-Tetrahydroxy-	2-(2,4-Dihydroxyphenyl)-3,5,7-trihydroxychromen-4-one
Flavone	Luteolin (V)	3′,4′,5,7-Tetrahydroxy-	2-(3,4-Dihydroxyphenyl)-5,7-dihydroxy-4-chromenone
	Apigenin (VI)	4′,5,7-Trihydroxy-	5,7-Dihydroxy-2-(4-hydroxyphenyl)-4*H*-1-benzopyran-4-one
	Chrysin (VII)	5,7-Dihydroxy-	5,7-Dihydroxy-2-phenyl-4*H*-chromen-4-one
	Baicalein (VIII)	5,6,7-Trihydroxy-	5,6,7-Trihydroxy-2-phenyl-chromen-4-one
Flavanone	Naringenin (IX)	4′,5,7-Trihydroxy-	5,7-Dihydroxy-2-(4-hydroxyphenyl)chroman-4-one
	Hesperetin (X)	4′-Methoxy-5′,5,7-trihydroxy-	(*S*)-2,3-Dihydro-5,7-dihydroxy-2-(3-hydroxy-4-methoxyphenyl)-4*H*-1-benzopyran-4-one
	Eriodictyol (XI)	3′,4′,5,7-Tetrahydroxy-	(2*S*)-2-(3,4-Dihydroxyphenyl)-5,7-dihydroxy-4-chromanone
	Taxifolin (XII)	4′,5′,5,7-Tetrahydroxy-	(2*R*,3*R*)-2-(3,4-Dihydroxyphenyl)-3,5,7-trihydroxy-2,3-dihydrochromen-4-one
Flavanol	(+)-Catechin (XIII)[a]	3′,4′,5,7-Tetrahydroxy-	(2*R*,3*S*)-2-(3,4-Dihydroxyphenyl)-3,4-dihydro-2*H*-chromene-3,5,7-triol
	(−)-Epicatechin (XIV)[a]	3′,4′,5,7-Tetrahydroxy-	(2*R*,3*S*)-2-(3,4-Dihydroxyphenyl)-3,4-dihydro-2*H*-chromene-3,5,7-triol

(Continued)

TABLE 1.1 (*Continued*)
Structural Substituents of the Examples of Flavonoids Described in Figure 1.2 as Follows: Flavonols (I–IV), Flavones (V–VIII), Flavanones (IX–XII), Flavanols (XIII–XVII), Anthocyanidins (XVIII–XXIII), and Isoflavones (XXIV–XXVI) Are Presented

Subclass	Examples	Substitutions	IUPAC Name
	(+)-Gallocatechin (XV)	3′,4′,5′,5,7-Pentahydroxy-	(2*R*,3*S*)-2-(3,4,5-Trihydroxyphenyl)-3,5,7-chromanetriol
	Epigallocatechin (XVI)	3′,4′,5,7-Tetrahydroxy-3-galoyl-	[(2*R*,3*R*)-2-(3,4-Dihydroxyphenyl)-5,7-dihydroxy-3,4-dihydro-2*H*-chromen-3-yl] 3,4,5-trihydroxybenzoate
	Epigallocatechin-3-gallate (XVII)	3′,4′,5′,5,7-Pentahydroxy-3-galoyl-	[(2*R*,3*R*)-5,7-Dihydroxy-2-(3,4,5-trihydroxyphenyl) chroman-3-yl] 3,4,5-trihydroxybenzoate
Anthocyanidin	Pelargonidin (XVIII)	4′,5,7-Trihydroxy-	2-(4-Hydroxyphenyl) chromenylium-3,5,7-triol
	Cyanidine (XIX)	3′,4′,5,7-Tetrahydroxy-	2-(3,4-Dihydroxyphenyl) chromenylium-3,5,7-triol
	Delphinidin (XX)	3′,4′,5′,5,7-Pentahydroxy-	2-(3,4,5-Trihydroxyphenyl) chromenylium-3,5,7-triol
	Peonidin (XXI)	3′-Methoxy-4′,5,7-trihydroxy-	2-(4-Hydroxy-3-methoxyphenyl) chromenylium-3,5,7-triol
	Petunidin (XXII)	3′,4′,5,7-Tetrahydroxy-5′-methoxy-	2-(3,4-Dihydroxy-5-methoxyphenyl)-3,5,7-trihydroxychromenylium
	Malvidin (XXIII)	3′,5′-Dimethoxy-4′,5,7-trihydroxy-	3,5,7-Trihydroxy-2-(4-hydroxy-3,5-dimethoxyphenyl)chromenium
Isoflavone	Genistein (XXIV)	4′,5,7-Trihydroxy-	5,7-Dihydroxy-3-(4-hydroxyphenyl) chromen-4-one
	Daidzein (XXV)	4′,7-Dihydroxy-	7-Hydroxy-3-(4-hydroxyphenyl) chromen-4-one
	Formononetin (XXVI)	4′-Methoxy-7-trihydroxy-	7-Hydroxy-3-(4-methoxyphenyl) chromen-4-one

[a] The chiral carbons 2 and 3 are in a *trans* configuration (compound XIII) and *cis* configuration (compound XIV), respectively.

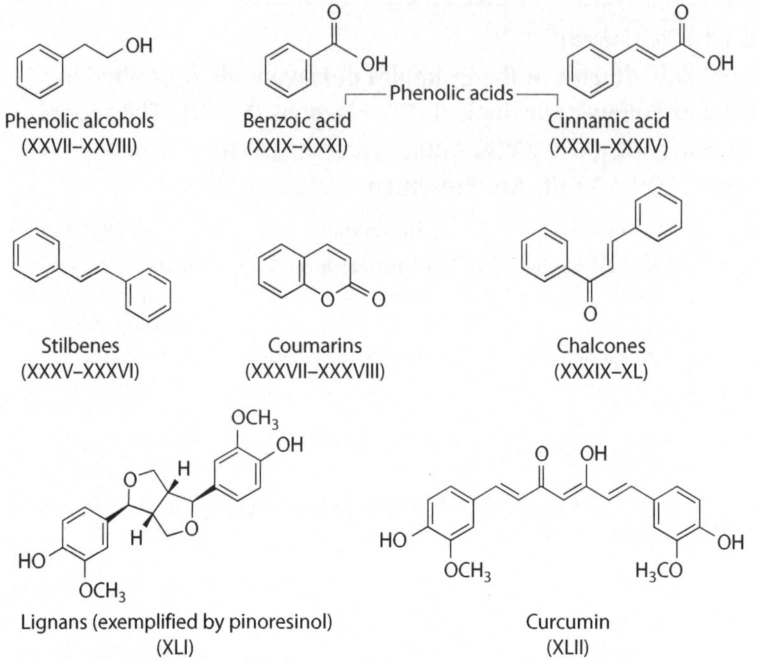

FIGURE 1.3 Basic structural formulae of the six major subclasses of nonflavonoid polyphenols. Under the name of each subclass, Roman-numbered examples, as quoted in the text, are given in parentheses (XXVII–XLI). The substituents of each of these examples are presented in Table 1.2. In addition, curcumin (XLII) is included as an example of another nonflavonoid compound.

(Chen et al., 2014). Other flavonoid subclasses that are comparatively minor dietary components are the dihydroflavonols, flavan-3,4-diols, and aurones.

The second class of polyphenols, defined as nonflavonoid phenolics, includes a more heterogeneous group of compounds that, like flavonoids, also comprises six subclasses (Figure 1.3 and Table 1.2): the first and second subclasses are the phenolic alcohols [e.g., tyrosol (XXVII), hydroxytyrosol (XXVIII)] and the phenolic acids, respectively. The phenolic acid subclass, which includes phenols possessing at least one carboxylic acid function, can be subdivided into the hydroxybenzoic acids (C6–C1) [e.g., gallic (XXIX), vanillic (XXX), and protocatechuic (XXXI)] and hydroxycinnamic acids (C6–C3) [e.g., caffeic (XXXII), *p*-coumaric (XXXIII), and ferulic (XXXIV)] derivatives. The third subclass is represented by the stilbenes (C6–C2–C6) [e.g., resveratrol (XXXV) and piceatannol (XXXVI)]. The fourth, fifth, and sixth subclasses are the coumarins (C6–C3) [e.g., esculetin (XXXVII) and scopoletin (XXXVIII)], the chalcones (C6–C3–C6) [e.g., butein (XXXIX) and phloretin (XL)], and the lignans (C6–C3–C3–C6) [e.g., pinoresinol (XLI)] (Croft, 1998; Heleno et al., 2015; Saibabu et al., 2015), respectively. As shown in Figure 1.3 and Table 1.2, curcumin (XLII) is also included as an example of another nonflavonoid compound.

TABLE 1.2
Structural Substituents of the Examples of the Six Subclasses of Nonflavonoid Polyphenols Described in Figure 1.3 as Follows: Benzoic Acid Derivatives (XXIX–XXXI), Cinnamic Acid Derivatives (XXXII–XXXIV), Stilbenes (XXXV–XXXVI), Coumarins (XXXVII–XXXVIII), Chalcones (XXXIX–XL), and Lignans (XLI) Are Presented

Subclass	Examples	Substitutions	IUPAC Name
Phenolic alcohols	Tyrosol (XXVII)	4-Hydroxy-	4-(2-Hydroxyethyl)phenol
	Hydroxytyrosol (XXVIII)	3,4-Dihydroxy-	4-(2-Hydroxyethyl)-1,2-benzenediol
Phenolic acids			
Benzoic acid derivatives	Gallic (XXIX)	3,4,5-Trihydroxy-	3,4,5-Trihydroxybenzoic acid
	Vanillic (XXX)	3-Methoxy-4-hydroxy-	4-Hydroxy-3-methoxybenzoic acid
	Protocatechuic (XXXI)	3,4-Dihydroxy-	3,4-Dihydroxybenzoic acid
Hydroxycinnamic derivatives	Caffeic (XXXII)	3,4-Dihydroxy-	3-(3,4-Dihydroxyphenyl)-2-propenoic acid
	p-Coumaric (XXXIII)	4-Hydroxy-	(*E*)-3-(4-Hydroxyphenyl)-2-propenoic acid
	Ferulic (XXXIV)	3-Methoxy-4-hydroxy-	(*E*)-3-(4-Hydroxy-3-methoxy-phenyl) prop-2-enoic acid
Stilbenes	Resveratrol (XXXV)	3,5,4′-Trihydroxy-trans-	Trans-3,5,4′-Trihydroxystilbene
	Piceatannol (XXXVI)	3′,4′,3,5-Tetrahydroxy-trans-	4-[(*E*)-2-(3,5-Dihydroxyphenyl) ethenyl]benzene-1,2-diol
Coumarins	Esculetin (XXXVII)	6,7-Dihydroxy-	6,7-Dihydroxy-2-chromenone
	Scopoletin (XXXVIII)	6-Methoxy-7-hydroxy-	7-Hydroxy-6-methoxychromen-2-one
Chalcones	Butein (XXXIX)	2′,4′,3,4-tetrahydroxy-	(*E*)-1-(2,4-Dihydroxyphenyl)-3-(3,4-dihydroxyphenyl) prop-2-en-1-one
	Phloretin (XL)	2′,4′,6′,4-Tetrahydroxy-dyhydro-	3-(4-Hydroxyphenyl)-1-(2,4,6-trihydroxyphenyl) propan-1-one

(*Continued*)

TABLE 1.2 (*Continued*)
Structural Substituents of the Examples of the Six Subclasses of Nonflavonoid Polyphenols Described in Figure 1.3 as Follows: Benzoic Acid Derivatives (XXIX–XXXI), Cinnamic Acid Derivatives (XXXII–XXXIV), Stilbenes (XXXV–XXXVI), Coumarins (XXXVII–XXXVIII), Chalcones (XXXIX–XL), and Lignans (XLI) Are Presented

Subclass	Examples	Substitutions	IUPAC Name
Lignans	Pinoresinol (XLI)	Full name/structure	4-[(3*S*,3*aR*,6*S*,6*aR*)-6-(4-hydroxy-3-methoxyphenyl)-1,3,3*a*,4,6,6*a*-hexahydrofuro[3,4-c]furan-3-yl]-2-methoxyphenol
Other nonflavonoids	Curcumin (XLII)	Full name/structure	(1*E*,6*E*)-1,7-Bis(4-hydroxy-3-methoxyphenyl)-1,6-heptadiene-3,5-dione

Note: In addition, curcumin (XLII) is included as an example of another nonflavonoid compound.

1.3 POLYPHENOLS AND THEIR MAJOR HEALTH-RELATED BIOACTIVITIES

For decades, polyphenols have been seen, primarily, as antioxidant molecules (Landete, 2012; Sureda et al., 2014; Virgili and Marino, 2008). This classical view emerged from the well-established *in vitro* ability of these compounds to scavenge reactive oxygen species (ROS), a term which comprises mainly oxygen- and nitrogen-centered free radicals (e.g., superoxide, hydroxyl, and nitric oxide) as well as other nonradical prooxidant species (e.g., hydrogen peroxide, peroxynitrite, and hypochlorite) (Carocho and Ferreira, 2013; Seifried et al., 2007; Sies, 1997). In addition, the contention that polyphenols are a synonym of antioxidant molecules was prompted by the broad recognition of the potential of these compounds to prevent the initiation, or at least ameliorate the rate of development, of diverse ROS-related non-transmissible chronic diseases (NTCD) (Cassidy et al., 2011; Grassi et al., 2010; Hooper et al., 2008; Huxley and Neil, 2003; Kris-Etherton et al., 2002; McCullough et al., 2012; Perez-Vizcaino and Duarte, 2010; Peterson et al., 2012; Rangel-Huerta et al., 2015; Rossi et al., 2010; Sarriá et al., 2015; Scalbert et al., 2005; Wang and Stoner, 2008; Wang and Yang, 2012; Wang et al., 2014). From a mechanistic point of view, the ROS-scavenging properties of polyphenols have been primarily associated with their ability to donate either a hydrogen atom or a single electron to the free radical species (Figure 1.4, exemplified for the 4′-hydroxyl or the 4′-phenoxyl group of the protonated and deprotonated forms of quercetin, respectively, and for a hydroxyl radical as example of ROS) (Amic et al., 2007;

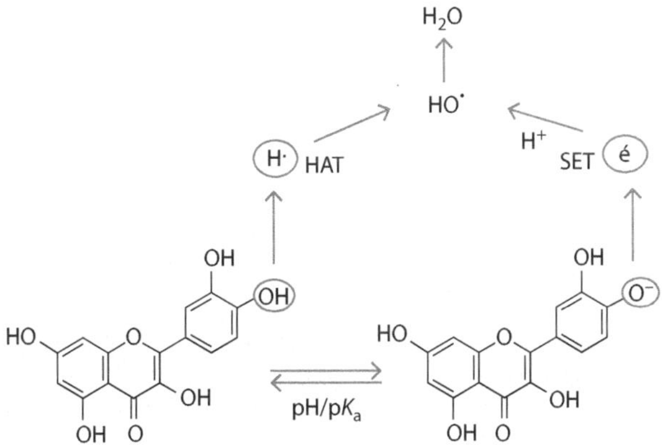

FIGURE 1.4 Schematic representation of the hydrogen atom transfer (HAT) and the single-electron transfer (SET) mechanisms by which a polyphenol (the case shown is for the quercetin molecule) stabilizes a free radical (depicted for a hydroxyl radical). The SET mechanism is favored under conditions in which the phenolate species predominates.

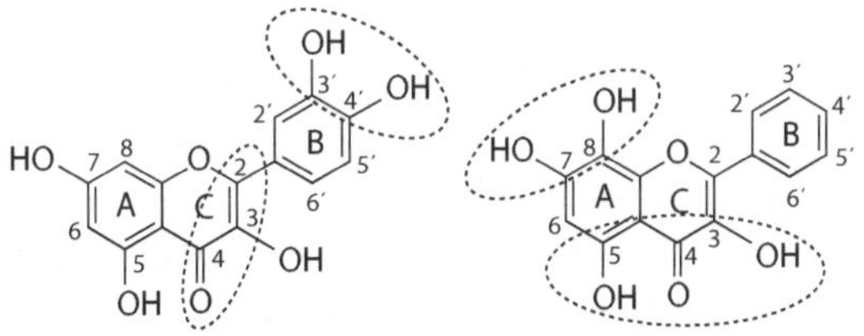

FIGURE 1.5 Summary of the main structural criteria that modulate the antioxidant activity of flavonoids. (Taken from Amic, D. et al., *Curr. Med. Chem.*, 14, 827, 2007.)

Lucic et al., 2008; Vauzour et al., 2010). Although the ROS-scavenging capacity of flavonoids (and that of most polyphenols) is, in general, proportional to the number of its hydrogen/electron-donating phenolic groups, the structural location of such groups can often significantly influence such activity (Amic et al., 2007; RiceEvans et al., 1996; Vauzour et al., 2010). As shown in Figure 1.5, in the case of most flavonoids, the ROS-scavenging capacity appears to greatly depend on the presence of the following structural features: (1) an *o*-dihydroxy (3′,4′-diOH, i.e., a catechol structure) group in the B ring, which serves as a radical target site and confers high stability to the flavonoid phenoxyl radicals formed via hydrogen bonding or by expanded electron delocalization; (2) a C2–C3 double bond in conjugation with a 4-keto function in the C ring, which is responsible for the electron delocalization arising from the B ring; and (3) an additional presence of a 3- and a 5-hydroxyl group for maximal radical scavenging capacity (shown in the right side of Figure 1.5). In the absence of *o*-dihydroxy structure in the B ring, hydroxyl substituents in a catechol structure on the A ring have been shown to be able to somehow compensate and become a large determinant of flavonoid antiradical activity (Amic et al., 2007; Lucic et al., 2008).

In certain cases, however, in addition to affecting the ROS-scavenging activity, the structural location of the phenolic groups can also confer iron- and copper-chelating properties to some polyphenols (Milaeva, 2011; Valko et al., 2006). Chelation can redox-inactivate these metals, thus preventing them from catalyzing the reductive conversion of molecular oxygen into superoxide anion and that of hydrogen peroxide into hydroxyl radicals. In the case of flavonoids, the *o*-dihydroxy structure, and the 4-keto and the 3- or 5-hydroxyl groups, is considered essential for their metal-chelating ability (Figure 1.5). The ability of polyphenols to exert an antioxidant action is, however, by no means limited to their "direct" ROS-scavenging and metal-chelating mechanisms. In fact, it extends to a larger and probably a biologically more relevant number of other indirect antioxidant actions (Dinkova-Kostova and Talalay, 2008; Pisoschi and Pop, 2015). Among these is the ability to induce the expression of genes coding the *de novo* synthesis of ROS-removing enzymes, which only some particular polyphenols have (Dinkova-Kostova and Talalay, 2008; Ji et al., 2015; Stepanic et al., 2015; Yang et al., 2011, 2014). Examples of such enzymes are superoxide dismutase (SOD), catalase, and glutathione peroxidase, which respectively contribute to the intracellular removal of superoxide anions, hydrogen peroxide, and lipid peroxides. Some polyphenols are, on the other hand, able to lower the intracellular formation of ROS by inhibiting the activity of ROS-forming enzymes like xanthine oxidase, NADPH oxidase, and lipoxygenase, and by repressing the expression of genes coding for the synthesis of such enzymes (Mann et al., 2009). In addition, some particular polyphenols can contribute to the removal of ROS by increasing the expression of genes that code for the synthesis of enzymes involved in the generation of endogenous antioxidants (Espinosa-Diez et al., 2015). A remarkable example of this is the enzyme gamma-glutamyl-cysteine synthase (GGS; glutamate cysteine ligase catalytic subunit), whose activity is rate limiting in the synthesis of the tripeptide-reduced glutathione (gamma-glutamyl-cysteinyl-glycine or GSH). GSH plays a fundamental antioxidant role within cells in the direct scavenging of ROS, and by acting as a cofactor of the enzyme glutathione peroxidase, by providing electrons for the reductive removal of hydrogen peroxide and lipid hydroperoxides (Na and Surh, 2008). During the last decade, researchers have recognized that an important mechanism by which certain polyphenols can induce the synthesis of antioxidant enzymes (and that of other cytoprotective enzymes) is through their interaction with the Keapl (Kelch-like ECH-associated protein)–Nrf2 (nuclear erythroid–related factor 2) system. Nrf2 is a redox-sensitive transcription factor that upon being activated by an oxidative signal within the cytoplasm, it undergoes translocation into the nucleus where it binds to DNA-ARE regions (antioxidant response elements) of gene promoters, inducing gene expression. Under basal or normal conditions, Nrf2 is retained in the cytoplasm through its binding to the cytoskeletal protein Keapl, facilitating its ubiquitination and degradation, and thereby keeping down the Nrf2 levels. Upon exposure of cells to certain polyphenols, Nrf2 is released; then translocating into the nucleus leads to an enhanced expression of various antioxidant and phase II detoxifying enzymes. Through these actions, some polyphenols have been shown to induce the synthesis of GGS, SOD, GSH-transferases, heme oxigenase-1, and NADPH-quinone oxidase (all linked to ARE-regulated genes), protecting cells against oxidative and lytic damage (Dinkova-Kostova and Talalay, 2008;

Espinosa-Diez et al., 2015; Ji et al., 2015; Mann et al., 2009; Na and Surh, 2008; Pisoschi and Pop, 2015; Stepanic et al., 2015; Yang et al., 2011, 2014). Low ROS levels (mild oxidative stress conditions) have also been shown to increase the expression of Nrf2 and facilitate the dissociation of Keap1–Nrf2 complex, inducing antioxidant responses. Interestingly, certain polyphenols can also exert these actions at concentrations that are considerably lower than those normally required for them to act *in vitro* as simple ROS-scavengers. The exact mechanism(s) by which these polyphenols are able to facilitate the Nrf2 dissociation remains to be elucidated. However, some evidence suggests that when exposed to a mild oxidative milieu, these polyphenols might undergo swift oxidation, promoting, thereby, the subsequent oxidation of some critical cysteine moieties that are present in the Nrf2/Keap1 complex (Naidu et al., 2015; Tebay et al., 2015).

Oxidative stress refers, in general terms, to a condition in which the antioxidant defense mechanisms (both the enzyme-based and the nonenzymatic ones) are surpassed by a noncontrolled rate of ROS production and ROS-induced oxidative modification of biological macromolecules (i.e., lipids, proteins, and nucleic acids) (Lichtenberg and Pinchuk, 2015; Lushchak, 2014; Rodriguez-Mateos et al., 2014; Sies, 2015). Oxidative stress, whether acting as a primary causal mechanism or as a disease-worsening factor, is largely recognized as a "common denominator" of a number of cardiovascular, tumoral, and neurodegenerative diseases (Montezano et al., 2015; Rangel-Huerta et al., 2015; Sies, 2015). Although a large body of literature supports the concept that many of the benefits of polyphenols against oxidative stress-bearing conditions are likely to stem from antioxidant actions exerted by these compounds (whether directly or indirectly), during the last decade an increasing number of studies have provided basic and clinical evidences that, in addition to their antioxidant actions, some polyphenols have the ability to exert many antioxidant-unrelated actions (Sandoval-Acuña et al., 2014). Among the most studied bioactivities is the ability of some polyphenols to exert (1) anti-inflammatory (Mladenka et al., 2010; Tangney and Rasmussen, 2013), (2) anti-atherogenic (Vauzour et al., 2010), (3) vasodilating (Galleano et al., 2009; Schewe et al., 2008; Schnorr et al., 2008), (4) antiplatelet aggregation (Ostertag et al., 2010; Santhakumar et al., 2014; Wright et al., 2013), and (5) anti-nutritional actions (interfering metal absorption and carbohydrate and lipid digestion/absorption) (Kim et al., 2011; Ma et al., 2011; Williamson, 2013). In addition, a number of selected polyphenols have been recognized for their ability to exert anti-mutagenic and anticarcinogenic actions, whether affecting cell regulation, differentiation, proliferation, or apoptosis (Kuno et al., 2012; Yang et al., 2013). Moreover, some polyphenols have been shown to exert the following actions: (1) antiallergic (Magrone and Jirillo, 2012), (2) antimicrobial (Daglia, 2012; Pastene et al., 2009a,b, 2010; Xie et al., 2015), (3) antiviral (Andrae-Marobela et al., 2013), (4) immune-modulating (Ghiringhelli et al., 2012), (5) lipid profile–normalizing (Kim et al., 2011), (6) estrogenic (Vitale et al., 2013), (7) neurosedating (Jäger and Saaby, 2011), (8) insulin-potentiating (Munir et al., 2013), and/or (9) anti-angiogenic (Mojzis et al., 2008). In contrast to the ROS-scavenging property of polyphenols, which is shared by each phenolic structure, the aforementioned antioxidant-unrelated bioactivities are restricted to only a very limited number of polyphenols (Sandoval-Acuña et al., 2014). Understanding the structural particularities (pharmacophores) that underlie the bio-selective actions of these polyphenols and assessing their clinical usefulness will be fundamental toward the goal of broadening—beyond their traditional antioxidant activity—the scope of health benefit potential of these compounds (Kishimoto et al., 2013). Nonetheless, despite the increasing pleiotropic character that some polyphenols appear to display, in the current epidemiological, preclinical, and clinical research scenario, the potential of polyphenols to promote health benefits is still dominated, or at least mostly associated with, their ability to exert antioxidant-related actions. Thus, a number of studies have been conducted in experimental models of oxidative stress-related disorders, where ROS have been clearly implied as an oxidative stress-causing factor; these studies indicate that the appropriate administration (i.e., doses, frequency, and mode) of polyphenols (whether these are given as pure compounds, as polyphenolic extracts, or as polyphenol-rich diets) effectively contributes to either prevent or ameliorate most of the metabolic consequences associated with such stress (Carrasco-Pozo et al., 2011a,b, 2012; Kishimoto et al., 2013). However, since in humans most NTCD are multifactorial in terms of their causative factors, targeting oxidative stress by the sole administration of polyphenols may not always prove to be sufficient to secure the total prevention and protection sought (Dyck and Schrauwen, 2015; Jain et al., 2015; Persson et al., 2014; Rodriguez-Mateos et al., 2014). Nonetheless, on the basis of the antioxidant property exhibited by all polyphenol molecules, it is not surprising to see that—beyond their structural differences—most of these compounds are often able to bring some degree of health benefit. Thus, beyond the potential of some selected polyphenols to act as pleiotropic molecules (Kishimoto et al., 2013; Rodriguez-Mateos et al., 2014), the sole ability of each polyphenol to exert antioxidant actions itself warrants the widespread and continuous interest in these compounds.

1.4 POLYPHENOLIC CONTENT AND ANTIOXIDANT CAPACITY ASSESSMENT

The total antioxidant capacity (TAC) of any *in vitro* tested plant or food material could be expected to result from the individual plus the interactive (additive, competitive, synergistic, and even antagonistic) contribution made by each of its extracted and assayable components (Ajila et al., 2011; Carocho and Ferreira, 2013; Fraga et al., 2014; Galleano et al., 2010; Zhang et al., 2014). In the case of most human edible plants, especially fruits and vegetables, the antioxidant capacity is mostly accounted for by the following types of molecules: (1) the so-called antioxidant vitamins, comprising ascorbic acid and α-tocopherol, preferentially present in citric fruits and seed oils, respectively; (2) the provitamin A carotenoids, comprises mainly β-carotene, α-carotene, and β-cryptoxanthin, largely contained in some orange and yellow fruits, and some vegetables; (3) other carotenoids, such as lycopene, lutein, and zeaxanthin, contained in tomatoes, spinach, and brussels sprouts; and finally (4) a large variety of flavonoid and nonflavonoid phenolic compounds. Polyphenols account for most of the total antioxidant activity of fruits and vegetables; the few exceptions are primarily defined by the type of food and, to a much lesser extent, the type of antioxidant assay applied (Cook and Samman, 1996; Duthie and Crozier, 2000; Proteggente et al., 2002; Rodriguez-Casado, 2016). Establishing how rich a foodstuff is in polyphenols constitutes an early and fundamental step toward assessing its potential to serve as a dietary source of polyphenolic/antioxidant compounds. Polyphenols have long remained compounds of interest for the food, pharmaceutical, and cosmetic industries, whether ingested as part of a polyphenol-rich natural, a processed, or a functional food (understood as any food matrix into which a polyphenol-rich extract has been added), or consumed in the form of a nutraceutical or a cosmeceutical product.

Different *in vitro* analytical methodologies have been developed to assay the polyphenolic richness of a food, a concept that can comprise assessing the concentration as well as the structural diversity of all its extractable/solubilized phenolic components. Among the most commonly employed methods to assess the polyphenolic richness are those directed to assess the (1) total phenolic content of a sample that corresponds to assessing the total content of the sum of all its flavonoid and nonflavonoid compounds, (2) total content of a structurally related phenolic subgroup (i.e., aimed at assessing the total amount of only the flavonoids), (3) total content of a particular polyphenol subclass (aimed at assessing, for instance, total anthocyanins, total flavones, or total flavan-3-ols present in a sample), and (4) exact content of a single polyphenol molecule within a sample (e.g., specifically the content of quercetin, epicatechin, or caffeic acid) (Bobo-Garcia et al., 2015; Diaz-Garcia et al., 2013; Ignat et al., 2011; Medina, 2011; Moon and Shibamoto, 2009; Motilva et al., 2013; Natale et al., 2015).

The most commonly used and widespread analytical approach to assess the total phenolic content of a sample is through the use of the Folin–Ciocalteu (F-C) reagent (Magalhães et al., 2008; Medina-Remon et al., 2009; Prior et al., 2005). The assay is based on the direct reaction between the phenolic compounds extracted from a sample and the redox-sensitive F-C reagent, a mixture of phosphomolybdate and phosphotungstate. The product of this redox reaction, a blue complex, is easily measured within the visible portion of the spectrum. Since the F-C assay is run under alkaline conditions (pH 10), virtually all the assayable phenolics are present in their nonprotonated form; thus this assay expresses, essentially, the ability of all F-C-reactive fenolate to donate an electron to the F-C reagent. Due to the simplicity and readiness of the F-C method, for decades this assay has remained as the single most common and widespread analytical approach to assess total phenolics. However, the F-C assay is not specific for phenolic compounds as it measures the ability of virtually any substance capable of reducing the F-C reagent (Huang et al., 2005; Magalhães et al., 2008; Margraf et al., 2015). Thus, unless properly controlled (for which no experimental approach has so far proven to be optimal) (Bridi et al., 2014; Folch-Cano et al., 2013), this assay will not discriminate between polyphenols and many other compounds capable of reducing the F-C reagent, like glucose, fructose, and ascorbic acid. These compounds could be naturally found in many polyphenol-rich samples (e.g., several sweet and citric fruits and its corresponding beverages) or in samples of processed foodstuff. Singleton and Rossi (1965), Ainsworth and Gillespie (2007), and Magalhães et al. (2010), among several other researchers, have discussed the F-C method and the potential confounding effects of various non-polyphenolic compounds; some authors have also proposed modified methods to correct or avoid the confounding factors and to reduce the time required to complete the running of the classical method (Bridi et al., 2014). Despite its potential limitations, the Folin–Ciocalteu assay (in its classical and in some of its modified versions) has remained and, most likely for some time, will continue to be the most commonly employed assay to quickly assess the total phenolic contents.

In addition to assessing their polyphenolic richness, food samples are often assessed in terms of their actual "total antioxidant capacity." Different *in vitro* methodologies have been developed to estimate the TAC of a sample, whether this is constituted by a single antioxidant molecule or by a complex mixture of such compounds. Most employed methodologies are based on strategies aimed at evaluating (1) the consumption of stable free radicals by the antioxidants present in the sample, (2) the capability of such antioxidants to reduce cupric or ferric ions, or (3) the ability of the sample's antioxidants to protect a target molecule from undergoing oxidation when exposed to a reactive free radical.

In assaying the ability to scavenge stable radicals, $DPPH^{\bullet}$ (2,2-diphenyl-1-picrylhydrazyl) and $ABTS^{\bullet+}$ (2,2-Azinobis-(3-ethylbenzothiazole-6-sulfonate)) are the two most commonly used free radicals, as these two reagents are highly stable and colored (Lopez-Alarcon and Denicola, 2013; Prior et al., 2005). While $DPPH^{\bullet}$ is commercially available, $ABTS^{\bullet+}$ must be generated through the oxidation of ABTS by compounds, such as $K_2S_2O_8$, MnO_2, or peroxyl radicals (see the following text). The $ABTS^{\bullet+}$-based assay is often named as the TEAC (Trolox-equivalent antioxidant capacity) assay. While $DPPH^{\bullet}$ is soluble in organic solvents (e.g., ethanol), $ABTS^{\bullet+}$ is soluble in both aqueous and hydroalcoholic media. The main shortcomings of the DPPH and TEAC assays are the complexity of their reaction mechanisms, the strong dependence of the results with the experimental conditions used, and the apparent biological irrelevance of using $DPPH^{\bullet}$ or $ABTS^{\bullet+}$ as ROS (Lopez-Alarcon and Denicola, 2013; Prior et al., 2005). In addition, $ABTS^{\bullet+}$ and $DPPH^{\bullet}$ are two sterically hindered nitrogen-centered radicals, and the reaction between these ROS and diverse antioxidants, rather than reflecting specific reactivity, appears to reflect primarily steric hindrance. In turn, the ROS involved in the oxidative damage of biological macromolecules and in the deterioration of food components are mainly oxygen-centered (i.e., superoxide, hydroxyl, lipoperoxy, and lipoalcoxy) radicals. Given this, and since polyphenols are known to react preferentially with oxygen-centered radicals, the $ABTS^{\bullet+}$ and $DPPH^{\bullet}$ radicals might not be the best models for assessing the biologically relevant potential antioxidant capacity.

In the case of the TAC methods based on the reduction of metal ions by polyphenols, the FRAP (ferric-reducing antioxidant power) and CUPRAC (cupric-reducing antioxidant capacity) assays evaluate the capacity of a sample to reduce either ferric ions or cupric ions, respectively, both in aqueous media. While the FRAP assay uses pyridyl triazine to complex ferric ions (Fe(III)-TPTZ), the CUPRAC assay employs neocuproine to complex cupric ions (Cu(II)–Nc) (Amorati and Valgimigli, 2015; Apak et al., 2007; Campos et al., 2009; Celik et al., 2010). In both assays, the ability of polyphenols (as well as that of other molecules) to reduce the formerly referred complexes is monitored spectrophotometrically along a defined period of time. The FRAP assay is run at acidic pH (3.6); at this pH, all phenolic groups are likely to be protonated. In turn, the CUPRAC assay is performed at pH 7.0; thus, depending on the pK_a of its phenolic hydroxyls, polyphenols of an identical sample will be present in the CUPRAC cocktail assay as a mixture of protonated and nonprotonated hydroxyl groups. Although both methods are experimentally easy to implement, the results obtained from applying FRAP and CUPRAC assays will depend largely on the incubation time that the samples are allowed in order to react with the corresponding iron and copper metal complexes (Apak et al., 2004; Benzie and Choi, 2014; Henriquez et al., 2011). Thus, as time can substantially affect the endpoint results, much caution should be taken when comparing FRAP or CUPRAC data from different laboratories (Apak et al., 2007).

ORAC and TRAP are two competitive assays that assess the ability of molecules to inhibit the oxidative consumption of an easy-to-quantify ROS-sensitive target molecule (Dorta et al., 2015). In the ORAC (oxygen radical absorbance capacity) assay, the hydrophilic and thermolabile azo compound AAPH (2,2′-azobis(2-amidinopropane)hydrochloride) is routinely used as a source of peroxyl radicals (ROO•); the photostable and low-cost reagent fluorescein is used as an ROS-sensitive fluorescent probe (Huang et al., 2005; Ou et al., 2013). Thermal decomposition of AAPH at neutral pH yields carbon-centered azo radicals, which swiftly react with molecular oxygen to give a steady flux of the highly reactive ROO• radicals. When the ORAC assay is aimed at assessing hydrophobic antioxidants, a cyclodextrin derivative has been used as a water-solubility enhancer (Huang et al., 2002). Instead of AAPH, the lipid-soluble 2,2′-azobis(2,4-dimethylvaleronitrile) (AMVN) has also been employed as peroxyl radical generator. Although ORAC is an *in vitro* assay, peroxyl radicals are considered to be of much closer biological relevance than those generated from DPPH or ABTS, as the former ROS are commonly formed during lipid and protein oxidation chain reactions. Such ROS have harmful effects not only on the health of most living organisms but are also associated with the quality of foods through the oxidative deterioration of its lipid and protein components. In the fluorescein-based ORAC assay, polyphenols as well as any other molecule capable of reacting at neutral pH with peroxyl radicals (such as vitamins C and E, and carotenes) will be evaluated. In the presence of these antioxidants, the rate of fluorescence decay is concentration-dependently slowed down, resulting in an increment of the area under the curve (AUC) associated with the time-dependent oxidation of fluorescein (Huang et al., 2005; Ou et al., 2013). The ORAC antioxidant capacity (also referred to as activity) is determined from competition kinetics and is estimated from the difference in the AUC described by the decay of the fluorescence's probe when it is exposed to peroxyl radicals in the presence and absence of the sample's antioxidants. Most commonly, the AUC values are compared with that induced by Trolox® (a hydrosoluble vitamin E analog), allowing for the determination of an ORAC value in terms of this reference compound, that is expressed as micromoles of Trolox equivalents per sample mass or volume unit. The ORAC assay takes into account the initial reaction rate and the total extent of inhibition, which includes the action of slow-reacting or secondary antioxidant products formed. In recent years, Lopez-Alarcon et al. (Ortiz et al., 2012) have proposed the use of pyrogallol red (PGR), as an ROO•-target molecule alternative to the fluorescein probe. Interestingly, these authors provided evidence that by means of PGR, the ORAC assay can distinguish between the ROS-scavenging contribution of ascorbic acid and all other ROO•-scavengers components present in a sample (Lopez-Alarcon and Lissi, 2006). Since PGR absorbs within the

UV–visible spectra, running the PGR-based ORAC assay demands only a simple spectrophotometer.

The fluorescein-based ORAC assay, developed during the 1990s by scientists working at the National Institutes of Health (NIH) and the U.S. Department of Agriculture (USDA), was used earlier to compile data on the antioxidant capacity of hundreds of foods and beverages consumed in the United States (Cao et al., 1993; Ou et al., 2001; Prior et al., 2003; Wang et al., 1996). Since then, the ORAC assay has been widely employed by many active research groups. Although the ORAC assay can be run under well-standardized *in vitro* conditions, assessing the total antioxidant activity with ORAC does not imply that the food components in the sample will become equally available to react with ROS under *in vivo* conditions (namely, within the organism after the sample ingestion). Several assumptions are required to use ORAC values as a quantitative indication of the *in vivo* antioxidant ability of polyphenols and other antioxidants. A major one is that each of the molecules present in the *in vitro* sample would have to be present (in the same proportion, concentrations, and time) in the biological milieu where these are meant to exert *in vivo* their antioxidant action. However, such a requirement may not be very likely to be fulfilled due to the complexity of the absorption, distribution, and biotransformation phenomena that often affect each of the ingested polyphenols. The ORAC values might represent the *in vivo* scenario somehow more accurately when the sample is subjected to no or minimal absorption (i.e., in the stomach or in the skin). Certainly, such type of considerations will apply not only to the ORAC but also to any other *in vitro* antioxidant capacity assay.

Less frequently employed than the ORAC is the TRAP (total radical-trapping antioxidant parameter) assay (Pellegrini et al., 2003; Queiros et al., 2013; Wayner et al., 1987). This method measures the decay of a fluorescent target during a controlled peroxidation reaction initiated by either peroxyl or hydroxyl radicals (the latter ROS are generated from a mixture of H_2O_2 plus Fe^{2+}; Fenton reaction). It was initially introduced for the determination of the antioxidant status of human plasma; yet, it was soon after adapted to be extensively employed in the characterization of the total antioxidant capacity of a large variety of plant food samples (Wayner et al., 1985). Different variations of the TRAP assay have been developed to assess natural products (Fraga et al., 2014; Lopez-Alarcon and Denicola, 2013); the TRAP values reflect mainly the number of $ROO^•$ trapped per antioxidant molecule; thus, it is an index exclusively related to the stoichiometry of the $ROO^•$-antioxidant reaction. Typically, the TRAP assay is run using AAPH as a source of ROS and dichlorofluorescein (DCFH), instead of the initially used β-phycoerythrin, as a fluorescent oxidizable probe (Huang et al., 2005). The oxidation of DCFH by $ROO^•$ yields the highly fluorescent product dichlorofluorescein. In the TRAP assay, the presence of antioxidant compounds competitively inhibits the decrease of the fluorescence signal. TRAP values are usually calculated from the length of the lag phase (namely, the time during which an antioxidant action is exerted before the target/probe oxidation process starts) caused by the addition of the antioxidants contained in the sample, and results are compared with those obtained after using Trolox as antioxidant standard. Unfortunately, the use of the lag phase in this assay can be a major shortcoming since not all antioxidant molecules possess a clearly defined lag phase and also because this method ignores the antioxidant capacity profile after the lag phase. As discussed by Prior et al. (2005), the application of different criteria to establish the length of the lag phase as an endpoint makes it difficult to compare data between laboratories.

Despite the limitations of the TRAP and other antioxidant assays discussed in the earlier text, these assays will likely to be continuously applied to samples of different nature, like natural extracts, complex foods, by-products of the food industry, or simply to isolated compounds (Fraga et al., 2014; Lopez-Alarcon and Denicola, 2013). While the FRAP, CUPRAC, ABTS, and the AAPH-based ORAC and TRAP assays have been employed to measure the antioxidant capacity of samples containing primarily hydrophilic compounds, the DDPH assay is the only method that has been routinely applied to both aqueous and lipophilic extracts of plant foods. When run under carefully standardized and controlled experimental conditions, each of these methods can give rise to accurate and repeatable values; however, conclusive data on the antioxidant ability of a given sample may substantially differ between one method and another. The ranking of antioxidant capacity of a group of samples determined by different assays usually follows the same trend irrespective of the method considered. However, since the assays are not always based on identical reaction mechanisms, often major differences in the ranking of particular samples can be found (Fraga et al., 2014; Perez-Jimenez et al., 2010a). Thus, to characterize the antioxidant capacity of a sample and elaborate rankings, more than one method would need to be used. In addition, whenever comparisons are sought, these should only be performed between values of antioxidant capacity obtained using identical sample preparation protocols, extraction conditions (e.g., solvent composition, pH, temperature, and maceration times), and methods of analysis. To provide comprehensive information on the total antioxidant capacity of a foodstuff, the pros and cons of each of the assays applied should be taken into account, and whenever possible, at least two, and preferably more TAC assays should be conducted. Detailed information and discussion on the main features, advantages, and drawbacks of the these assays, and on other methodologies employed in the determination of the antioxidant capacity of pure polyphenolic compounds or mixtures of these in complex matrices, can be found in several valuable articles and reviews (Fraga et al., 2014; Magalhães et al., 2008). In addition, when comparing results obtained from two or more of the such antioxidant capacity assays, particular attention should be placed on the fact that antioxidant molecules participate differently in some of these assays, depending on their reaction mechanisms. While in the case of the ORAC and TRAP assays the antioxidant molecules are believed to be preferentially involved in a hydrogen atom transfer (HAT) type of reaction (see Figure 1.4), in the FRAP, CUPRAC, ABTS, and DPPH assays the antioxidants appear to react overall through an electron transfer (ET) reaction type. Generally, the non-protonated form of most phenolics (i.e., the phenolate form) is believed to be more reactive toward ROS in the form of a single-electron transfer (SET) mechanism of action. When the antioxidants under assay are polyphenols (as exemplified in Figure 1.4 for quercetin), a hydrogen atom belonging to at least one of its phenol moieties is transferred to a hydroxyl free radical stabilizing the latter in the form of a water molecule. In such a reaction, the phenol is converted into an aryloxy radical which subsequently undergoes disproportionation to produce a phenolic quinone (not shown). Due to the ability of phenoxy radicals to undergo resonance delocalization throughout the phenolic ring, the aryloxy radicals generally have much higher stability than the radicals formed during the attack of ROS to biological substrates (i.e., lipids, proteins, and nucleic acids) (Choe and Min, 2006; Perez-Jimenez et al., 2008). To be effective, phenolic antioxidants should react much faster with the free radicals than with the biomolecules sought to protect. However, depending on the biological targets and the reaction circumstances, some phenolic compounds could also behave as prooxidants (Chobot et al., 2013); this aspect is discussed in Chapter 2.

1.5 FRUITS AND VEGETABLES AS A MAJOR SOURCE OF POLYPHENOLS

During the last two decades, considerable evidence has accumulated to substantiate the claim that a high and regular consumption of fruits and vegetables is directly associated with several major health benefits (Arts and Hollman, 2005; Boeing et al., 2012; Nagle et al., 2015; Quinones et al., 2013; Scalbert et al., 2005). Such benefits are expressed, at the epidemiological level, as a reduction in the relative risk of developing several cardiovascular (Rangel-Huerta et al., 2015) and tumoral diseases (Luo et al., 2015; Turati et al., 2015; Vieira et al., 2016; Wang et al., 2015). Although fruits and vegetables contain a large number of health-promoting phytochemicals, experimental evidence supports the concept that at least a significant part of the benefits associated with the regular consumption of these foods can be attributed to molecules that display antioxidant properties, especially polyphenols (Herrera et al., 2009; Mates et al., 2011). In fact, an increasing number of clinical studies now support the existence of a direct relationship between the ingestion of certain polyphenol-rich foods and the improvement of several biochemical and/or physiological markers of health status. Examples of recent studies are the human health benefits associated with the ingestion of epicatechin-rich cocoa (Basu et al., 2015; Dower et al., 2015; Ponzo et al., 2015; Sansone et al., 2015), anthocyanin-rich berries (Davinelli et al., 2015; McAnulty et al., 2014), and hydroxytyrosol-rich olive oil (Bulotta et al., 2014; Granados-Principal et al., 2010). Like other phytochemicals, polyphenols are not present in isolation in foods and therefore, their antioxidant capacities are likely to depend strongly on the occurrence of complex interactions between these compounds and other components present in the food matrix. Nonetheless, assaying their total content in diverse plant food matrices has long been considered a practical, yet indirect and partial, form of estimating the polyphenol-dependent antioxidant richness of foods. Measuring total phenolics and the fluorescein-based ORAC activity, our laboratory (Speisky et al., 2012) systematically characterized the antioxidant richness of a large number of species/varieties of fruits and vegetables grown in Chile and in other west and southern neighboring countries of South America. From these studies, the digital database http://www.portalantioxidantes.com/ was launched, which compiles the results of analyzing nearly 150 fruit species and varieties (Speisky et al., 2012). Noteworthy, rather than compiling data from papers published by many different research centers (see the following text); in our study, all fruit analyses were conducted by a single ISO/IEC 17025-certified laboratory (of antioxidants at the Nutrition and Food Technology of the University of Chile). In addition, the database includes the antioxidant richness of several south Andes native fruits. Some native berry species, like maqui (*Aristotelia chilensis*), murtilla (*Ugni molinae*), and calafate (*Berberis microphylla*) largely outscored the antioxidant capacity of traditional berry fruits—blueberries, raspberries, strawberries, and blackberries—also grown and consumed in Chile. In this database, major differences in total phenolics and ORAC values were also observed as a function of the fruit variety; those differences were particularly large among varieties of berries, avocado, cherries, and apples. As shown in the database and reported by Speisky et al. (2012), in some of the whole fruit studied (pears, apples, apricots, quinces, plums, and peaches), a significant part of the ORAC values was found to be accounted for by antioxidants present in the peel. Recently (in 2014), this database was updated with ORAC and total phenolics values of over a hundred species and varieties of vegetables regularly grown and consumed in the South American region. As seen with certain fruits, some native varieties of potatoes, beans, onions, and other species were found to be considerably richer in ORAC and phenolics, compared with their corresponding nonnative or introduced varieties. This warrants further exploration of such antioxidant-rich native varieties in terms of their chemical composition, nutritional functionality, and nutraceutical potential. The www.portalantioxidantes.com database is likely to be a useful tool to estimate the intake of total phenolics and, through the ORAC data, the antioxidant-related dietary contribution of some native varieties of fruits and vegetables commonly consumed by populations from Chile and some of its west and southern neighboring countries. As with other databases, it might also serve to establish possible associations between the intake of antioxidants through fruit and vegetables (i.e., total polyphenols and ORAC-associated antioxidants) and the prevalence of some of those diseases whose etiologies appear to be primarily linked to oxidative stress. The release of a database on antioxidants from fruits and vegetables from the west-southern region of South America was inspired by the ORAC and total phenolics database launched earlier (in 2010) by scientists from the USDA on several hundred foodstuffs commonly consumed by the American population (http://www.ars.usda.gov/Services/docs.htm?docid=15866) (2015b). Despite the broad use of the aforementioned database, recently the USDA's Nutrient Data Laboratory (NDL) decided to remove its ORAC Database for Selected Foods from the NDL website (http://www.ars.usda.gov/Services/docs.htm?docid=15866) (2015b). That decision was partly prompted by the repeated misuse of the ORAC values contained in the database by some food- and dietary supplement–manufacturing companies, which indiscriminately promoted their products and grossly guided consumer choices on food and dietary supplements. This decision, however, was unfortunate since the misuse of any scientific information should not be a reason to censor it (Prior, 2015). Rather than assuming that its data on ORAC and total phenolics were not valid or relevant, the governmental emphasis should have been placed on procuring, via regulations and public education, the proper interpretation and marketing use of that information. This withdrawal of information had no impact on the scientific community, as papers published since have continued reporting applications of the ORAC methodology. It should be stressed though, that ORAC values are only an index of the ability of a given sample to scavenge peroxyl radicals under *in vitro* conditions. This assay suffers the same limitations of any other "in test tube" assay to analyze nutrients or food ingredients (Prior, 2015).

Although all polyphenol molecules are ORAC-assayable antioxidants, some specific polyphenols also have the potential to modulate other bioactivities not accounted for by the ORAC value. Therefore, various research laboratories as well as health and food science organizations assessed and compiled data on the concentration of individual phenolic compounds in foodstuffs. Early work by scientists from the USDA led, in 2003, to the online publication of a database describing the flavonoid contents of 225 food items, which was followed by a second release in 2006 on flavonoid composition of 392 food items, a third release in 2011 for 500 food items, a 3.1 release in 2013, and more recently (September 2015), a 3.2 release for 506 food items (http://www.ars.usda.gov/SP2UserFiles/Place/80400525/Data/Flav/Flav3.2.pdf). The latter release comprised detailed data on 26 predominant dietary flavonoids, grouped into five subclasses: flavonols, flavones, flavanones, flavan-3-ols, and anthocyanidins (http://www.ars.usda.gov/nutrientdata/flav); isoflavones were not included in this database. A separate database, "USDA Iowa State University Database on the Isoflavones Content of Foods" first released in 1999, was updated in 2008 and published on the Nutrient Data Laboratory's (NDL) website: http://www.ars.usda.gov/nutrientdata. Similarly, data on proanthocyanidins, which were also not included in the original database, are available in the "USDA Database for the Proanthocyanidin Content of Selected Foods" released on NDL's website in August 2004 (see Bhagwat et al., 2013).

A totally separate but highly comprehensive database on specific polyphenols in foodstuffs, the Phenol-Explorer database (http://phenol-explorer.eu/), was launched in 2009 by Scalbert and coworkers (Neveu et al., 2010; Perez-Jimenez et al., 2010b). This database offers data on each of the major individual phenolics known to be present in most commonly consumed foods. It initially compiled over 37,000 original data points collected from several hundred articles published in peer-reviewed journals (Neveu et al., 2010). Interestingly, the Phenol-Explorer web database also details 383 polyphenol metabolites identified in human and animal biofluids (data from examining 221 publications). Recently, Rothwell and coworkers (2015) exploited such data to characterize the so-called polyphenol metabolome, an expression which refers to the complete set of polyphenols or polyphenol metabolites that are either naturally present in foods or that can be found in biological specimens (fluids and tissues) following their ingestion as a result of their endogenous metabolism, including the one accounted for by the microbiota (Zamora-Ros et al., 2014). Although the Phenol-Explorer was initially conceived as a repository of polyphenol food composition data, it has extended to now include, whenever proper literature is available, pharmacokinetic data on parent and metabolites of polyphenols (Neveu et al., 2010; Rothwell et al., 2013, 2015). Food processing can often constitute an important cause of losses in the polyphenol contents, whether caused by commonly applied thermal processes, by skin or seed removal, by metal or enzyme-catalyzed oxidation, or simply by leaching of these compounds into processing oil or water (Rickman et al., 2007). Thus, the Phenol-Explorer database has also recently enhanced its scope to also include some data on the possible effects of processing on phenolic contents of foods (Rothwell et al., 2013, 2015).

In addition to occurring in fruits and vegetables, phenolic compounds can also be found naturally, and occasionally in relatively high concentrations, in certain fermented beverages (like red wine and beer), some cereals (e.g., wheat, oat) and grains (like cocoa and coffee), certain spices (e.g., oregano, rosemary, cinnamon), and in many herbal infusion plant leaves (e.g., *Camellia sinensis*, *Peumus boldus*, *Haplopappus baylahuen*) (Speisky et al., 2006). Considering the broad number of polyphenol-rich foods we are exposed to, it is not surprising to find reports indicating that the average total intake of polyphenols in Western populations can reach as high as 2 g/day per person (Chun et al., 2007; Del Rio et al., 2013; Perez-Jimenez et al., 2011). However, it is generally believed that the daily intake of phenolic compounds range from 100 mg to 1 g, depending on the food ingested, on how rich in polyphenols these are, and on the frequency and portion of their consumption (Ovaskainen et al., 2008; Ross and Kasum, 2002; Saura-Calixto et al., 2007; Thu et al., 2004; Zamora-Ros et al., 2015). Hence, at its upper daily intake level, the consumption of polyphenols can be up to 10-fold higher than that of dietary vitamin C and the provitamin A carotenoids, over 20 times greater than dietary vitamin E, and near 100-fold higher than a variety of antioxidant terpenoids. Despite the considerable clinical evidence for a diversity of health benefits of some phenolics, no Dietary Reference Intake (DRI) currently exists for these compounds. Yet, during the last two decades, scientist and authorities of the United States and Canada have made important collaborative efforts in that direction (Atkinson, 2011). Unfortunately, however, it still appears that insufficient data are available as to warrant the inclusion of polyphenols in the Dietary Reference Intake guidance. The European Union, on the other hand, has also been taking steps to standardize dietary recommendations for polyphenols across Europe (Fairweather-Tait, 2011). As indicated by Williamson and Holst (2008): "we are heading in the right direction but bioavailability and dose–effects, including toxic levels, need to be established before DRIs can be considered." Certainly, to obtain such data, particularly through human studies, an enormous undertaking that would require significant public governmental funding would be needed. Compared to the vast epidemiological evidence supporting the contention of increasing the consumption of various polyphenol-rich foods as a health-promoting action, relatively less evidence exists for the health benefits of ingesting specific polyphenolic compounds. Future studies focusing on human clinical trials of selected polyphenols and their combinations should be carried out. In addition to assessing their eventual clinical efficacy, these compounds will also need to be investigated in terms of risk assessment, and their safety evaluation should include the observation of the possible undesirable associated effects (Bhullar and Rupasinghe, 2013). In terms of dietary polyphenols recommendations, based on the currently available information, it would seem that, so far, the best dietary advice remains to increase the consumption of these compounds in the form of a large variety of plant-based foods.

REFERENCES

Ainsworth, E.A. and Gillespie, K.M. 2007. Estimation of total phenolic content and other oxidation substrates in plant tissues using Folin-Ciocalteu reagent. *Nature Protocols* 2:875–877.

Ajila, C.M., Brar, S.K., Verma, M., Tyagi, R.D., Godbout, S., and Valero, J.R. 2011. Extraction and analysis of polyphenols: Recent trends. *Critical Reviews in Biotechnology* 31:227–249.

Amic, D., Davidovic-Amic, D., Beslo, D., Rastija, V., Lucic, B., and Trinajstic, N. 2007. SAR and QSAR of the antioxidant activity of flavonoids. *Current Medicinal Chemistry* 14:827–845.

Amorati, R. and Valgimigli, L. 2015. Advantages and limitations of common testing methods for antioxidants. *Free Radical Research* 49:633–649.

Andrae-Marobela, K., Ghislain, F.W., Okatch, H., and Majinda, R.R.T. 2013. Polyphenols: A diverse class of multi-target anti-HIV-1 agents. *Current Drug Metabolism* 14:392–413.

Apak, R., Guclu, K., Demirata, B., Ozyurek, M., Celik, S.E., Bektasoglu, B., Berker, K.I., and Ozyurt, D. 2007. Comparative evaluation of various total antioxidant capacity assays applied to phenolic compounds with the CUPRAC assay. *Molecules* 12:1496–1547.

Apak, R., Guclu, K., Ozyurek, M., and Karademir, S.E. 2004. Novel total antioxidant capacity index for dietary polyphenols and vitamins C and E, using their cupric ion reducing capability in the presence of neocuproine: CUPRAC method. *Journal of Agricultural and Food Chemistry* 52:7970–7981.

Arts, I.C.W. and Hollman, P.C.H. 2005. Polyphenols and disease risk in epidemiologic studies. *American Journal of Clinical Nutrition* 81:317S–325S.

Atkinson, S.A. 2011. Defining the process of dietary reference intakes: Framework for the United States and Canada. *American Journal of Clinical Nutrition* 94:655S–657S.

Basu, A., Betts, N.M., Leyva, M.J., Fu, D.X., Aston, C.E., and Lyons, T.J. 2015. Acute cocoa supplementation increases postprandial HDL cholesterol and insulin in obese adults with type 2 diabetes after consumption of a high-fat breakfast. *Journal of Nutrition* 145:2325–2332.

Beecher, G.R. 2003. Overview of dietary flavonoids: Nomenclature, occurrence and intake. *Journal of Nutrition* 133:3248S–3254S.

Benzie, I.F. and Choi, S.W. 2014. Antioxidants in food: Content, measurement, significance, action, cautions, caveats, and research needs. *Advances in Food and Nutrition Research* 71:1–53.

Bhagwat, S. and Haytowitz, D.B. 2015. USDA database for the flavonoid content of selected foods, release 3.2. U.S. Department of Agriculture, Agricultural Research Service, Washington, DC. Nutrient Data Laboratory Home Page: http://www.ars.usda.gov/nutrientdata/flav.

Bhagwat, S., Haytowitz, D.B., and Holden, J.M. (Ret.). 2013. USDA database for the flavonoid content of selected foods, release 3.1. U.S. Department of Agriculture, Agricultural Research Service, Washington, DC. Nutrient Data Laboratory Home Page: http://www.ars.usda.gov/nutrientdata/flav.

Bhullar, K.S. and Rupasinghe, H.P.V. 2013. Polyphenols: Multipotent therapeutic agents in neurodegenerative diseases. *Oxidative Medicine and Cellular Longevity* 2013:891748.

Bobo-Garcia, G., Davidov-Pardo, G., Arroqui, C., Virseda, P., Marin-Arroyo, M.R., and Navarro, M. 2015. Intra-laboratory validation of microplate methods for total phenolic content and antioxidant activity on polyphenolic extracts, and comparison with conventional spectrophotometric methods. *Journal of the Science of Food and Agriculture* 95:204–209.

Boeing, H., Bechthold, A., Bub, A. et al. 2012. Critical review: Vegetables and fruit in the prevention of chronic diseases. *European Journal of Nutrition* 51:637–663.

Bridi, R., Troncoso, M.J., Folch-Cano, C., Fuentes, J., Speisky, H., and Lopez-Alarcon, C. 2014. A polyvinylpolypyrrolidone (PVPP)-assisted folin-ciocalteu assay to assess total phenol content of commercial beverages. *Food Analytical Methods* 7:2075–2083.

Bulotta, S., Celano, M., Lepore, S.M., Montalcini, T., Pujia, A., and Russo, D. 2014. Beneficial effects of the olive oil phenolic components oleuropein and hydroxytyrosol: Focus on protection against cardiovascular and metabolic diseases. *Journal of Translational Medicine* 12:219.

Campos, C., Guzman, R., Lopez-Fernandez, E., and Casado, A. 2009. Evaluation of the copper(II) reduction assay using bathocuproinedisulfonic acid disodium salt for the total antioxidant capacity assessment: The CUPRAC-BCS assay. *Analytical Biochemistry* 392:37–44.

Cao, G.H., Alessio, H.M., and Cutler, R.G. 1993. Oxygen-radical absorbance capacity assay for antioxidants. *Free Radical Biology and Medicine* 14:303–311.

Carocho, M. and Ferreira, I.C.F.R. 2013. A review on antioxidants, prooxidants and related controversy: Natural and synthetic compounds, screening and analysis methodologies and future perspectives. *Food and Chemical Toxicology* 51:15–25.

Carrasco-Pozo, C., Gotteland, M., and Speisky, H. 2011a. Apple peel polyphenol extract protects against indomethacin-induced damage in Caco-2 cells by preventing mitochondrial complex I inhibition. *Journal of Agricultural and Food Chemistry* 59:11501–11508.

Carrasco-Pozo, C., Mizgier, M.L., Speisky, H., and Gotteland, M. 2012. Differential protective effects of quercetin, resveratrol, rutin and epigallocatechin gallate against mitochondrial dysfunction induced by indomethacin in Caco-2 cells. *Chemico-Biological Interactions* 195:199–205.

Carrasco-Pozo, C., Speisky, H., Brunser, O., Pastene, E., and Gotteland, M. 2011b. Apple peel polyphenols protect against gastrointestinal mucosa alterations induced by indomethacin in rats. *Journal of Agricultural and Food Chemistry* 59:6459–6466.

Cassidy, A., O'Reilly, E.J., Kay, C., Sampson, L., Franz, M., Forman, J.P., Curhan, G., and Rimm, E.B. 2011. Habitual intake of flavonoid subclasses and incident hypertension in adults. *American Journal of Clinical Nutrition* 93:338–347.

Celik, S.E., Ozyurek, M., Guclu, K., and Apak, R. 2010. Solvent effects on the antioxidant capacity of lipophilic and hydrophilic antioxidants measured by CUPRAC, ABTS/persulphate and FRAP methods. *Talanta* 81:1300–1309.

Chen, X., Mukwaya, E., Wong, M.S., and Zhang, Y. 2014. A systematic review on biological activities of prenylated flavonoids. *Pharmaceutical Biology* 52:655–660.

Chobot, V., Kubicova, L., Bachmann, G., and Hadacek, F. 2013. Versatile redox chemistry complicates antioxidant capacity assessment: Flavonoids as milieu-dependent anti- and pro-oxidants. *International Journal of Molecular Sciences* 14:11830–11841.

Choe, E. and Min, D.B. 2006. Chemistry and reactions of reactive oxygen species in foods. *Critical Reviews in Food Science and Nutrition* 46:1–22.

Chun, O.K., Chung, S.J., and Song, W.O. 2007. Estimated dietary flavonoid intake and major food sources of US adults. *Journal of Nutrition* 137:1244–1252.

Cook, N.C. and Samman, S. 1996. Flavonoids: Chemistry, metabolism, cardioprotective effects, and dietary sources. *Journal of Nutritional Biochemistry* 7:66–76.

Corcoran, M.P., Mckay, D.L., and Blumberg, J.B. 2012. Flavonoid basics: chemistry, sources, mechanisms of action, and safety. *Journal of Nutrition in Gerontology and Geriatrics* 31:176–189.

Croft, K.D. 1998. The chemistry and biological effects of flavonoids and phenolic acids. *Annals of the New York Academy of Sciences* 854:435–442.

Daglia, M. 2012. Polyphenols as antimicrobial agents. *Current Opinion in Biotechnology* 23:174–181.

Davinelli, S., Bertoglio, J.C., Zarrelli, A., Pina, R., and Scapagnini, G. 2015. A randomized clinical trial evaluating the efficacy of an anthocyanin-maqui berry extract (Delphinol (R)) on oxidative stress biomarkers. *Journal of the American College of Nutrition* 34:28–33.

Del Rio, D., Rodriguez-Mateos, A., Spencer, J.P.E., Tognolini, M., Borges, G., and Crozier, A. 2013. Dietary (poly)phenolics in human health: Structures, bioavailability, and evidence of protective effects against chronic diseases. *Antioxidants and Redox Signaling* 18:1818–1892.

Diaz-Garcia, M.C., Obon, J.M., Castellar, M.R., Collado, J., and Alacid, M. 2013. Quantification by UHPLC of total individual polyphenols in fruit juices. *Food Chemistry* 138:938–949.

Dinkova-Kostova, A.T. and Talalay, P. 2008. Direct and indirect antioxidant properties of inducers of cytoprotective proteins. *Molecular Nutrition and Food Research* 52:S128–S138.

Dorta, E., Fuentes-Lemus, E., Aspee, A., Atala, E., Speisky, H., Bridi, R., Lissi, E., and Lopez-Alarcon, C. 2015. The ORAC (oxygen radical absorbance capacity) index does not reflect the capacity of antioxidants to trap peroxyl radicals. *RSC Advances* 5:39899–39902.

Dower, J.I., Geleijnse, J.M., Gijsbers, L., Zock, P.L., Kromhout, D., and Hollman, P.C.H. 2015. Effects of the pure flavonoids epicatechin and quercetin on vascular function and cardiometabolic health: A randomized, double-blind, placebo-controlled, crossover trial. *American Journal of Clinical Nutrition* 101:914–921.

Duthie, G. and Crozier, A. 2000. Plant-derived phenolic antioxidants. *Current Opinion in Lipidology* 11:43–47.

Dyck, J.R.B. and Schrauwen, P. 2015. Resveratrol: Challenges in translating preclinical findings to improved patient outcomes. *Biochimica et Biophysica Acta* 1852:1069–1070.

Espinosa-Diez, C., Miguel, V., Mennerich, D., Kietzmann, T., Sánchez-Pérez, P., Cadenas, S., and Lamas, S. 2015. Antioxidant responses and cellular adjustments to oxidative stress. *Redox Biology* 6:183–197.

Fairweather-Tait, S.J. 2011. Contribution made by biomarkers of status to an FP6 Network of Excellence, EURopean micronutrient RECommendations Aligned (EURRECA). *American Journal of Clinical Nutrition* 94:651S–654S.

Folch-Cano, C., Olea-Azar, C., and Speisky, H. 2013. Structural and thermodynamic factors on the adsorption process of phenolic compounds onto polyvinylpolypyrrolidone. *Colloids and Surfaces A: Physicochemical and Engineering Aspect* 418:105–111.

Fraga, C.G., Oteiza, P.I., and Galleano, M. 2014. In vitro measurements and interpretation of total antioxidant capacity. *Biochimica et Biophysica Acta* 1840:931–934.

Galleano, M., Oteiza, P.I., and Fraga, C.G. 2009. Cocoa, chocolate, and cardiovascular disease. *Journal of Cardiovascular Pharmacology* 54:483–490.

Galleano, M., Verstraeten, S.V., Oteiza, P.I., and Fraga, C.G. 2010. Antioxidant actions of flavonoids: Thermodynamic and kinetic analysis. *Archives of Biochemistry and Biophysics* 501:23–30.

Ghiringhelli, F., Rebe, C., Hichami, A., and Delmas, D. 2012. Immunomodulation and anti-inflammatory roles of polyphenols as anticancer agents. *Anti-Cancer Agents in Medicinal Chemistry* 12:852–873.

Granados-Principal, S., Quiles, J.L., Ramirez-Tortosa, C.L., Sanchez-Rovira, P., and Ramirez-Tortosa, M.C. 2010. Hydroxytyrosol: From laboratory investigations to future clinical trials. *Nutrition Reviews* 68:191–206.

Grassi, D., Desideri, G., and Ferri, C. 2010. Flavonoids: Antioxidants against atherosclerosis. *Nutrients* 2:889–902.

Heleno, S.A., Martins, A., Queiroz, M.J.R.P., and Ferreira, I.C.F.R. 2015. Bioactivity of phenolic acids: Metabolites versus parent compounds: A review. *Food Chemistry* 173:501–513.

Henriquez, C., Lopez-Alarcon, C., Gomez, M., Lutz, M., and Speisky, H. 2011. Time-dependence of ferric reducing antioxidant power (FRAP) index in Chilean apples and berries. *Archivos Latinoamericanos de Nutrici6n* 61:323–332.

Herrera, E., Jimenez, R., Aruoma, O.I., Hercberg, S., Sanchez-Garcia, I., and Fraga, C. 2009. Aspects of antioxidant foods and supplements in health and disease. *Nutrition Reviews* 67:S140–S144.

Hooper, L., Kroon, P.A., Rimm, E.B., Cohn, J.S., Harvey, I., Le Cornu, K.A., Ryder, J.J., Hall, W.L., and Cassidy, A. 2008. Flavonoids, flavonoid-rich foods, and cardiovascular risk: A meta-analysis of randomized controlled trials. *American Journal of Clinical Nutrition* 88:38–50.

Huang, D.J., Ou, B.X., Hampsch-Woodill, M., Flanagan, J.A., and Deemer, E.K. 2002. Development and validation of oxygen radical absorbance capacity assay for lipophilic antioxidants using randomly methylated beta-cyclodextrin as the solubility enhancer. *Journal of Agricultural and Food Chemistry* 50:1815–1821.

Huang, D.J., Ou, B.X., and Prior, R.L. 2005. The chemistry behind antioxidant capacity assays. *Journal of Agricultural and Food Chemistry* 53:1841–1856.

Huxley, R.R. and Neil, H.A.W. 2003. The relation between dietary flavonol intake and coronary heart disease mortality: A meta-analysis of prospective cohort studies. *European Journal of Clinical Nutrition* 57:904–908.

Ignat, I., Volf, I., and Popa, V.I. 2011. A critical review of methods for characterisation of polyphenolic compounds in fruits and vegetables. *Food Chemistry* 126:1821–1835.

Jäger, A.K. and Saaby, L. 2011. Flavonoids and the CNS. *Molecules* 16:1471–1485.

Jain, A.K., Mehra, N,K., and Swarnakar, N.K. 2015. Role of antioxidants for the treatment of cardiovascular diseases: Challenges and opportunities. *Current Pharmaceutical Design* 21:4441–4455.

Ji, L.L., Sheng, Y.C., Zheng, Z.Y., Shi, L., and Wang, Z.T. 2015. The involvement of p62-Keapl-Nrf2 antioxidative signaling pathway and JNK in the protection of natural flavonoid quercetin against hepatotoxicity. *Free Radical Biology and Medicine* 85:12–23.

Kim, A., Chiu, A., Barone, M.K., Avino, D., Wang, F., Coleman, C.I., Phung, O.J. 2011. Green tea catechins decrease total and low-density lipoprotein cholesterol: A systematic review and meta-analysis. *Journal of the American Dietetic Association* 111:1720–1729.

Kim, E.Y., Pai, T.K., and Han, O. 2011. Effect of bioactive dietary polyphenols on zinc transport across the intestinal Caco-2 cell monolayers. *Journal of Agricultural and Food Chemistry* 59:3606–3612.

Kishimoto, Y., Tani, M., and Kondo, K. 2013. Pleiotropic preventive effects of dietary polyphenols in cardiovascular diseases. *European Journal of Clinical Nutrition* 67:532–535.

Kris-Etherton, P.M., Hecker, K.D., Bonanome, A., Coval, S.M., Binkoski, A.E., Hilpert, K.F., Griel, A.E., and Etherton, T.D. 2002. Bioactive compounds in foods: Their role in the prevention of cardiovascular disease and cancer. *American Journal of Medicine* 113:71–88.

Kühnau, J. 1976. The flavonoids. A class of semi-essential food components: Their role in human nutrition. *World Review of Nutrition and Dietetics* 24:117–191.

Kuno, T., Tsukamoto, T., Hara, A., and Tanaka, T. 2012. Cancer chemoprevention through the induction of apoptosis by natural compounds. *Journal of Biophysical Chemistry* 3:156–173.

Landete, J.M. 2012. Updated knowledge about polyphenols: Functions, bioavailability, metabolism, and health. *Critical Reviews in Food Science and Nutrition* 52:936–948.

Lichtenberg, D. and Pinchuk, I. 2015. Oxidative stress, the term and the concept. *Biochemical and Biophysical Research Communications* 461:441–444.

Liu, R.H. 2013. Dietary bioactive compounds and their health implications. *Journal of Food Science* 78:A18–A25.

Lopez-Alarcon, C. and Denicola, A. 2013. Evaluating the antioxidant capacity of natural products: A review on chemical and cellular-based assays. *Analytica Chimica Acta* 763:1–10.

Lopez-Alarcon, C. and Lissi, E. 2006. A novel and simple ORAC methodology based on the interaction of Pyrogallol Red with peroxyl radicals. *Free Radical Research* 40:979–985.

Lucic, B., Amic, D., and Trinajstic, N. 2008. Antioxidant QSAR modeling as exemplified on polyphenols. *Methods in Molecular Biology* 477:207–218.

Luo, W.P., Fang, Y.J., Lu, M.S., Zhong, X., Chen, Y.M., and Zhang, C.X. 2015. High consumption of vegetable and fruit colour groups is inversely associated with the risk of colorectal cancer: A case-control study. *British Journal of Nutrition* 113:1129–1138.

Lushchak, V.I. 2014. Free radicals, reactive oxygen species, oxidative stress and its classification. *Chemico-Biological Interactions* 224:164–175.

Ma, Q.Y., Kim, E.Y., Lindsay, E.A., and Han, O. 2011. Bioactive dietary polyphenols inhibit heme iron absorption in a dose-dependent manner in human intestinal Caco-2 cells. *Journal of Food Science* 76:H143–H150.

Magalhães, L.M., Santos, F., Segundo, M.A., Reis, S., and Lima, J.L.F.C. 2010. Rapid microplate high-throughput methodology for assessment of Folin-Ciocalteu reducing capacity. *Talanta* 83:441–447.

Magalhães, L.M., Segundo, M.A., Reis, S., and Lima, J.L.F.C. 2008. Methodological aspects about in vitro evaluation of antioxidant properties. *Analytica Chimica Acta* 613:1–19.

Magrone, T. and Jirillo, E. 2012. Influence of polyphenols on allergic immune reactions: Mechanisms of action. *Proceedings of the Nutrition Society* 71:316–321.

Mann, G.E., Bonacasa, B., Ishii, T., and Siow, R.C.M. 2009. Targeting the redox sensitive Nrf2-Keapl defense pathway in cardiovascular disease: Protection afforded by dietary isoflavones. *Current Opinion in Pharmacology* 9:139–145.

Margraf, T., Karnopp, A.R., Rosso, N.D., and Granato, D. 2015. Comparison between Folin-Ciocalteu and Prussian blue assays to estimate the total phenolic content of juices and teas using 96-well microplates. *Journal of Food Science* 80(11):C2397–C2403.

Mates, J.M., Segura, J.A., Alonso, F.J., and Marquez, J. 2011. Anticancer antioxidant regulatory functions of phytochemicals. *Current Medicinal Chemistry* 18:2315–2338.

McAnulty, L.S., Collier, S.R., Landram, M.J., Whittaker, D.S., Isaacs, S.E., Klemka, J.M., Cheek, S.L., Arms, J.C., and McAnulty, S.R. 2014. Six weeks daily ingestion of whole blueberry powder increases natural killer cell counts and reduces arterial stiffness in sedentary males and females. *Nutrition Research* 34:577–584.

McCullough, M.L., Peterson, J.J., Patel, R., Jacques, P.F., Shah, R., and Dwyer, J.T. 2012. Flavonoid intake and cardiovascular disease mortality in a prospective cohort of US adults. *American Journal of Clinical Nutrition* 95:454–464.

Medina, M.B. 2011. Simple and rapid method for the analysis of phenolic compounds in beverages and grains. *Journal of Agricultural and Food Chemistry* 59:1565–1571.

Medina-Remon, A., Barrionuevo-Gonzalez, A., Zamora-Ros, R., Andres-Lacueva, C., Estruch, R., Martinez-Gonzalez, M.A., Diez-Espino, J., and Lamuela-Raventos, R.M. 2009. Rapid Folin-Ciocalteu method using microtiter 96-well plate cartridges for solid phase extraction to assess urinary total phenolic compounds, as a biomarker of total polyphenols intake. *Analytica Chimica Acta* 634:54–60.

Mierziak, J., Kostyn, K., and Kulma, A. 2014. Flavonoids as important molecules of plant interactions with the environment. *Molecules* 19:16240–16265.

Milaeva, E.R. 2011. Metal-based antioxidants: Potential therapeutic candidates for prevention the oxidative stress-related carcinogenesis: Mini-review. *Current Topics in Medicinal Chemistry* 11:2703–2713.

Mladenka, P., Zatloukalova, L., Filipsky, T., and Hrdina, R. 2010. Cardiovascular effects of flavonoids are not caused only by direct antioxidant activity. *Free Radical Biology and Medicine* 49:963–975.

Mojzis, J., Varinska, L., Mojzisova, G., Kostova, I., and Mirossay, L. 2008. Antiangiogenic effects of flavonoids and chalcones. *Pharmacological Research* 57:259–265.

Montezano, A.C., Dulak-Lis, M., Tsiropoulou, S., Harvey, A., Briones, A.M., and Touyz, R.M.2015. Oxidative stress and human hypertension: Vascular mechanisms, biomarkers, and novel therapies. *Canadian Journal of Cardiology* 31:631–641.

Moon, J.K. and Shibamoto, T. 2009. Antioxidant assays for plant and food components. *Journal of Agricultural and Food Chemistry* 57:1655–1666.

Motilva, M.J., Serra, A., and Macia, A. 2013. Analysis of food polyphenols by ultra high-performance liquid chromatography coupled to mass spectrometry: An overview. *Journal of Chromatography A* 1292:66–82.

Munir, K.M., Chandrasekaran, S., Gao, F., and Quon, M.J. 2013. Mechanisms for food polyphenols to ameliorate insulin resistance and endothelial dysfunction: Therapeutic implications for diabetes and its cardiovascular complications. *American Journal of Physiology: Endocrinology and Metabolism* 305:E679–E686.

Na, H.K. and Surh, Y.J. 2008. Modulation of Nrf2-mediated antioxidant and detoxifying enzyme induction by the green tea polyphenol EGCG. *Food and Chemical Toxicology* 46:1271–1278.

Nagle, C.M., Wilson, L.F., Hughes, M.C.B., Ibiebele, T.I., Miura, K., Bain, C.J., Whiteman, D.C., and Webb, P.M. 2015. Cancers in Australia in 2010 attributable to inadequate consumption of fruit, non-starchy vegetables and dietary fibre. *Australian and New Zealand Journal of Public Health* 39:422–428.

Naidu, S.D., Kostov, R.V., and Dinkova-Kostova, A.T. 2015. Transcription factors Hsf1 and Nrf2 engage in crosstalk for cytoprotection. *Trends in Pharmacological Sciences* 36:6–14.

Natale, A., Nardiello, D., Palermo, C., Muscarella, M., Quinto, M., and Centonze, D. 2015. Development of an analytical method for the determination of polyphenolic compounds in vegetable origin samples by liquid chromatography and pulsed amperometric detection at a glassy carbon electrode. *Journal of Chromatography A* 1420:66–73.

Neveu, V., Perez-Jimenez, J., Vos, F. et al. 2010. Phenol-explorer: An online comprehensive database on polyphenol contents in foods. *Database (Oxford)* 2010:bap024.

Nutrient Data Laboratory. 2010. USDA database for the oxygen radical absorbance capacity (ORAC) of selected foods, release 2. U.S. Department of Agriculture, Agricultural Research Service, Washington, DC. http://www.ars.usda.gov/Services/docs.htm?docid=1586 (accessed December 15, 2011).

Ortiz, R., Antilen, M., Speisky, H., Aliaga, M.E., Lopez-Alarcon, C., and Baugh, S. 2012. Application of a microplate-based ORAC-pyrogallol red assay for the estimation of antioxidant capacity: First action 2012.03. *Journal of AOAC International* 95:1558–1561.

Ostertag, L.M., O'Kennedy, N., Kroon, P.A., Duthie, G.G., and de Roos, B. 2010. Impact of dietary polyphenols on human platelet function: A critical review of controlled dietary intervention studies. *Molecular Nutrition and Food Research* 54:60–81.

Ou, B.X., Chang, T., Huang, D.J., and Prior, R.L. 2013. Determination of total antioxidant capacity by oxygen radical absorbance capacity (ORAC) using fluorescein as the fluorescence probe: First action 2012.23. *Journal of AOAC International* 96:1372–1376.

Ou, B.X., Hampsch-Woodill, M., and Prior, R.L. 2001. Development and validation of an improved oxygen radical absorbance capacity assay using fluorescein as the fluorescent probe. *Journal of Agricultural and Food Chemistry* 49:4619–4626.

Ovaskainen, M.L., Torronen, R., Koponen, J.M., Sinkko, H., Hellstrom, J., Reinivuo, H., and Mattila, P. 2008. Dietary intake and major food sources of polyphenols in Finnish adults. *Journal of Nutrition* 138:562–566.

Pastene, E., Speisky, F., Garcia, A., Moreno, J., Troncoso, M., and Figueroa, G. 2010. In vitro and in vivo effects of apple peel polyphenols against *Helicobacter pylori*. *Journal of Agricultural and Food Chemistry* 58:7172–7179.

Pastene, E., Speisky, H., Troncoso, M., Alarcon, J., and Figueroa, G. 2009a. In vitro inhibitory effect of apple peel extract on the growth of helicobacter pylori and respiratory burst induced on human neutrophils. *Journal of Agricultural and Food Chemistry* 57:7743–7749.

Pastene, E., Troncoso, M., Figueroa, G., Alarcon, J., and Speisky, H. 2009b. Association between polymerization degree of apple peel polyphenols and inhibition of helicobacter pylori urease. *Journal of Agricultural and Food Chemistry* 57:416–424.

Pellegrini, N., Serafini, M., Colombi, B., Del Rio, D., Salvatore, S., Bianchi, M., and Brighenti, F. 2003. Total antioxidant capacity of plant foods, beverages and oils consumed in Italy assessed by three different in vitro assays. *Journal of Nutrition* 133:2812–2819.

Perez-Jimenez, J., Arranz, S., Tabernero, M., Diaz-Rubio, M.E., Serrano, J., Goni, I., and Saura-Calixto, F. 2008. Updated methodology to determine antioxidant capacity in plant foods, oils and beverages: Extraction, measurement and expression of results. *Food Research International* 41:274–285.

Perez-Jimenez, J., Fezeu, L., Touvier, M., Arnault, N., Manach, C., Hercberg, S., Galan, P., and Scalbert, A. 2011. Dietary intake of 337 polyphenols in French adults. *American Journal of Clinical Nutrition* 93:1220–1228.

Perez-Jimenez, J., Neveu, V., Vos, F., and Scalbert, A. 2010a. Identification of the 100 richest dietary sources of polyphenols: An application of the Phenol-Explorer database. *European Journal of Clinical Nutrition* 64:S112–S120.

Perez-Jimenez, J., Neveu, V., Vos, F., and Scalbert, A. 2010b. Systematic analysis of the content of 502 polyphenols in 452 foods and beverages: An application of the phenol-explorer database. *Journal of Agricultural and Food Chemistry* 58:4959–4969.

Perez-Vizcaino, F. and Duarte, J. 2010. Flavonols and cardiovascular disease. *Molecular Aspects of Medicine* 31:478–494.

Persson, T., Popescu, B.O., and Cedazo-Minguez, A. 2014. Oxidative stress in Alzheimer's disease: Why did antioxidant therapy fail? *Oxidative Medicine and Cellular Longevity* 2014:427318.

Peterson, J.J., Dwyer, J.T., Jacques, P.F., and McCullough, M.L. 2012. Associations between flavonoids and cardiovascular disease incidence or mortality in European and US populations. *Nutrition Reviews* 70:491–508.

Pisoschi, A.M. and Pop, A. 2015. The role of antioxidants in the chemistry of oxidative stress: A review. *European Journal of Medicinal Chemistry* 96:55–74.

Ponzo, V., Goitre, I., Fadda, M., Gambino, R., De Francesco, A., Soldati, L., Gentile, L., Magistroni, P., Cassader, M., and Bo, S. 2015. Dietary flavonoid intake and cardiovascular risk: A population-based cohort study. *Journal of Translational Medicine* 13:218.

Prior, R.L. Antioxidant food databases? Valuable or not? Brunswick Laboratories website. http://www.brunswicklabs.com/Portals/153979/docs/A Response to the USDA ORAC Statement.pdf (accessed October 16, 2015).

Prior, R.L., Hoang, H., Gu, L.W., Wu, X.L., Bacchiocca, M., Howard, L., Hampsch-Woodill, M., Huang, D.J., Ou, B.X., and Jacob, R. 2003. Assays for hydrophilic and lipophilic antioxidant capacity (oxygen radical absorbance capacity [ORAC(FL)] of plasma and other biological and food samples. *Journal of Agricultural and Food Chemistry* 51:3273–3279.

Prior, R.L., Wu, X.L., and Schaich, K. 2005. Standardized methods for the determination of antioxidant capacity and phenolics in foods and dietary supplements. *Journal of Agricultural and Food Chemistry* 53:4290–4302.

Proteggente, A.R., Pannala, A.S., Paganga, G., Van Buren, L., Wagner, E., Wiseman, S., Van De Put, F., Dacombe, C., and Rice-Evans, C.A. 2002. The antioxidant activity of regularly consumed fruit and vegetables reflects their phenolic and vitamin C composition. *Free Radical Research* 36:217–233.

Queiros, R.B., Tafulo, P.A.R., and Sales, M.G.F. 2013. Assessing and comparing the total antioxidant capacity of commercial beverages: Application to beers, wines, waters and soft drinks using TRAP, TEAC and FRAP methods. *Combinatorial Chemistry and High Throughput Screening* 16:22–31.

Quinones, M., Miguel, M., and Aleixandre, A. 2013. Beneficial effects of polyphenols on cardiovascular disease. *Pharmacological Research* 68:125–131.

Rangel-Huerta, O.D., Pastor-Villaescusa, B., Aguilera, C.M., and Gil, A. 2015. A Systematic review of the efficacy of bioactive compounds in cardiovascular disease: Phenolic compounds. *Nutrients* 7:5177–5216.

RiceEvans, C.A., Miller, N.J., and Paganga, G. 1996. Structure-antioxidant activity relationships of flavonoids and phenolic acids. *Free Radical Biology and Medicine* 20:933–956.

Rickman, J.C., Barrett, D.M., Bruhn, CM. 2007. Nutritional comparison of fresh, frozen and canned fruits and vegetables. Part 1. Vitamins C and B and phenolic compounds. *Journal of the Science of Food and Agriculture* 87:930–944.

Rodriguez-Casado, A. 2016. The health potential of fruits and vegetables phytochemicals: Notable examples. *Critical Reviews in Food Science and Nutrition* 56(7):1097–107.

Rodriguez-Mateos, A., Vauzour, D., Krueger, C.G., Shanmuganayagam, D., Reed, J., Calani, L., Mena, P., Del Rio, D., and Crozier, A. 2014. Bioavailability, bioactivity and impact on health of dietary flavonoids and related compounds: An update. *Archives of Toxicology* 88:1803–1853.

Ross, J.A. and Kasum, C.M. 2002. Dietary flavonoids: Bioavailability, metabolic effects, and safety. *Annual Review of Nutrition* 22:19–34.

Rossi, M., Bosetti, C., Negri, E., Lagiou, P., and La Vecchia, C. 2010. Flavonoids, proanthocyanidins, and cancer risk: A network of case-control studies from Italy. *Nutrition and Cancer: An International Journal* 62:871–877.

Rothwell, J.A., Medina-Remon, A., Perez-Jimenez, J., Neveu, V., Knaze, V., Slimani, N., and Scalbert, A. 2015. Effects of food processing on polyphenol contents: A systematic analysis using Phenol-Explorer data. *Molecular Nutrition and Food Research* 59:160–170.

Rothwell, J.A., Perez-Jimenez, J., Neveu, V. et al. 2013. Phenol-Explorer 3.0: A major update of the Phenol-Explorer database to incorporate data on the effects of food processing on polyphenol content. *Database (Oxford)* 2013:bat070.

Saibabu, V., Fatima, Z., Khan, L., and Hameed, S. 2015. Therapeutic potential of dietary phenolic acids. *Advances in Pharmacological Sciences* 2015:823539.

Sandoval-Acuña, C., Ferreira, J., and Speisky, H. 2014. Polyphenols and mitochondria: An update on their increasingly emerging ROS-scavenging independent actions. *Archives of Biochemistry and Biophysics* 559:75–90.

Sansone, R., Rodriguez-Mateos, A., Heuel, J. et al. 2015. Cocoa flavanol intake improves endothelial function and Framingham Risk Score in healthy men and women: A randomised, controlled, double-masked trial: The Flaviola Health Study. *British Journal of Nutrition* 114:1246–1255.

Santhakumar, B., Bulmer, C., and Singh, I. 2014. A review of the mechanisms and effectiveness of dietary polyphenols in reducing oxidative stress and thrombotic risk. *Journal of Nutrition and Dietetics* 27:1–21.

Sarriá, B., Martinez-Lopez, S., Sierra-Cinos, J.L., Garcia-Diaz, L., Goya, L., Mateos, R., and Bravo, L. 2015. Effects of bioactive constituents in functional cocoa products on cardiovascular health in humans. *Food Chemistry* 174:214–218.

Saura-Calixto, F., Serrano, J., and Goni, I. 2007. Intake and bioaccessibility of total polyphenols in a whole diet. *Food Chemistry* 101:492–501.

Scalbert, A., Manach, C., Morand, C., Remesy, C., and Jimenez, L. 2005. Dietary polyphenols and the prevention of diseases. *Critical Reviews in Food Science and Nutrition* 45:287–306.

Scalbert, A. and Williamson, G. 2000. Dietary intake and bioavailability of polyphenols. *Journal of Nutrition* 30:2073S–2085S.

Schewe, T., Steffen, Y., and Sies, H. 2008. How do dietary flavanols improve vascular function? A position paper. *Archives of Biochemistry and Biophysics* 476:102–106.

Schnorr, O., Brossette, T., Momma, T.Y., Kleinbongard, P., Keen, C.L., Schroeter, H., and Sies, H. 2008. Cocoa flavanols lower vascular arginase activity in human endothelial cells in vitro and in erythrocytes in vivo. *Archives of Biochemistry and Biophysics* 476:211–215.

Seifried, H.E., Anderson, D.E., Fisher, E.I., and Milner, J.A. 2007. A review of the interaction among dietary antioxidants and reactive oxygen species. *Journal of Nutritional Biochemistry* 18:567–579.

Sies, H. 1997. Oxidative stress: Oxidants and antioxidants. *Experimental Physiology* 82:291–295.

Sies, H. 2015. Oxidative stress: A concept in redox biology and medicine. *Redox Biology* 4:180–183.

Singleton, V.L. and Rossi, J.A. 1965. Colorimetry of total phenolics with phosphomolybdic-phosphotungstic acid reagents. *American Journal of Enology and Viticulture* 16:144–158.

Speisky, H., Lopez-Alarcon, C., Gomez, M., Fuentes, J., and Sandoval-Acuña, C. 2012. First web-based database on total phenolics and oxygen radical absorbance capacity (ORAC) of fruits produced and consumed within the south Andes region of South America. *Journal of Agricultural and Food Chemistry* 60:8851–8859.

Speisky, H., Rocco, C., Carrasco, C., Lissi, E.A., and Lopez-Alarcon, C. 2006. Antioxidant screening of medicinal herbal teas. *Phytotherapy Research* 20:462–467.

Stepanic, V., Gasparovic, A.C., Troselj, K.G., Amic, D., and Zarkovic, N. 2015. Selected attributes of polyphenols in targeting oxidative stress in cancer. *Current Topics in Medicinal Chemistry* 15:496–509.

Strack, D. 1997. Phenolic metabolism. In *Plant Biochemistry*, eds. P. Dey and J. Harborne, pp. 387–416. San Diego, CA: Academic Press.

Strack, D. and Wray, V. 1992. Anthocyanins. In *The Flavonoids: Advances in Research Since 1986*, ed. J.B. Harborne, pp. 1–22. London, UK: Chapman & Hall.

Sureda, A., Tejada, S., Bibiloni, M.M., Tur, J.A., and Pons, A. 2014. Polyphenols: Well beyond the antioxidant capacity: Polyphenol supplementation and exercise-induced oxidative stress and inflammation. *Current Pharmaceutical Biotechnology* 15:373–379.

Tangney, C.C. and Rasmussen, H.E. 2013. Polyphenols, inflammation, and cardiovascular disease. *Current Atherosclerosis Reports* 15:324.

Tebay, L.E., Robertson, H., Durant, S.T., Vitale, S.R., Penning, T.M., Dinkova-Kostova, A.T., and Hayes, J.D. 2015. Mechanisms of activation of the transcription factor Nrf2 by redox stressors, nutrient cues, and energy status and the pathways through which it attenuates degenerative disease. *Free Radical Biology and Medicine* 88(Pt B):108–146.

Thu, N.N., Sakurai, C., Uto, H., Van Chuyen, N., Lien, D.T.K., Yamamoto, S., Ohmori, R., and Kondo, K. 2004. The polyphenol content and antioxidant activities of the main edible vegetables in northern Vietnam. *Journal of Nutritional Science and Vitaminology* 50:203–210.

Turati, F., Rossi, M., Pelucchi, C., Levi, F., and La Vecchia, C. 2015. Fruit and vegetables and cancer risk: A review of southern European studies. *British Journal of Nutrition* 113:S102–S110.

Valko, M., Rhodes, C.J., Moncol, J., Izakovic, M., and Mazur, M. 2006. Free radicals, metals and antioxidants in oxidative stress-induced cancer. *Chemico-Biological Interactions* 160:1–40.

Vauzour, D., Rodriguez-Mateos, A., Corona, G., Oruna-Concha, M.J., and Spencer, J.P.E. 2010. Polyphenols and human health: Prevention of disease and mechanisms of action. *Nutrients* 2:1106–1131.

Vieira, A.R., Abar, L., Vingeliene, S., Chan, D.S., Aune, D., Navarro-Rosenblatt, D., Stevens, C., Greenwood, D., and Norat, T. 2016. Fruits, vegetables and lung cancer risk: A systematic review and meta-analysis. *Annals of Oncology* 27(1):81–96.

Virgili, F. and Marino, M. 2008. Regulation of cellular signals from nutritional molecules: A specific role for phytochemicals, beyond antioxidant activity. *Free Radical Biology and Medicine* 45:1205–1216.

Visioli, F., De La Lastra, C.A., Andres-Lacueva, C. et al. 2011. Polyphenols and human health: A prospectus. *Critical Reviews in Food Science and Nutrition* 51:524–546.

Vitale, D.C., Piazza, C., Melilli, B., Drago, F., and Salomone, S. 2013. Isoflavones: Estrogenic activity, biological effect and bioavailability. *European Journal of Drug Metabolism* 38:15–25.

Wang, H., Cao, G.H., and Prior, R.L. 1996. Total antioxidant capacity of fruits. *Journal of Agricultural and Food Chemistry* 44:701–705.

Wang, L.S. and Stoner, G.D. 2008. Anthocyanins and their role in cancer prevention. *Cancer Letters* 269:281–290.

Wang, L.S. and Yang, Z.J. 2012. Flavonols intake and the risk of coronary heart disease: A meta-analysis of cohort studies. *Atherosclerosis* 222:270–273.

Wang, M., Qin, S., Zhang, T., Song, X., and Zhang, S. 2015. The effect of fruit and vegetable intake on the development of lung cancer: A meta-analysis of 32 publications and 20414 cases. *European Journal of Clinical Nutrition* 69(11):1184–1192.

Wang, X., Ouyang, Y.Y., Liu, J., and Zhao, G. 2014. Flavonoid intake and risk of CVD: A systematic review and meta-analysis of prospective cohort studies. *British Journal of Nutrition* 111:1–11.

Wang, Z.M., Nie, Z.L., Zhou, B., Lian, X.Q., Zhao, H., Gao, W., Wang, Y.S., Jia, E.Z., Wang, L.S., and Yang, Z.J. 2012. Flavonols intake and the risk of coronary heart disease: A meta-analysis of cohort studies. *Atherosclerosis* 222:270–273.

Wasson, A.P., Pellerone, F.I., and Mathesius, U. 2006. Silencing the flavonoid pathway in *Medicago truncatula* inhibits root nodule formation and prevents auxin transport regulation by rhizobia. *Plant Cell* 18:1617–1629.

Wayner, D.D., Burton, G.W., Ingold, K.U., Barclay, L.R., and Locke, S.J. 1987. The relative contributions of vitamin E, urate, ascorbate and proteins to the total peroxyl radical-trapping antioxidant activity of human blood plasma. *Biochimica et Biophysica Acta* 924:408–419.

Wayner, D.D., Burton, G.W., Ingold, K.U., and Locke, S. 1985. Quantitative measurement of the total, peroxyl radical-trapping antioxidant capability of human blood plasma by controlled peroxidation. The important contribution made by plasma proteins. *FEBS Letters* 187:33–37.

Williamson, G. 2013. Possible effects of dietary polyphenols on sugar absorption and digestion. *Molecular Nutrition and Food Research* 57:48–57.

Williamson, G. and Holst, B. 2008. Dietary reference intake (DRI) value for dietary polyphenols: Are we heading in the right direction? *British Journal of Nutrition* 99:S55–S58.

Wright, B., Spencer, J.P.E., Lovegrove, J.A., and Gibbins, J.M. 2013. Insights into dietary flavonoids as molecular templates for the design of anti-platelet drugs. *Cardiovascular Research* 97:13–22.

Xie, Y.X., Yang, W.J., Tang, F., Chen, X.Q., and Ren, L.C, 2015. Antibacterial activities of flavonoids: Structure-activity relationship and mechanism. *Current Medicinal Chemistry* 22:132–149.

Yang, C.S., Li, G.X., Yang, Z.H., Guan, F., Chen, A., and Ju, J. 2013. Cancer prevention by tocopherols and tea polyphenols. *Cancer Letters* 334:79–85.

Yang, J.H., Shin, B.Y., Han, J.Y., Kim, M.G., Wi, J.E., Kim, Y.W., Cho, I.J., Kim, S.C., Shin, S.M., and Ki, S.H. 2014. Isorhamnetin protects against oxidative stress by activating Nrf2 and inducing the expression of its target genes. *Toxicology and Applied Pharmacology* 274:293–301.

Yang, Y.C., Lii, C.K., Lin, A.H., Yeh, Y.W., Yao, H.T., Li, C.C., Liu, K.L., and Chen, H.W. 2011. Induction of glutathione synthesis and heme oxygenase 1 by the flavonoids butein and phloretin is mediated through the ERK/Nrf2 pathway and protects against oxidative stress. *Free Radical Biology and Medicine* 51:2073–2081.

Yao, L.H., Jiang, Y.M., Shi, J., Tomas-Barberan, F.A., Datta, N., Singanusong, R., and Chen, S.S. 2004. Flavonoids in food and their health benefits. *Plant Foods for Human Nutrition* 59:113–122.

Zamora-Ros, R., Knaze, V., Rothwell, J. et al. 2015. Dietary polyphenol intake in Europe: The European Prospective Investigation into Cancer and Nutrition (EPIC) study. *European Journal of Nutrition*. [Epub ahead of print]. DOI 10.1007/s00394-015-0950-x.

Zamora-Ros, R., Touillaud, M., Rothwell, J.A., Romieu, I., and Scalbert, A. 2014. Measuring exposure to the polyphenol metabolome in observational epidemiologic studies: Current tools and applications and their limits. *American Journal of Clinical Nutrition* 100:11–26.

Zhang, H., Yu, D.D., Sun, J., Liu, X.T., Jiang, L., Guo, H.Y., and Ren, F.Z. 2014. Interaction of plant phenols with food macronutrients: Characterisation and nutritional-physiological consequences. *Nutrition Research Reviews* 27:1–15.

2 Key Aspects of Polyphenols and Health

Metabolic Fate, Mechanisms of Action, and Influence on Gut Microbiota

Josep Lluís Torres, Sara Ramos-Romero, and Jara Pérez-Jiménez

CONTENTS

2.1 INTRODUCTION

It is generally accepted by the scientific community that polyphenols are beneficial for humans. They are mainly regarded as life-extending antioxidants because they efficiently scavenge oxidant species, notably free radicals. This assumption is based on the free radical theory of aging proposed by Denham Harman in his 1956 paper (Harman, 1956). He wrote: "Aging and the degenerative diseases associated with it are attributed basically to the deleterious side attacks of free radicals on cell constituents and on the connective tissue." Free radicals and other oxidative species such as hydrogen peroxide trigger oxidation of biomolecules such as DNA, proteins, and lipids; Harman suggested later that "dietary manipulations [are] expected to lower the rate of production of free radical reaction damage" (Harman, 1981). To Harman, these manipulations would essentially consist of reducing intake, but to many others, it followed logically from his theory that any agent with a capacity to scavenge free radicals would, in principle, prevent degenerative diseases and delay aging through an "antioxidant" action. Naturally occurring phenolic compounds arose as one of the main candidates for antioxidants. Already in 1936, Rusznyák and Szent-Györgyi had described flavonols as active health-promoting agents in the absence of ascorbic acid (vitamin C) and called them vitamin P (Rusznyák and Szent-Györgyi, 1936). Later, the association of the free radical scavenging activity of flavonols and other phenolics with the promotion of health aroused scientific interest for polyphenols as chemopreventive agents (Herrmann, 1973; Huang et al., 1992). This has now expanded from the scientific community to the general population.

More recently, nonredox polyphenol–protein interactions (e.g., enzyme–substrate, receptor–ligand) have been recognized as physiologically relevant. Actually, before polymeric phenolics were associated with free radical scavenging and the promotion of health, their capacity to interact with collagen was long known and applied to the manufacturing of leather (Gustavson, 1949). Polyphenols have been shown to interact with many proteins other than collagen, including salivary proline-rich proteins, fibronectin, and fibrinogen (de Freitas and Mateus, 2001; Hung et al., 2005). It has even been suggested that these polyphenol/protein interactions may be more directly related to disease prevention than the redox behavior itself (Sang et al., 2005).

Polyphenols evolved in plants to serve a variety of functions that include photoprotection, oxidative-stress regulation, biological communication with the microorganisms of the rhizosphere, developmental regulation, and defense against parasites (Mouradov and Spangenberg, 2014). Given their protein-binding action, they were first believed to be "quantitative" defenses naturally selected to impair active proteins in a rather unspecific fashion. A closer look at the extraordinary structural variety of the thousands of tannin molecules suggests more selective "qualitative" interactions that would be the product of an evolutionary arms race between plants and parasites (Zucker, 1983). This structurally diverse collection of putatively active ligands and substrates appears to help maintain the homeostasis of animals that feed on plants.

The health-promoting effects of polyphenols have been extensively described in the scientific literature (Crozier et al., 2009; Khurana et al., 2013; Okuda, 2005; Scalbert et al., 2005; Singh et al., 2011) and partially backed by clinical trials, mainly centered on risk factors of cardiovascular diseases (CVDs) (Hooper et al., 2008; Taubert et al., 2007). Nevertheless, the efficacy of polyphenols in humans has been contested. Main concerns about the activity of dietary polyphenols include the poor absorption of oligomers and the fast metabolization of the absorbed fraction. The latter results in low bioavailability of the native structures, lack of activity of the metabolites, and an inability to keep the redox homeostasis of a healthy organism or overcoming the excess of free radicals, when the defense mechanisms are seriously compromised (Gutteridge and Halliwell, 2010; Halliwell et al., 2005; Spencer et al., 2001). The picture that seems to emerge from this controversial information is that polyphenols exert a beneficial influence on human homeostasis which is difficult to explain and substantiate.

The main applied interest of polyphenols appears to be their use as dietary ingredients or functional food components. Some health claims involving phenolics have received favorable opinions by the European Food Safety Authority (EFSA), namely, protection of LDL particles from oxidative stress by polyphenols in olive oil standardized by the content of hydroxytyrosol and its derivatives (e.g., oleuropein complex) (EFSA Panel on Dietetic Products, 2011) and maintenance of normal endothelium-dependent vasodilation by cocoa flavanols (EFSA Panel on Dietetic Products, 2012, 2014). In both cases, in accordance with the earlier discussion, the experts recognize a cause and effect relationship between the consumption of polyphenols and maintenance of a normal physiological function, but they point out that the mechanism(s) of action are unknown. In view of the open controversy over the functionality of plant polyphenols, this chapter will critically address some key aspects of the cross talk between these ubiquitous plant secondary metabolites and the mammalian organism. The efficacy of polyphenols *in vivo* and their possible mechanisms of action will also be discussed.

To illustrate the variety of possible interactions of polyphenols with the living organism, Figure 2.1 highlights the structural features that make them such a versatile family of natural products. Polyphenols are amphiphilic structures that are able to (1) establish hydrophilic and hydrophobic interactions, (2) chelate metal ions, (3) absorb UV light, and (4) donate hydrogen atoms and/or transfer electrons. Their putative biological activities are connected to one or more of these chemical and physicochemical properties.

2.2 ABSORPTION AND METABOLIC FATE

As polyphenols are metabolized fast and extensively, their biological activity may be attributed, at least in part, to their metabolites. A general overview of the metabolic fate of polyphenols is shown in Figure 2.2 (Cardona et al., 2013; Monagas et al., 2010).

2.2.1 Metabolic Transformations

Different transformations can take place in different parts of the organism.

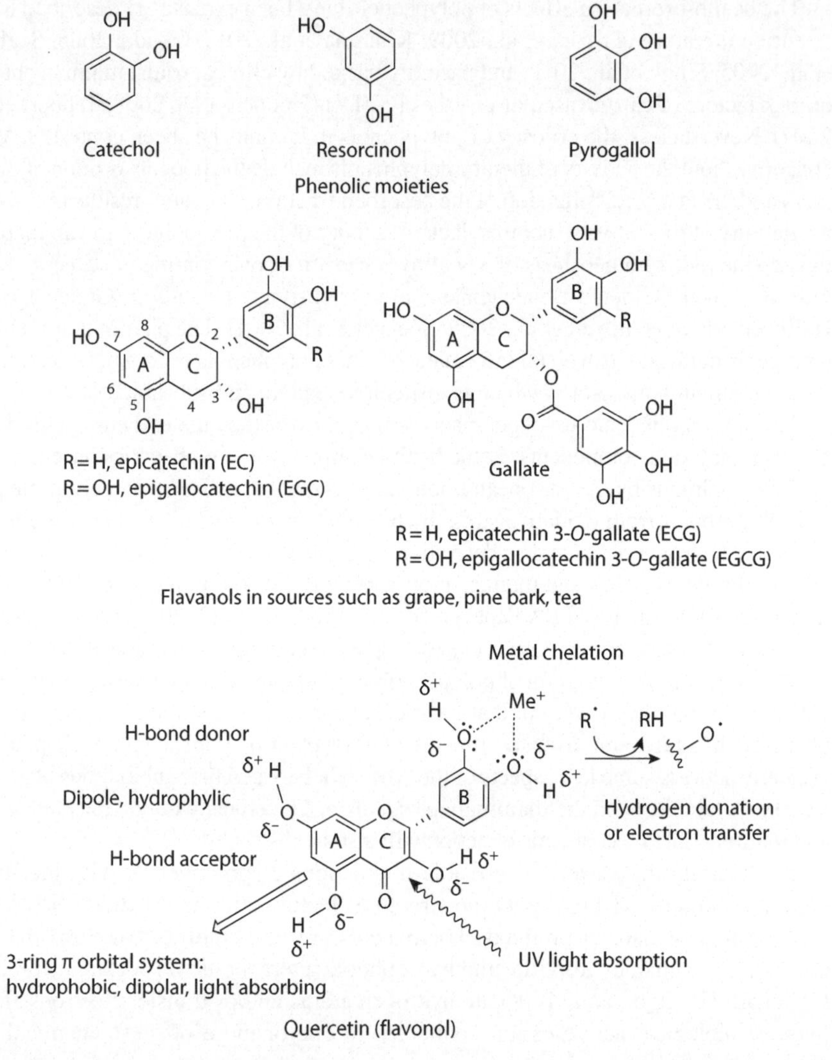

FIGURE 2.1 Representative polyphenol structures and their chemical and physicochemical properties.

2.2.1.1 The Oral Cavity and Stomach

Some modifications take place in these organs, traditionally considered as minor sites of polyphenol metabolization. Esterases present in the oral cavity may release galloyl substituents from proanthocyanidins (Yang et al., 1999). The interactions of proanthocyanidins with salivary proteins have been widely reported but little is yet known about how this may affect their bioavailability (Ferruzzi et al., 2012). Later, the acidic environment in the stomach may partially hydrolyze some of these proanthocyanidins (Spencer et al., 2000). A small fraction of some

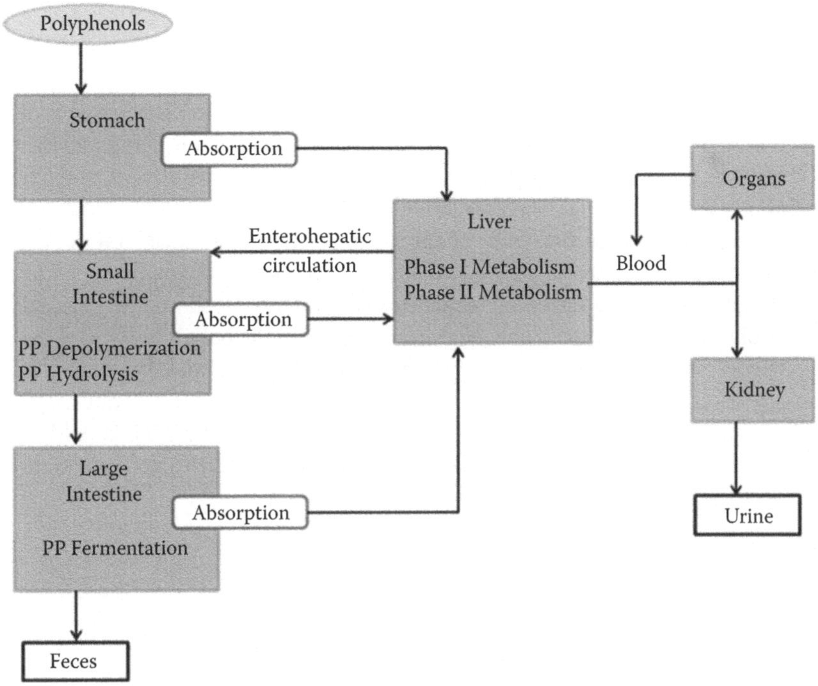

FIGURE 2.2 Scheme of the metabolic transformations of polyphenols after intake.

low-molecular-weight polyphenols (e.g., chlorogenic acids) may be absorbed in the stomach (Lafay et al., 2006).

2.2.1.2 The Small Intestine

Many polyphenols are absorbed in the small intestine. The aglycones (i.e., species without any non-phenolic-associated compound) are directly absorbed. In general, polyphenols present as glycosides or esters of organic acids are poorly absorbed and are first cleaved to yield the aglycone. Anthocyanins are the only polyphenols that are directly and extensively absorbed as glycosides. The cleavage of phenolic conjugates may take place in the enterocytes of the brush border or in the lumen of the digestive tube. In enterocytes, lactase-phlorizin hydrolase is mainly responsible for the transformations; in the lumen, digestive enzymes such as pancreatic esterases hydrolyze ester bonds. For instance, chlorogenic acids from coffee yield quinic acid and free phenolic acids (Amarante da Encarnação et al., 2015).

Polyphenols of higher molecular weight, such as hydrolyzable tannins or oligomers and polymers of proanthocyanidins, are not absorbed as such in the small intestine but may be partially depolymerized as suggested by the detection of (epi)catechin monomers and metabolites in the plasma and urine of rats given polymeric proanthocyanidins (Mateos-Martín et al., 2012; Touriño et al., 2011).

2.2.1.3 The Large Intestine

Those polyphenols reaching the colon (most oligomers and polymers and also smaller species) may follow three different processes: (1) direct absorption by the colonocytes, (2) excretion in the feces, and (3) transformation by the microbiota into metabolites that may be either absorbed or excreted. Indeed, the latter is being increasingly recognized as one of the key aspects to explain the biological effects of polyphenols (see Section 2.4). Microbial transformations originate a wide variety of metabolites; some of them (e.g., urolithins from ellagitannins) are specific for certain classes of phenolics, while others (e.g., phenolic acids from flavanols, flavonols, or anthocyanins) are generated by a variety of structures (Figure 2.3). It should be stressed that even insoluble and nonextractable high-molecular-weight polyphenols that used to be considered inert are also object of colonic transformation and therefore result in a variety of absorbable metabolites (Mateos-Martín et al., 2012). Transformation in the colon involves a wide variety of chemical reactions that include hydrolysis, ring fission, reduction, decarboxylation, demethylation, and dehydroxylation (Selma et al., 2009).

2.2.1.4 The Liver

Polyphenols and metabolites absorbed in the stomach, the small intestine, and the large intestine will reach the liver, where they may be subjected to phase I and, most commonly, phase II transformations (Donovan et al., 2001; Scalbert and

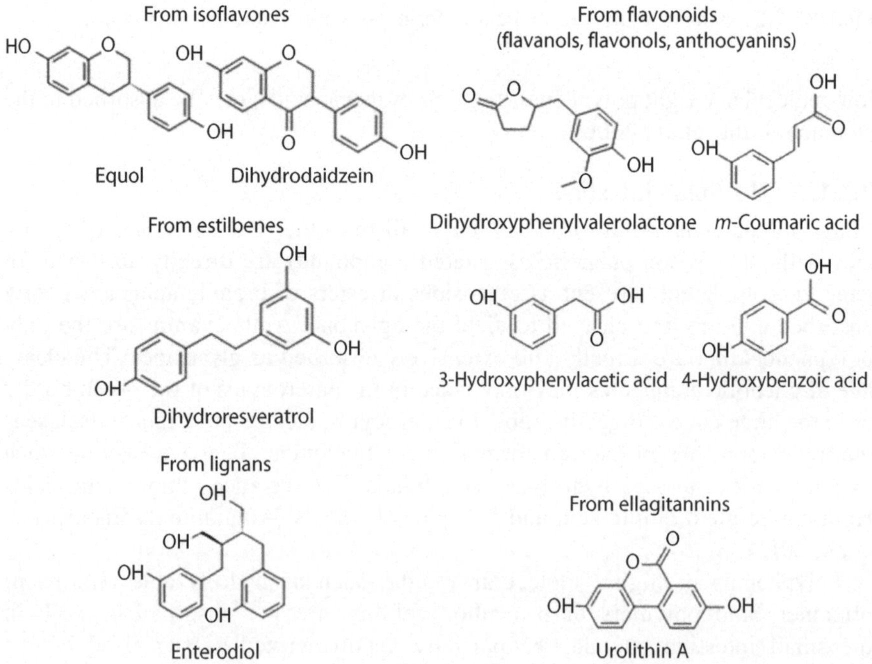

FIGURE 2.3 Examples of metabolites derived from colonic fermentation of different types of polyphenols.

Williamson, 2000). The most common transformations are conjugations with one or several substituents, mainly sulfate, methyl, and glucuronidyl groups, catalyzed by sulfotransferases, catechol-*O*-methyltransferases, and UDP-glucuronosyltransferases, respectively; some of them may also take place in the small intestine. Derivatizations modify the hydrophilic/hydrophobic balance of the original molecules to facilitate their transport and excretion. Then the conjugates may (1) return to the digestive tube by enterohepatic circulation, (2) be directly excreted through urine, and (3) enter the blood stream, circulate, and eventually be excreted through urine.

The metabolites that enter the blood stream may exert an effect in different organs, either immediately or after accumulation. Little is known to date about the formation of deposits of polyphenol metabolites in the organs, but some studies have reported interesting data about this. For instance, the accumulation of metabolites in brown adipose tissue, a tissue of increasing relevance in relation to cardiometabolic pathologies, was reported after the supplementation to rats with grape proanthocyanidins at a dose that could be easily incorporated into a real diet (Serra et al., 2013).

2.2.2 Biological Activity of Polyphenol Metabolites

Increasing evidence supports the idea that circulating metabolites are major contributors to the *in vivo* effects of polyphenols (Williamson et al., 2005; Williamson and Clifford, 2010). The levels of certain polyphenol metabolites have been related to the amelioration of some cardiometabolic risk factors; for instance, inverse associations between urinary enterolactone, a microbial metabolite from lignans; and abdominal obesity, high serum C-reactive proteins or high serum triglycerides have been observed (Frankenfeld, 2014).

These findings indicate a need for *in vitro* studies that investigate the possible mechanisms of polyphenol action made with metabolites instead of with the original compounds. An increasing number of papers have already explored the biological activities of polyphenol metabolites (Heleno et al., 2015). The action mechanisms of these compounds are discussed in other sections of this chapter, but an important aspect that should be considered is that different modifications of the original polyphenols may have different effects on their biological activities that are difficult to predict. For instance, glucuronidated and sulfated metabolites of quercetin, which prevent endothelial dysfunction, did not exhibit the vasorelaxant effect in rat aorta attributed to intact quercetin (Lodi et al., 2009), whereas methylation of hesperetin improved the anti-inflammatory activities of the parent compound (Yang et al., 2012).

2.2.3 Metabolites as Markers of Polyphenol Intake

A proper evaluation of intake is a prerequisite for the validation of clinical trials and epidemiological studies. The well-known weaknesses of food questionnaires and the limitations of the existing databases on polyphenol content make it difficult to obtain reliable data on consumption. Thus, the use of biomarkers of polyphenol intake, based on determinations of metabolites in biological fluids, is emerging as an alternative approach (Andrés-Lacueva and Zamora-Ros, 2010). A systematic review of more than 150 human studies that measured urinary metabolites, after supplementation with polyphenol-rich products or with pure compounds, detected three categories of polyphenols that may be characterized by their urinary metabolites, namely, (1) polyphenols that exhibit a dose–response effect and medium urinary recovery (10%–50%), for example, isoflavones and hydroxytyrosol; (2) polyphenols that exhibit a dose–response effect and very low urinary recovery (<1%), for example, anthocyanins; and (3) polyphenols with low urinary recovery (<10%) and no dose–response effect, for which urinary metabolites cannot be used as biomarkers of intake, for example, flavanones, flavonols, and catechins (Perez-Jimenez et al., 2010).

2.3 ANTIOXIDANT ACTIVITY OF POLYPHENOLS

The alleged antioxidant activity of polyphenols may be attributed to their capacity to directly scavenge free radicals or to other more complex physiological mechanisms.

2.3.1 Free Radical Scavenging

Polyphenols participate in redox reactions because their catechol and pyrogallol moieties (germinal di- and trihydroxyl groups, respectively) are easily oxidized (Figure 2.4).

Thus, they are able to react with reactive oxygen species (ROS) primarily generated by the mitochondria or incorporated from the environment. ROS, which include $O_2^{\bullet-}$, $HO^\bullet$, $^\bullet NO$, H_2O_2, $^1O_2$, HOCl, as well as oxide and peroxide radicals ($RO^\bullet$ and $ROO^\bullet$) from biomolecules, are putatively deleterious agents (Halliwell, 2006; Halliwell and Gutteridge, 1984) that cause oxidative stress when the endogenous defense mechanisms are overwhelmed (Halliwell and Whiteman, 2004; Sies, 1991). Polyphenols are good ROS scavengers because they react faster with them than many biomolecules, and the radicals they form (semiquinones) do not react or react very slowly with these biomolecules. The capacity of polyphenols to directly scavenge ROS comes from their tendency to donate hydrogen atoms or to transfer electrons (Wright et al., 2001) (Figure 2.4). Hydrogen atom transfer (HAT) seems to be the predominant mechanism in nonpolar environments (Foti et al., 2008) and may be the predominant mechanism against lipid peroxides ($LOO^\bullet$) generated during lipid (LH) autoxidation. In this case, polyphenols act as chain-breaking antioxidants because they stop the chain of radical-mediated peroxidations along the lipid molecule. The bond dissociation energy (BDE) is the most important parameter determining the reactivity of polyphenols by HAT (Wright et al., 2001). The second mechanism is the single-electron transfer (SET). In this case, at least two variants have been proposed depending on what particle (proton or electron) is released first: sequential proton loss electron transfer (SPLET) (Foti et al., 2004; Litwinienko and Ingold, 2003) and proton-coupled electron transfer (PCET) (Sjödin et al., 2006). The ionization potential (IP) of the phenolic hydroxyls is the most significant parameter in this kind of mechanism; the lower the IP value, the easier the electron abstraction and the reaction with free radicals. Electron transfer mechanisms are favored in polar environments such as blood and the cellular cytosol (Carreras et al., 2009; Litwinienko and Ingold, 2007). Different structural features contribute to the stabilization of the

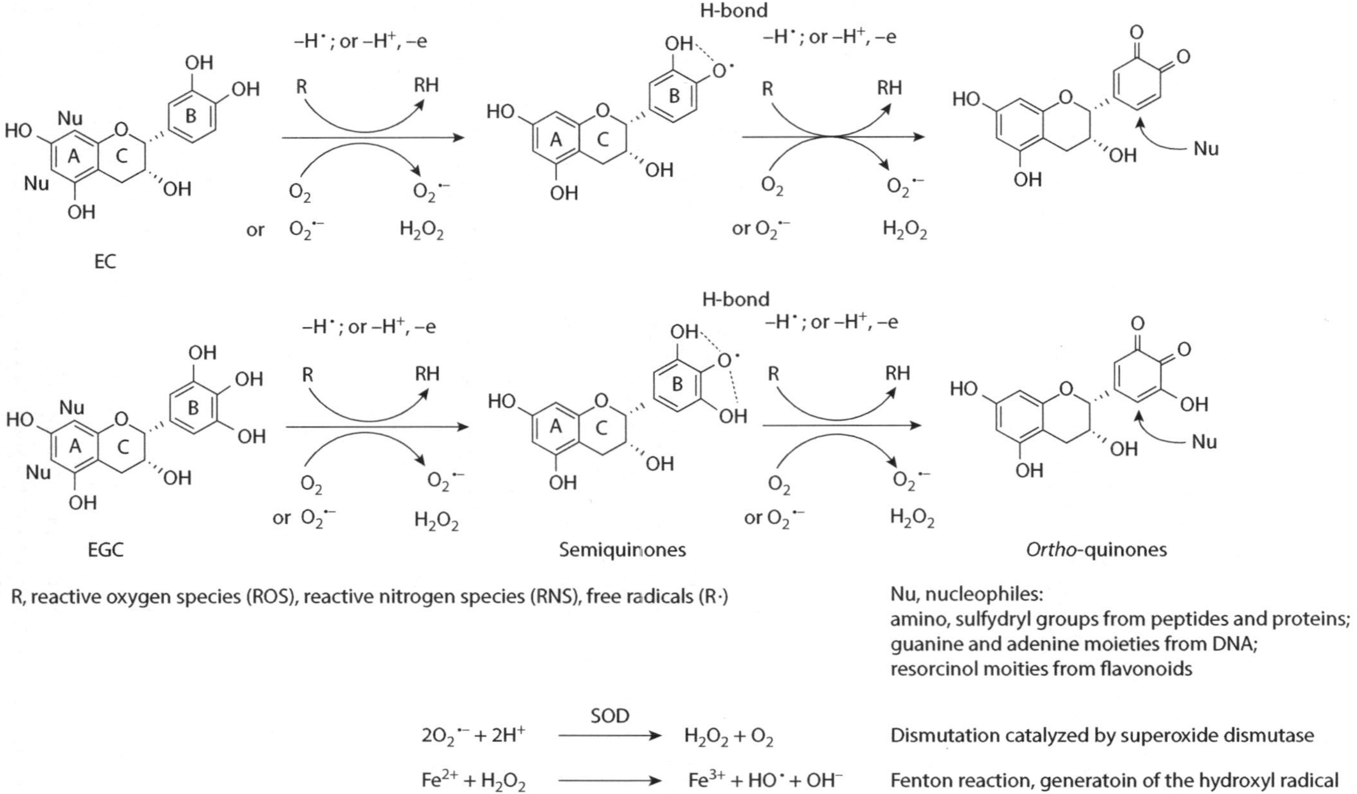

FIGURE 2.4 Chemical reactivity of flavanols as representative polyphenols.

phenoxy radical resulting from the first hydrogen or electron transfer (Figure 2.4, first step in flavanol oxidation). These include the adjacent hydroxyl groups that form hydrogen bonds with the phenoxy radical and substituents in ortho and para positions that stabilize the phenoxy radical by resonance or hyperconjugation effects (Leopoldini et al., 2004). The reactivity of the catechol and pyrogallol groups is lost when conjugated by phase II metabolization (see Section 2.1).

Another possible antioxidant action of polyphenols comes from their ability to chelate divalent ions such as Fe^{2+} (ferrous ion) (Figure 2.1). The ferrous ion contributes to generate the highly aggressive hydroxy radical ($HO^{\bullet}$) by reacting with hydrogen peroxide (Fenton reaction, Figure 2.4). The catechol and pyrogallol moieties of polyphenols may reduce the incidence of the Fenton reaction by chelating Fe^{2+} (Andjelković et al., 2006; Rice-Evans et al., 1996). Again, phase II conjugation with glucuronide, methyl, or sulfate groups will preclude this reaction from occurring.

2.3.2 Prooxidant Activity

Formation of the phenoxy radical is followed by the transfer of a second electron that results in the formation of an ortho-quinone or hydroxy-ortho-quinone after rearrangement of electrons within the phenolic system. These ortho-quinones are reactive molecules because they include electrophilic positions that may engage in reactions with nucleophiles (Figure 2.4). In the living organism, ortho-quinones are reactive against nucleophiles such as the amino and sulfhydryl groups of proteins (Liberato et al., 1981), guanine and adenine moieties in DNA (Cavalieri et al., 1997), and the resorcinol moiety (ring A) of flavonoids (Dangles et al., 2000; Guyot et al., 1996). Quinones are potentially toxic and carcinogenic because they may alter the structure of proteins and DNA (Awad et al., 2001; Galati et al., 2001; Cavalieri et al., 2002). At high concentrations, they may also result in the production of ROS through redox cycling catalyzed by nicotinamide adenine dinucleotide phosphate (NADPH)-dependent reductases (Awad et al., 2001; Liehr and Roy, 1990). Cell defense mechanisms include removal of quinones by methylation catalyzed by catechol-*O*-methyltransferase or by other conjugation reactions, such as glutathione, glucuronide, or sulfate formation as described for estrogens and neurotransmitters (Antonio et al., 2003; Butterworth et al., 1997; Roy et al., 1990). On the other hand, reaction of the ortho-quinones with the nucleophilic positions on the resorcinol moiety of polyphenols such as (−)-epicatechin (EC) regenerates the catechol group and increases the number of hydroxyls available for ROS scavenging (Bors et al., 2000; Dangles et al., 2000; Torres et al., 2002).

Polyphenols can also directly generate free radicals. The superoxide radical ($O_2^{\bullet-}$) may be formed from the catechol and pyrogallol moieties or the semiquinone radicals resulting from the first oxidation step by reaction with molecular oxygen (Figure 2.4) (Metodiewa et al., 1999; Miura et al., 1998). Pyrogallol and other tryhydroxybenzenes are more likely to generate $O_2^{\bullet-}$ than catechols (Alanko et al., 1999). Catechol and pyrogallol can also react with $O_2^{\bullet-}$ to give H_2O_2 (Deeble et al., 1988; Miura et al., 1998) (Figure 2.4). As stated earlier, hydrogen peroxide reacts rapidly with Fe^{2+} to give the hydroxyl radical ($HO^{\bullet}$) by the Fenton reaction (Figure 2.4). Both H_2O_2 and $HO^{\bullet}$ are potentially toxic. Hydrogen peroxide may have different deleterious cellular effects, depending on its concentration and the cell type, both directly or through the Fe^{2+}-mediated generation of $HO^{\bullet}$ (Long et al., 2000). The hydroxyl radical $HO^{\bullet}$ rapidly reacts with DNA and other biomolecules and may induce cancer and premature aging (Halliwell and Gutteridge, 1999). However, H_2O_2 and free radicals are involved in cell signaling and the normal functioning of the organism (Halliwell, 2012; Mittler et al., 2011). Because of this dual role of ROS as both deleterious and beneficial agents in homeostasis, their concentrations are carefully controlled by various defense and regulatory mechanisms (Imlay, 2008; Rhee and Woo, 2010).

2.3.3 Antioxidants or Prooxidants?

Are polyphenols predominantly ROS scavengers (antioxidants) or ROS producers (prooxidants)? What is the final overall redox effect of polyphenols on the life of animal organisms? And, as ROS participate in cell signaling, how do polyphenols, as both scavengers and producers, influence homeostasis? Do they? As phenolics are extensively metabolized into redox unreactive species, it is quite unlikely that they have any systemic scavenging activity. On the other hand, the skin and the intestinal tract are two organs where high local concentrations of intact polyphenols may exert an effective ROS-scavenging action. Polyphenols topically applied to skin may remove potentially damaging free radicals from exogenous sources, such as pollution and radiation, and preclude their formation by absorbing radiation (photoprotection) (Afaq and Katiyar, 2011; Hsu, 2005; Matito et al., 2011). Protection against UV radiation is a significant activity of flavonoids that is independent of their phase II metabolization because the highly conjugated (3–8 double bonds that alternate with single bonds) UVA and UVB absorbing π-system (see quercetin in Figure 2.1) is not altered by derivatization of the hydroxyl functions (Basu-Modak et al., 2003; Hsu, 2005). The gut is another organ where intact phenolics may have significant scavenging activity before or while they are absorbed through the enterocytes (Halliwell et al., 2000). High-molecular-weight nonextractable proanthocyanidins may be particularly efficient in the gut as they gradually release scavenging moieties, such as EC, while transiting along the intestinal tract (Mateos-Martín et al., 2012; Touriño et al., 2011). But again, as the catechol, pyrogallol, and gallate moieties of polyphenols also generate ROS, what is their ultimate redox action in the skin, gut, or anywhere else? It seems evident that the animals that feed on plants regard polyphenols as toxic agents because they readily metabolize them into inactive conjugates. As polyphenols have the capacity to modify the delicately balanced redox homeostasis, the organism reacts by activating its antioxidant defenses as shown in the case of prolonged exposure to grape skin polymeric proanthocyanidins, which triggered the overexpression of peroxiredoxins (Lizárraga et al., 2011). Paradoxically, the ultimate antioxidant effect of phenolics may very well come from their ROS-generating ability, an effect that would fit in the concept of hormesis, defined as a "process in which a low dose of a stressful stimulus activates an adaptive response that increases the resistance of the cell or organism to a moderate to severe level of stress" (Calabrese et al., 2007). Mitohormesis refers to the hormetic effect triggered by oxidative stress generated by the mitochondrial activity. Indeed, increased oxidative stress and increased longevity have been directly related (Ristow and Schmeisser, 2011;

Viña et al., 2013). Physical exercise and caloric restriction are two factors that trigger mild oxidative stress and have been proved to promote health and longevity (Lanza et al., 2008; Ristow and Zarse, 2010). Actually, caloric restriction is efficient at extending life span in yeast, worms, fish, rats, mice, and it may be the only nutrition intervention to do so in humans so far (Redman and Ravussin, 2011). The life span–extending action of caloric restriction has been directly linked to mild oxidative stress in nematodes and mice (Schulz et al., 2007; Weimer et al., 2014).

The putative cancer prevention activity of some polyphenols such as tea (−)-epigallocatechin-3-*O*-gallate (EGCG) may also be attributed, in part, to some kind of hormetic effect (Lambert and Elias, 2010). Apart from promoting antioxidant defenses, ROS generated by some polyphenols may be responsible for the induction of apoptosis in tumor cells, probably mediated by hydrogen peroxide (Lozano et al., 2006; Yang et al., 2000). Care should be taken not to generalize this to the situation *in vivo* where the availability of both oxygen and intact phenolic hydroxyl groups may not account for any proapoptotic effect (Halliwell, 2008; Long et al., 2000).

The redox reactivity of polyphenol antioxidants may help to promote health, prevent cancer, and promote longevity by acting either as ROS scavengers that directly eliminate oxidant species or as ROS producers that activate endogenous antioxidant defenses and regulatory systems.

2.4 INTERACTION WITH SIGNALING PATHWAYS AND ANTI-INFLAMMATORY ACTIVITY

Polyphenolic structures appear to influence signaling pathways involved in a wide variety of cell processes, from inflammation to apoptosis. This explains why they have been related to the prevention of several chronic diseases of different types, from neurological disorders to certain kinds of cancer (Ramos, 2008; Spencer et al., 2012). The mechanisms of action may involve redox reactions of different kinds and/or protein–ligand interactions. The latter are more structure specific than the redox activity and mainly attributed *in vivo* to the circulating metabolites either at the doses achieved with the diet or at higher concentrations achieved by supplementation (Fraga and Oteiza, 2011; Rodríguez-Mateos et al., 2014; Spencer et al., 2003).

2.4.1 Mechanisms Involved in the Effect of Polyphenols on Signaling Pathways

The effects of polyphenols on cell signaling cascades have been observed in two different locations of the cell: the cell nucleus and the cell membrane (Fraga and Oteiza, 2011). For instance, procyanidin dimers B1 and B2 have been shown to interact inside the cell *in vitro* with transcription factor NF-κB (nuclear factor kappa-light-chain enhancer of activated B cells), the endpoint of several signaling pathways involved in the activation of inflammation. This interaction, through bonds between certain hydroxyl groups of procyanidins and the arginine residues of NF-κB, inhibits its ability to bind to DNA (Mackenzie et al., 2009). The observation that the more rigid type A procyanidin dimers did not combine with NF-κB emphasizes the structural selectivity of polyphenol/peptide interactions. On the cell surface, polyphenols interact with proteins in the lipid rafts, specialized areas of the cell membrane that are rich in sphingolipids, cholesterol, and proteins involved in cell signaling (e.g., NADH oxidase, endothelial nitric oxide synthase, heme-oxygenase-1). It has been shown that EC increases the levels of nitric oxide, a potent vasodilator, in endothelial cells, through the inhibition of NADPH oxidase (Steffen et al., 2007), and that EGCG inhibits the activation of the epidermal growth factor, mediated by a raft-located receptor, in HT29 colon cancer cells (Adachi et al., 2007). High molecular weight proanthocyanidins, unable to enter the cell, may exert an effect in signaling pathways by interacting with these membrane proteins on epithelial cells (Erlejman et al., 2008).

2.4.1.1 Main Molecular Targets of Polyphenols

The molecular targets of polyphenols for which more cumulative evidence exists are summarized in the following text. Polyphenols have shown either activating or inhibitory effects depending on the compound and/or the dose (Ramos, 2008; Spencer et al., 2003); this is relevant because the putative health-promoting effects of phenolics may require activation or inhibition depending on the kind of target cell (e.g., cancer versus healthy cells). For this same reason, caution is needed so as not to generalize about the effects of dietary polyphenols as a whole:

- *MAPK (mitogen-activated protein kinase).* MAPKs include extracellular signal–regulated kinase (ERK 1/2), c-Jun N-amino terminal kinase (JNK 1/2/3), and p38 kinase (p38αβ), which are key factors in the transduction of extracellular signals into cellular responses. They are involved in the activation of many genes, such as those associated with (1) inflammation through an increased expression of iNOS—inducible nitric oxide synthase—and cytokines, (2) cell adhesion and transendothelial migration, and (3) apoptosis (Claude et al., 2014; Cordero-Herrera et al., 2013b; Spencer et al., 2012). The activity of polyphenols, particularly flavonoids, on these pathways may be explained by their structural homology to PD98059 (2′-amino-3′-methoxyflavone), a well-known MAPK inhibitor (Spencer et al., 2012). Different MAPKs have different effects on cell survival; ERK seems to promote survival, while JNK stimulates the expression of the proapoptotic factors. Several phenolic acids have been shown to reduce cell death by protecting pancreatic beta cells against tert-butyl hydroperoxide–induced toxicity and via an activation of the ERK pathway (Fernández-Millán et al., 2014). Interestingly, these compounds derive from the colonic fermentation of procyanidins, which emphasizes the relevance of phenolic metabolites. Other flavanol metabolites such as EC conjugates with methyl and glucuronyl moieties reduced monocyte adhesion at physiologically significant concentrations by inhibiting p38 phosphorylation (Claude et al., 2014). In contrast, (−)-epicatechin-3-*O*-gallate (ECG)-induced apoptosis in human colon cancer cells by activating the p53-p38/JNK cascade (Cordero-Herrera et al., 2013b).
- *PI3K/Akt (phosphatidylinositol-3-kinase/protein kinase B or Akt) pathway.* This is an apoptosis-related pathway that when activated in healthy cells promotes normal cell growth and differentiation, a pursued activity for neural

cells; when activated in cancer cells, it will lead to cell death by apoptosis. Certain flavonoids may show both activities depending on dose and type of conjugation; interestingly, *O*-methylated quercetin metabolites were less effective than the aglycone and glucuronidated conjugates were inactive, probably due to differences in the capacity of the different structures to enter neurons (Spencer et al., 2003). The citrus flavanone hesperetin, at physiological nanomolar concentrations, was able to activate this pathway by inhibiting proapoptotic proteins such as apoptosis signal-regulating kinase 1 (ASK1) and caspases 3 and 9 in cortical neurons (Vauzour et al., 2007). Other flavonoids exert a proapoptotic effect in different cell models of cancer by activating the expression of the proapoptotic proteins such as Bax or caspases 3, 7, and 9 (Ramos, 2008). The activity of flavonoids such as quercetin and myricetin on this pathway may be explained by their structural homology to the PI3K inhibitor LY294002. These structures would compete for the ATP-binding pocket of the enzyme (Walker et al., 2000). Polyphenols may also exert other physiological effects related to the PI3K/Akt pathway. Cocoa flavanols modulate hepatic glucose production via downregulation of the key gluconeogenic enzyme PEPCK (phosphoenolpyruvate carboxykinase) *in vitro* at physiological concentrations (Cordero-Herrera et al., 2013a); this activity would be compatible with the putative beneficial effects of flavanols on type 2 diabetes.

- *PPARγ (peroxisome-proliferator-activated receptor γ).* The activation of PPARγ is also involved in many different cellular processes. For instance, it increases the expression of ABCA1 (ATP-binding cassette transporter), which causes an increase in the efflux of cholesterol from macrophages in a dose-dependent manner. This prevents the accumulation of cholesterol in senescent macrophages present in atherosclerosis, cancer, and macular degeneration. The anthocyanins cyanidin 3-*O*-β-glucoside and peonidin 3-*O*-β-D-glucoside upregulate this signaling pathway in a dose-dependent manner in macrophage and macrophage-derived foam cells so that they increase cholesterol efflux (Xia et al., 2005). The ellagitannin punicalagin and gallic acid induce, via activation of the PPARγ pathway, the expression of the antiatherosclerotic enzyme paraoxonase/arylesterase 1 (PON1) in hepatocytes (Khateeb et al., 2010).
- *Nrf2 (nuclear factor erythroid 2-related factor 2).* This transcription factor, after being transposed to the cell nucleus, stimulates the transcription of genes that code for protective proteins (e.g., glutathione *S*-transferase, glutathione peroxidase, glutamate cysteine ligase) active against oxidative and electrophilic stress. Different phenolics promote the translocation of Nrf2 via activation of the ERK and PI3k/Akt pathways described earlier, as shown for hydroxytyrosol and hepatic cells; direct interactions with Nrf2 that stimulate its translocation may also occur (Fraga and Oteiza, 2011; Martín et al., 2010).
- *NF-κB (nuclear factor kappa-light-chain enhancer of activated B cells).* This ubiquitous transcription factor may be activated, among other mechanisms, by the MAPK signaling pathway described earlier. Once activated, it induces the expression of genes mainly related to inflammation.

Structurally diverse polyphenols such as chrysin and ellagic acid modulate NF-κB activation pathway in human intestinal Caco-2 cells at physiological concentrations (Romier et al., 2008). Anthocyanins reduce the expression of inflammatory enzymes such as inducible form of nitric oxide synthase and COX2 (cyclooxygenase-2) via inhibition of NF-κB as shown in macrophages and microglia cells (De Pascual-Teresa, 2014).

- *Activator protein 1 (AP-1).* AP-1 is a transcription factor involved in the regulation of the expression of genes that are activated in response to certain cytokines, growing factors, or oxidative stress; therefore, they control processes such as inflammation or proliferation. High-molecular-weight phenolic compounds from cranberry inhibited AP-1 in human gingival epithelial cells, thus reducing proinflammatory IL-6 (interleukin 6) production (Tipton et al., 2014).

2.4.2 Anti-Inflammatory Activity

Inflammation is a complex biological response of the vascular tissues to harmful stimuli (e.g., pathogens, damage/injury, and tissue stress) with the aim of removing the injurious agent and repairing the affected tissues in a process that lasts for a limited period of time, usually a few days (Medzhitov, 2008). The recognition of the inducer leads to a coordinated production of inflammatory mediators by activated immune cells such as mastocytes, neutrophils, macrophages, and platelets or derived from inactive plasmatic precursor proteins. These mediators include chemokines, complement fragments, cytokines (e.g., TNF-α, IL-1, IL-6), vasoactive compounds (•NO, histamine), peptides (serotonin, substance P), lipid mediators (eicosanoids), and inflammatory enzymes (caspases), among others (Medzhitov, 2008). The compounds released induce vasodilatation, selective extravasation of plasma proteins, and recruitment of leukocytes to the site of the injury, principally neutrophils that produce ROS and reactive nitrogen species (RNS), which eliminate the inducer and may also cause damage to cells of the surrounding tissue (Son et al., 2008). After this, recruited macrophages, tissue-resident cells, and cells involved in the adaptive immune response (i.e., B and T lymphocytes) eliminate the pathogen completely (Van Dyke and Kornman, 2008).

Pathological inflammation appears when the process becomes chronic due to persistence of the harmful condition and/or when tissue homeostasis is not reached for any other reason. Most chronic diseases such as arthritis, inflammatory bowel disease (IBD), metabolic syndrome (MetS), and CVD involve unresolved inflammatory responses. The modulation of inflammatory effectors by dietary compounds such as polyphenols is an interesting strategy to mitigate chronic inflammation.

2.4.2.1 Effects of Polyphenols on Chronic Inflammation

Polyphenols are active in models of rheumatoid arthritis. Their action mechanisms involve many varied activities, such as (1) inhibition of osteoclast/macrophage differentiation and function by the citrus flavonoid nobiletin (Murakami et al., 2007) and rutoside (Kauss et al., 2008); (2) estrogen modulation by genistein (Wang et al., 2008) and nobiletin (Murakami et al., 2007); (3) downregulation of "nuclear factor of activated T-cells" by EGCG (Morinobu et al., 2008); (4) inhibition of "a disintegrin-like and metalloprotease with thrombospondin type 1 motif 2/5" (ADAMTS) and protease expression in synovial fibroblasts and chondrocytes by nobiletin (Imada et al., 2008; Ishiwa et al., 2000; Lin et al., 2003); (5) antiproliferation (Sekine et al., 2008) and proapoptotosis (Byun et al., 2008) on synovial fibroblasts by flavopiridol and resveratrol, respectively; and (6) general immune depression by hesperidin (Li et al., 2008). The effect of hesperidin has also been tested in patients with rheumatoid arthritis with positive results (Kometani et al., 2008).

Dietary polyphenols, readily bioaccesible in the gut, and gut-associated lymphoid tissue (GALT) shortly after ingestion, have been shown to prevent or ameliorate inflammatory colitis in rodents probably through the inhibition of NF-κB-dependent gene expression and the induction of phase II antioxidant and detoxifying proteins (Shapiro et al., 2007). Glycosylation seems to significantly influence the anti-inflammatory activity of some flavonoids along the intestinal tract. While quercetin is absorbed in the small intestine, its glycosylated conjugates rutin and quercitrin reach the colon where quercetin is released by the microbiota (Comalada et al., 2005; Kwon et al., 2005). Thus, the aglycone quercetin ameliorates dextran sulfate sodium (DSS) colitis *in vivo*, only after being released in the colon from its glycosylated conjugate rutin (Kwon et al., 2005). On the other hand, some phytoestrogens (isoflavones genistein and daidzein) enhanced acute colitis in a rat model by increasing the expression of cyclooxygenase-2 (COX-2), a key enzyme in the prostaglandin biosynthesis from arachidonic acid (Seibel et al., 2008).

CVD are inflammatory conditions initiated by the adhesion of leucocytes on arterial walls that then migrate into the subendothelial space where they contribute to the formation of atherogenic plaques. This process is induced by NF-κB-mediated activation of the proinflammatory cytokines TNF and IL-1β (Lu et al., 2004). There is evidence that polyphenol-rich foods influence CVD biomarkers, but knowledge about the role of the different components and their metabolites is limited (Manach et al., 2005). Catechin and red wine extract have been shown to reduce thrombogenicity in mice (Soulat et al., 2006) and cocoa procyanidins decreased platelet function in humans while increasing plasma levels of catechin and epicatechin (Murphy et al., 2003). The effects of resveratrol, another polyphenol present in grape and wine, on the lipid profile and atherosclerosis have been extensively studied, but the results very much depend on the model used. Resveratrol is active in the ApoE-deficient mice model by reducing the levels of ICAM-1 and VCAM-1 and consequently blood lipid levels and the atherosclerotic plaque (Do et al., 2008). On the contrary, resveratrol did not modify the lipid levels of neither hypercholesterolemic rabbits (Wilson et al., 1996) nor normal rats (Turrens et al., 1997). Other *in vivo* effects of resveratrol are the reduction of cellular infiltration, fibrosis, and expression of proinflammatory cytokines in a model of autoimmune myocarditis (Yoshida et al., 2007). Morin, myricetin, and fisetin may play a role in ameliorating atherosclerosis by inhibiting LDL oxidation and blocking oxidized LDL uptake by macrophages (Lian et al., 2008). A similar LDL protection mechanism has been described for ECG in atherosclerotic lesions (Kawai et al., 2008b), kaempferol and rhamnocitrin (Tu et al., 2007), quercetin (Terao et al., 2008), and even quercetin-3-glucuronide

(Kawai et al., 2008a). Metabolization may modify the anti-inflammatory activities of flavonoids to different degrees as observed in the inhibition of eicosanoid production from human leukocytes by quercetin and its conjugates with methyl, sulfate, or glucuronyl groups (Loke et al., 2008).

MetS is a cluster of risk factors for CVDs and type 2 diabetes including insulin resistance (IR), central obesity, hyperlipidemia, and arterial hypertension that is also characterized by low-grade chronic inflammation (Kassi et al., 2011). Polyphenols such as catechins, anthocyanins, resveratrol, and curcumin have been shown to modulate physiological and molecular pathways that are involved in energy metabolism, adiposity, and obesity in animal models (Meydani and Hasan, 2010). Orally administered quercetin ameliorates hypertension, dyslipidemia, and hyperinsulinemia and lower TNF and iNOS expression in visceral fat of Zucker obese rats, a leptin-insensitive model of IR (Rivera et al., 2008). Quercetin and kaempferol may improve insulin sensitivity by acting as PPAR-γ partial agonists (Fang et al., 2008). Resveratrol also improves dyslipidemia, hypertension, hyperinsulinemia, liver fat, and proinflammatory TNF while increasing adiponectin and eNOS in visceral fat (Rivera et al., 2009). The action of resveratrol on IR and MetS may also derive, at least in part, from PPAR-γ activation (Kennedy et al., 2009).

2.5 EFFECTS ON SYMBIOTIC MICROBIOTA

Gut microbiota plays a crucial role in the metabolism of food components. The concept of "gut microbiome" defines a complex ecosystem made up of a large amount of bacteria and other microorganisms that coexist within the intestine. The physiological importance of this ecosystem resides in the huge metabolic capacity (about 100-fold greater than that of the host's liver [Zhu et al., 2010]) of the 1000–1150 prevalent bacterial species in the human gut (Qin et al., 2010). Most species in gut microbiota (98%) belong to only four phyla: Firmicutes (64%), Bacteroidetes (23%), Proteobacteria (8%), and Actinobacteria (3%) (Hold et al., 2002; Matsuki et al., 2004; Zhu et al., 2010). The microbial composition depends on the host's genetics and environmental factors such as vaginal or cesarean delivery at birth, neonatal breast or formula feeding, stress, age, hygiene, bacterial infections, antibiotic exposure, and diet. Particularly, the quality and quantity of the carbohydrates ingested by the host modulate the composition of the gut microbiota, as they are the main source of carbon and energy (Sanz, 2010). Unbalanced diets may disrupt the equilibrium between the host and its intestinal microbiota, leading to dysbiosis. This usually favors overgrowth of pathogens and weakens the host's immune system, possibly via alteration of the mucosal immunity (Noverr and Huffnagel, 2004). Moreover, the individual composition of the gut microbiota defines the susceptibility of the host to several diseases (Sekirov et al., 2010) and may play a crucial role in the onset and progression of several conditions including cancers, diabetes, nonalcoholic fatty liver disease, inflammatory bowel disease, as well as obesity and its related alterations (Jia et al., 2008; Sekirov et al., 2010). The Firmicutes/Bacteroidetes ratio is higher in obese than in lean humans and decreases with weight loss (Ley et al., 2006). This indicates that energy-dense diets predispose to obesity through microbiota deregulation and not only by an excess of energy intake.

2.5.1 Gut Microbiota and Polyphenols

Different food components such as polyphenols can influence the composition of gut microbiota and therefore they may modify their overall metabolic activity (Duda-Chodak, 2012). Polyphenols seem to favor the viability of potentially beneficial bacteria such as *Bifidobacterium* spp. and *Lactobacillus* spp. over potentially pathogenic ones, such as *Bacteroides* spp. and *Clostridium* spp. as observed for tea catechins and metabolites (Lee et al., 2006). Changes within the *Clostridium* genus appear to be particularly significant as some of its subgroups (e.g., *C. coccoides* and *C. leptum* groups) are beneficial to the host, whereas other species are pathogenic (e.g., *C. botulinum*, *C. tetani*, and *C. perfringens* that cause botulism, tetanus, and gas gangrene, respectively). Moreover, some *Clostridium* bacteria can be both beneficial and pathogenic depending on their proportion. For instance, *C. difficile* is a commensal bacteria that may also cause gastrointestinal problems with a wide spectrum of severity, from mild diarrhea to pseudomembranous colitis, toxic megacolon, sepsis, and even death (Dobson et al., 2003). Interestingly, catechin and its metabolites (5-(3′,4′-dihydroxyphenyl)-gamma-valerolactone, 5-phenyl-gamma-valerolactone, and phenylpropionic acid) selectively influence the growth on *Clostridium* species. They inhibit the growth of *C. histolyticum* (Tzounis et al., 2008), *C. perfringens*, and *C. difficile* (Lee et al., 2006) and promote that of the *C. coccoides–Eubacterium rectale* group (Tzounis et al., 2008). Similar to tea flavanols, red wine procyanidins also reduce the growth of *Clostridium* spp. while increasing that of other subgroups such as *Bacteroides* and *Lactobacillus* in rats (Dolara et al., 2005). This selective effect has also been observed in humans after a daily intake of red wine for 1 month; other species affected by red wine intake were *Enterococcus*, *Prevotella*, *Bifidobacterium*, *Eggerthella lenta*, and *Blautia coccoides–E. rectale* (Queipo-Ortuño et al., 2012). Citric polyphenols also differentially affect the growth and adhesion of some gut bacteria such as *Lactobacillus rhamnosus*, *Escherichia coli*, *Staphylococcus aureus*, and *Salmonella typhimurium in vitro* at doses similar to those active in the gastrointestinal tract (Parkar et al., 2008). Naringenin and phloridzin effectively inhibit adhesion of the pathogenic bacteria *S. typhimurium* to enterocytes, while caffeic acid, chlorogenic acid, *o*- and *p*-coumaric acid, catechin, EC, phloridzin, daidzein, genistein, and quercetin are the most active on *S. aureus*; phloridzin and rutin enhance the adherence of the probiotic *L. rhamnosus* (Parkar et al., 2008).

2.5.2 Mechanism of Action of Polyphenols on Gut Microbiota

Putatively beneficial microbiota plays a protective role against pathogens principally by two strategies: (1) occupying the gut lumen and (2) secreting antimicrobial compounds. Polyphenols, particularly flavonoids, may affect bacterial adhesion and/or viability (Duda-Chodak, 2012). The chemical functions (e.g., carbonyl, glycosyl) on the central ring of flavonoids (see Figure 2.1) seem to critically influence their growth inhibitory activity on gut microbiota, as exemplified by the differences detected with quercetin, naringenin, hesperetin, and their glycosylated conjugates (rutin, naringin, hesperidin, respectively) (Duda-Chodak, 2012; Duda-Chodak et al., 2015). The anti-adhesion activity of some polyphenols against potentially pathogenic bacteria has also been related to their structure. Cranberry proanthocyanidins that present A-type linkages (see Figure 6.3) appear to inhibit the adhesion of P-fimbriated *E. coli* to the urinary tract more efficiently than the less structurally restricted oligomers that include only B-type linkages (Foo et al., 2000).

By modifying the composition of gut microbiota, polyphenols may also influence their own metabolization as different types of bacteria may generate different metabolites. For instance, the soy isoflavone daidzein, a nonsteroidal estrogen with effects on menopause symptoms and preventive activity against coronary heart disease and breast, prostate, and colon cancers, is metabolized to equol only by some types of bacteria that may or may not be present in the colon depending on factors such as fermentation, intestinal transit, and redox balance, which can be modified by phenolics themselves (Yuan et al., 2007).

Some putatively deleterious outcomes resulting from the interaction between polyphenols and gut microbiota have also been described. Intestinal microbiota can transform quercetin and rutin into 3,4-dihydrophenylacetic acid (DOPAC), a metabolite of dopamine (Serra et al., 2012). Apart from its known anticancer, anti-inflammatory, cardioprotective, and neuroprotective properties, DOPAC may lead to mitochondrial dysfunction in the presence of the radical nitric oxide (•NO), a mechanism involved in the neurodegeneration associated with Parkinson's disease (Nunes et al., 2005). Other putatively detrimental effects of polyphenols directly or indirectly linked to a host's microbiota are prooxidant activity, modification of phase I and II metabolizing enzymes, drug interaction, and carcinogen interaction (Galati and O'Brien, 2004).

2.6 CONCLUDING REMARKS

Polyphenols interact with a great variety of physiological pathways in live organisms where they engage in redox reactions and/or establish protein–ligand interactions. Polyphenols are fast and extensively modified after ingestion so that their activity may be due to the overall contribution of the original structures and their metabolites. Gut microbiota plays a crucial role in the metabolization of polyphenols and at the same time may be modified by them. The cross talk between polyphenols and gut microbiota is being increasingly recognized as one of the keys to understanding their health benefits. The stimulation of endogenous defenses may also significantly contribute to the global effect of polyphenols. Traditionally thought of as antioxidants because they scavenge oxidant species, polyphenols may have earned this name by providing mild prooxidant cues. The action of plant polyphenols in the human organism appears to be the result of myriad phenomena that contribute to the maintenance of homeostasis.

REFERENCES

Adachi, S., Nagao, T., Ingolfsson, H.I., Maxfield, F.R., Andersen, O.S., Kopelovich, L., and Bernard Weinstein, I. 2007. The inhibitory effect of (–)-epigallocatechin gallate on activation of the epidermal growth factor receptor is associated with altered lipid order in HT29 colon cancer cells. *Cancer Research* 67:6493–6501.

Afaq, F. and Katiyar, S.K. 2011. Polyphenols: Skin photoprotection and inhibition of photocarcinogenesis. *Mini-Reviews in Medicinal Chemistry* 11:1200–1215.

Alanko, J., Riutta, A., Holm, P., Mucha, I., Vapaatalo, H., and Metsä-Ketelä, T. 1999. Modulation of arachidonic acid metabolism by phenols: Relation to their structure and antioxidant/prooxidant properties. *Free Radical Biology and Medicine* 26:193–201.

Amarante da Encarnação, J., Farrell, T.L., Ryder, A., Kraut, N.U., and Williamson, G. 2015. In vitro enzymic hydrolysis of chlorogenic acids in coffee. *Molecular Nutrition and Food Research* 59:231–239.

Andjelković, M., Camp, J.V., Meulenaer, B.D., Depaemelaere, G., Socaciu, C., Verloo, M., and Verhe, R. 2006. Iron-chelation properties of phenolic acids bearing catechol and galloyl groups. *Food Chemistry* 98:23–31.

Andrés-Lacueva, C. and Zamora-Ros, R. 2010. Wanted: Specific nutritional biomarkers for food consumption for the study of its protective role in health. *British Journal of Nutrition* 103:307–308.

Antonio, L., Xu, J., Little, J.M., Burchell, B., Magdalou, J., and Radominska-Pandya, A. 2003. Glucuronidation of catechols by human hepatic, gastric, and intestinal microsomal UDP-glucuronosyltransferases (UGT) and recombinant UGT1A6, UGT1A9, and UGT2B7. *Archives of Biochemistry and Biophysics* 411:251–261.

Awad, H.M., Boersma, M.G., Boeren, S., van Bladeren, P.V., Vervoort, J., and Rietjens, I.M.C.M. 2001. Structure-activity study on the quinone/quinone methide chemistry of flavonoids. *Chemical Research in Toxicology* 14:398–408.

Basu-Modak, S., Gordon, M.J., Dobson, L.H., Spencer, J.P., Rice-Evans, C., and Tyrrell, R.M. 2003. Epicatechin and its methylated metabolite attenuate UVA-induced oxidative damage to human skin fibroblasts. *Free Radical Biology and Medicine* 35:910–921.

Bors, W., Michel, C., and Stettmaier, K. 2000. Electron paramagnetic resonance studies of radical species of proanthocyanidins and gallate esters. *Archives of Biochemistry and Biophysics* 374:347–355.

Butterworth, M., Lau, S.S., and Monks, T.J. 1997. Formation of catechol estrogen glutathione conjugates and gamma-glutamyl transpeptidase-dependent nephrotoxicity of 17beta-estradiol in the golden Syrian hamster. *Carcinogenesis* 18:561–567.

Byun, H.S., Song, J.K., Kim, Y.R. et al. 2008. Caspase-8 has an essential role in resveratrol-induced apoptosis of rheumatoid fibroblast-like synoviocytes. *Rheumatology* 47:301–308.

Calabrese, E.J., Bachmann, K.A., John Bailer, A. et al. 2007. Biological stress response terminology: Integrating the concepts of adaptive response and preconditioning stress within a hormetic dose-response framework. *Toxicology and Applied Pharmacology* 222:122–128.

Cardona, F., Andrés-Lacueva, C., Tulipani, S., Tinahones, F.J., and Queipo-Ortuño, M.I. 2013. Benefits of polyphenols on gut microbiota and implications in human health. *Journal of Nutritional Biochemistry* 24:1415–1422.

Carreras, A., Esparbé, I., Brillas, E., Rius, J., Torres, J.L., and Juliá, L. 2009. Oxidant activity of tris(2,4,6-trichloro-3,5-dinitrophenyl)methyl radical with catechol and pyrogallol. Mechanistic considerations. *Journal of Organic Chemistry* 74:2368–2373.

Cavalieri, E.L., Li, K.-M., Balu, N., Saeed, M., Devanesan, P., Higginbotham, S., Zhao, J., Gross, M.L., and Rogan, E.G. 2002. Catechol ortho-quinones: The electrophilic compounds that form depurinating DNA adducts and could initiate cancer and other diseases. *Carcinogenesis* 23:1071–1077.

Cavalieri, E.L., Stack, D.E., Devanesan, P.D. et al. 1997. Molecular origin of cancer: Catechol estrogen-3,4-quinones as endogenous tumorinitiators. *Proceedings of the National Academy of Sciences of the United States of America* 94:10937–10942.

Claude, S., Boby, C., Rodríguez-Mateos, A., Spencer, J.P.E., Gérard, N., Morand, C., and Milenkovic, D. 2014. Flavanol metabolites reduce monocyte adhesion to endothelial cells through modulation of expression of genes via p38-MAPK and p65-Nf-kB pathways. *Molecular Nutrition and Food Research* 58:1016–1027.

Comalada, M., Camuesco, D., Sierra, S., Ballester, I., Xaus, J., Gálvez, J., and Zarzuelo, A. 2005. In vivo quercitrin anti-inflammatory effect involves release of quercetin, which inhibits inflammation through down-regulation of the NF-kappa B pathway. *European Journal of Immunology* 35:584–592.

Cordero-Herrera, I., Martín, M.A., Bravo, L., Goya, L., and Ramos, S. 2013a. Cocoa flavonoids improve insulin signalling and modulate glucose production via AKT and AMPK in HepG2 cells. *Molecular Nutrition and Food Research* 57:974–985.

Cordero-Herrera, I., Martín, M.A., Bravo, L., Goya, L., and Ramos, S. 2013b. Epicatechin gallate induces cell death via p53 activation and stimulation of p38 and JNK in human colon cancer SW480 cells. *Nutrition and Cancer* 65:718–728.

Crozier, A., Jaganath, I.B., and Clifford, M.N. 2009. Dietary phenolics: Chemistry, bioavailability and effects on health. *Natural Products Reports* 26:1001–1043.

Dangles, O., Fargeix, G., and Dufour, C. 2000. Antioxidant properties of anthocyanins and tannins: A mechanistic investigation with catechin and the 3′,4′,7-trihydroxyflavylium ion. *Journal of the Chemical Society, Perkin Transactions* 2:1653–1666.

de Freitas, V. and Mateus, N. 2001. Structural features of procyanidin interactions with salivary proteins. *Journal of Agricultural and Food Chemistry* 49:940–945.

de Pascual-Teresa, S. 2014. Molecular mechanisms involved in the cardiovascular and neuroprotective effects of anthocyanins. *Archives of Biochemistry and Biophysics* 559:68–74.

Deeble, D.J., Parsons, B.J., Phillips, G.O., Schuchmann, H.-P., and Von Sonntag, C. 1988. Superoxide radical reactions in aqueous solutions of pyrogallol and N-propyl gallate: The involvement of phenoxyl radicals. A pulse radiolysis study. *International Journal of Radiation Biology* 54:179–193.

Do, G.-M., Kwon, E.-Y., Kim, H.-J., Jeon, S.-M., Ha, T.-Y., Park, T., and Choi, M.-S. 2008. Long-term effects of resveratrol supplementation on suppression of atherogenic lesion formation and cholesterol synthesis in apo E-deficient mice. *Biochemical and Biophysical Research Communications* 374:55–59.

Dobson, G., Hickey, C., and Trinder, J. 2003. *Clostridium difficile* colitis causing toxic megacolon, severe sepsis and multiple organ dysfunction syndrome. *Intensive Care Medicine* 29:1030–1030.

Dolara, P., Luceri, C., De Filippo, C., Femia, A.P., Giovannelli, L., Caderni, G., Cecchini, C., Silvi, S., Orpianesi, C., and Cresci, A. 2005. Red wine polyphenols influence carcinogenesis, intestinal microflora, oxidative damage and gene expression profiles of colonic mucosa in F344 rats. *Mutation Research: Fundamental and Molecular Mechanisms of Mutagenesis* 591:237–246.

Donovan, J.L., Crespy, V., Manach, C., Morand, C., Besson, C., Scalbert, A., and Rémésy, C. 2001. Catechin is metabolized by both the small intestine and liver of rats. *Journal of Nutrition* 131:1753–1757.

Duda-Chodak, A. 2012. The inhibitory effect of polyphenols on human gut microbiota. *Journal of Physiology and Pharmacology* 63:497–503.

Duda-Chodak, A., Tarko, T., Satora, P., and Sroka, P. 2015. Interaction of dietary compounds, especially polyphenols, with the intestinal microbiota: A review. *European Journal of Nutrition* 54:325–341.

EFSA Panel on Dietetic Products, Nutrition and Allergies (NDA). 2011. Scientific opinion on the substantiation of health claims related to polyphenols in olive and protection of LDL particles from oxidative damage (ID 1333, 1638, 1639, 1696, 2865), maintenance of normal blood HDL cholesterol concentrations (ID 1639), maintenance of normal blood pressure (ID 3781), "anti-inflammatory properties" (ID 1882), "contributes to the upper respiratory tract health" (ID 3468), "can help to maintain a normal function of gastrointestinal tract" (3779), and "contributes to body defences against external agents" (ID 3467) pursuant to Article 13(1) of Regulation (EC) No 1924/2006 *EFSA Journal* 94:2033, 25pp.

EFSA Panel on Dietetic Products, Nutrition and Allergies (NDA). 2012. Scientific opinion on the substantiation of a health claim related to cocoa flavanols and maintenance of normal endothelium-dependent vasodilation pursuant to Article 13(5) of Regulation (EC) No 1924/2006. *EFSA Journal* 10:2809, 21pp.

EFSA Panel on Dietetic Products, Nutrition and Allergies (NDA). 2014. Scientific opinion on the modification of the authorisation of a health claim related to cocoa flavanols and maintenance of normal endothelium-dependent vasodilation pursuant to Article 13(5) of Regulation (EC) No 1924/2006 following a request in accordance with Article 19 of Regulation (EC) No 1924/2006. *EFSA Journal* 12:3654, 13pp.

Erlejman, A.G., Jaggers, G., Fraga, C.G., and Oteiza, P.I. 2008. TNFα-induced NF-κB activation and cell oxidant production are modulated by hexameric procyanidins in Caco-2 cells. *Archives of Biochemistry and Biophysics* 476:186–195.

Fang, X.-K., Gao, J., and Zhu, D.-N. 2008. Kaempferol and quercetin isolated from *Euonymus alatus* improve glucose uptake of 3T3-L1 cells without adipogenesis activity. *Life Sciences* 82:615–622.

Fernández-Millán, E., Ramos, S., Alvarez, C., Bravo, L., Goya, L., and Martín, M.A. 2014. Microbial phenolic metabolites improve glucose-stimulated insulin secretion and protect pancreatic beta cells against tert-butyl hydroperoxide-induced toxicity via ERKs and PKC pathways. *Food and Chemical Toxicology* 66:245–253.

Ferruzzi, M.G., Bordenave, N., and Hamaker, B.R. 2012. Does flavor impact function? Potential consequences of polyphenol-protein interactions in delivery and bioactivity of flavan-3-ols from foods. *Physiology and Behavior* 107:591–597.

Foo, L.Y., Lu, Y.R., Howell, A.B., and Vorsa, N. 2000. The structure of cranberry proanthocyanidins which inhibit adherence of uropathogenic P-fimbriated *Escherichia coli* in vitro. *Phytochemistry* 54:173–181.

Foti, M.C., Daquino, C., and Geraci, C. 2004. Electron-transfer reaction of cinnamic acids and their methyl esters with the DPPH center dot radical in alcoholic solutions. *Journal of Organic Chemistry* 69:2309–2314.

Foti, M.C., Daquino, C., Mackie, I.D., DiLabio, G.A., and Ingold, K.U. 2008. Reaction of phenols with the 2,2-diphenyl-1-picrylhydrazyl radical. Kinetics and DFT calculations applied to determine ArO-H bond dissociation enthalpies and reaction mechanism. *Journal of Organic Chemistry* 73:9270–9282.

Fraga, C.G. and Oteiza, P.I. 2011. Dietary flavonoids: Role of (−)-epicatechin and related procyanidins in cell signaling. *Free Radical Biology and Medicine* 51:813–823.

Frankenfeld, C.L. 2014. Cardiometabolic risk factors are associated with high urinary enterolactone concentration, independent of urinary enterodiol concentration and dietary fiber intake in adults. *Journal of Nutrition* 144:1445–1453.

Galati, G., Moridani, M.Y., Chan, T.S., and O'Brien, P.J. 2001. Peroxidative metabolism of apigenin and naringenin versus luteolin and quercetin: Glutathione oxidation and conjugation. *Free Radical Biology and Medicine* 30:370–382.

Galati, G. and O'Brien, P.J. 2004. Potential toxicity of flavonoids and other dietary phenolics: Significance for their chemopreventive and anticancer properties. *Free Radical Biology and Medicine* 37:287–303.

Gustavson, K.H. 1949. Some protein-chemical aspects of tanning processes. In *Advances in Protein Chemistry*, eds. M.L. Anson and T. Edsau John, pp. 353–421. San Diego, CA: Academic Press.

Gutteridge, J.M.C. and Halliwell, B. 2010. Antioxidants: Molecules, medicines, and myths. *Biochemical and Biophysical Research Communications* 393:561–564.

Guyot, S., Vercauteren, J., and Cheynier, V. 1996. Structural determination of colourless and yellow dimers resulting from (+)-catechin coupling catalysed by grape polyphenoloxidase. *Phytochemistry* 42:1279–1288.

Halliwell, B. 2006. Reactive species and antioxidants: Redox biology is a fundamental theme of aerobic life. *Plant Physiology* 141:312–322.

Halliwell, B. 2008. Are polyphenols antioxidants or pro-oxidants? What do we learn from cell culture and in vivo studies? *Archives of Biochemistry Biophysics* 476:107–112.

Halliwell, B. 2012. Free radicals and antioxidants: Updating a personal view. *Nutrition Reviews* 70:257–265.

Halliwell, B. and Gutteridge, J.M.C. 1984. Oxygen toxicity, oxygen radicals, transition metals and disease. *Biochemical Journal* 219:1–14.

Halliwell, B. and Gutteridge, J.M.C. 1999. *Free Radicals in Biology and Medicine.* Oxford, UK: Oxford Science.

Halliwell, B., Rafter, J., and Jenner, A. 2005. Health promotion by flavonoids, tocopherols, tocotrienols, and other phenols: Direct or indirect effects? Antioxidant or not? *American Journal of Clinical Nutrition* 81:268S–276S.

Halliwell, B. and Whiteman, M. 2004. Measuring reactive species and oxidative damage in vivo and in cell culture: How should you do it and what do the results mean? *British Journal of Pharmacology* 142:231–255.

Halliwell, B., Zhao, K., and Whiteman, M. 2000. The gastrointestinal tract: A major site of antioxidant action? *Free Radical Research* 33:819–830.

Harman, D. 1956. Aging: A theory based on free radical and radiation chemistry. *Journal of Gerontology* 11:298–300.

Harman, D. 1981. The aging process. *Proceedings of the National Academy of Sciences of the United States of America* 78:7124–7128.

Heleno, S.A., Martins, A., Queiroz, M.J.R.P., and Ferreira, I. 2015. Bioactivity of phenolic acids: Metabolites versus parent compounds: A review. *Food Chemistry* 173:501–513.

Herrmann, K. 1973. Plant phenolics as naturally occurring antioxidants. *Fette Seifen Anstrichmittel* 75:499–504.

Hold, G.L., Pryde, S.E., Russell, V.J., Furrie, E., and Flint, H.J. 2002. Assessment of microbial diversity in human colonic samples by 16S rDNA sequence analysis. *FEMS Microbiology Ecology* 39:33–39.

Hooper, L., Kroon, P.A., Rimm, E.B., Cohn, J.S., Harvey, I., Le Cornu, K.A., Ryder, J.J., Hall, W.L., and Cassidy, A. 2008. Flavonoids, flavonoid-rich foods, and cardiovascular risk: A meta-analysis of randomized controlled trials. *American Journal of Clinical Nutrition* 88:38–50.

Hsu, S. 2005. Green tea and the skin. *Journal of the America Academy of Dermatology* 52:1049–1059.

Huang, M.-T., Ho, C.-T., and Li, C.Y., eds. 1992. *Phenolic Compounds in Food and Their Effects on Health II. Antioxidants and Cancer Prevention*, Vol. 507, ACS Symposium Series. New York: American Chemical Society.

Hung, C.-F., Huang, T.-F., Chiang, H.-S., and Wu, W.-B. 2005. (−)-Epigallocatechin-3-gallate, a polyphenolic compound from green tea, inhibits fibroblast adhesion and migration through multiple mechanisms. *Journal of Cell Biochemistry* 96:183–197.

Imada, K., Lin, N., Liu, C., Lu, A., Chen, W., Yano, M., Sato, T., and Ito, A. 2008. Nobiletin, a citrus polymethoxy flavonoid, suppresses gene expression and production of aggrecanases-1 and -2 in collagen-induced arthritic mice. *Biochemical and Biophysical Research Communications* 373:181–185.

Imlay, J.A. 2008. Cellular defenses against superoxide and hydrogen peroxide. *Annual Review of Biochemistry* 77:755–776.

Ishiwa, J., Sato, T., Mimaki, Y., Sashida, Y., Yano, M., and Ito, A. 2000. A citrus flavonoid, nobiletin, suppresses production and gene expression of matrix metalloproteinase 9/gelatinase B in rabbit synovial fibroblasts. *Journal of Rheumatology* 27:20–25.

Jia, W., Li, H., Zhao, L., and Nicholson, J.K. 2008. Gut microbiota: A potential new territory for drug targeting. *Nature Reviews Drug Discovery* 7:123–129.

Kassi, E., Pervanidou, P., Kaltsas, G., and Chrousos, G. 2011. Metabolic syndrome: Definitions and controversies. *BMC Medicine* 9:48, 13pp.

Kauss, T., Moynet, D., Rambert, J., Al-Kharrat, A., Brajot, S., Thiolat, D., Ennemany, R., Fawaz, F., and Mossalayi, M.D. 2008. Rutoside decreases human macrophage-derived inflammatory mediators and improves clinical signs in adjuvant-induced arthritis. *Arthritis Research and Therapy* 10:R19, 9pp.

Kawai, Y., Nishikawa, T., Shiba, Y., Saito, S., Murota, K., Shibata, N., Kobayashi, M., Kanayama, M., Uchida, K., and Terao, J. 2008a. Macrophage as a target of quercetin glucuronides in human atherosclerotic arteries: Implication in the anti-atherosclerotic mechanism of dietary flavonoids. *Journal of Biological Chemistry* 283:9424–9434.

Kawai, Y., Tanaka, H., Murota, K., Naito, M., and Terao. J. 2008b. (–)-Epicatechin gallate accumulates in foamy macrophages in human atherosclerotic aorta: Implication in the anti-atherosclerotic actions of tea catechins. *Biochemical and Biophysical Research Communications* 374:527–532.

Kennedy, A., Overman, A., LaPoint, K., Hopkins, R., West, T., Chuang, C.-C., Martinez, K., Bell, D., and McIntosh, M. 2009. Conjugated linoleic acid-mediated inflammation and insulin resistance in human adipocytes are attenuated by resveratrol. *Journal of Lipid Research* 50:225–232.

Khateeb, J., Gantman, A., Kreitenberg, A.J., Aviram, M., and Fuhrman, B. 2010. Paraoxonase 1 (PON1) expression in hepatocytes is upregulated by pomegranate polyphenols: A role for PPAR-γ pathway. *Atherosclerosis* 208:119–125.

Khurana, S., Venkataraman, K., Hollingsworth, A., Piche, M., and Tai, T.C. 2013. Polyphenols: Benefits to the cardiovascular system in health and in aging. *Nutrients* 5:3779–3827.

Kometani, T., Fukuda, T., Kakuma, T., Kawaguchi, K., Tamura, W., Kumazawa, Y., and Nagata, K. 2008. Effects of alpha-glucosylhesperidin, a bioactive food material, on collagen-induced arthritis in mice and rheumatoid arthritis in humans. *Immunopharmacology and Immunotoxicology* 30:117–134.

Kwon, K.H., Murakami, A., Tanaka, T., and Ohigashi, H. 2005. Dietary rutin, but not its aglycone quercetin, ameliorates dextran sulfate sodium-induced experimental colitis in mice: Attenuation of pro-inflammatory gene expression. *Biochemical Pharmacology* 69:395–406.

Lafay, S., Gil-Izquierdo, A., Manach, C., Morand, C., Besson, C., and Scalbert, A. 2006. Chlorogenic acid is absorbed in its intact form in the stomach of rats. *Journal of Nutrition* 136:1192–1197.

Lambert, J.D. and Elias, R.J. 2010. The antioxidant and pro-oxidant activities of green tea polyphenols: A role in cancer prevention. *Archives of Biochemistry and Biophysics* 501:65–72.

Lanza, I.R., Short, D.K., Short, K.R., Raghavakaimal, S., Basu, R., Joyner, M.J., McConnell, J.P., and Sreekumaran Nair, K. 2008. Endurance exercise as a countermeasure for aging. *Diabetes* 57:2933–2942.

Lee, H.C., Jenner, A.M., Low, C.S., and Lee, Y.K. 2006. Effect of tea phenolics and their aromatic fecal bacterial metabolites on intestinal microbiota. *Research in Microbiology* 157:876–884.

Leopoldini, M., Marino, T., Russo, N., and Toscano, M. 2004. Antioxidant properties of phenolic compounds: H-Atom versus electron transfer mechanism. *Journal of Physical Chemistry A* 108:4916–4922.

Ley, R.E., Turnbaugh, P.J., Klein, S., and Gordon, J.I. 2006. Microbial ecology: Human gut microbes associated with obesity. *Nature* 444:1022–1023.

Li, R., Li, J., Cai, L., Hu, C.-M., and Zhang, L. 2008. Suppression of adjuvant arthritis by hesperidin in rats and its mechanisms. *Journal of Pharmacy and Pharmacology* 60:221–228.

Lian, T.-W., Wang, L., Lo, Y.-H., Jen Huang, I., and Wu, M.-J. 2008. Fisetin, morin and myricetin attenuate CD36 expression and oxLDL uptake in U937-derived macrophages. *Biochimica et Biophysica Acta-Molecular and Cell Biology of Lipids* 1781:601–609.

Liberato, D.J., Byers, V.S., Dennick, R.G., and Castagnoli, N. 1981. Regiospecific attack of nitrogen and sulfur nucleophiles on quinones derived from poison oak/ivy catechols (urushiols) and analogs as models for urushiol-protein conjugate formation. *Journal of Medicinal Chemistry* 24:28–33.

Liehr, J.G. and Roy, D. 1990. Free radical generation by redox cycling of estrogens. *Free Radical Biology and Medicine* 8:415–423.

Lin, N., Sato, T., Takayama, Y., Mimaki, Y., Sashida, Y., Yano, M., and Ito, A. 2003. Novel anti-inflammatory actions of nobiletin, a citrus polymethoxy flavonoid, on human synovial fibroblasts and mouse macrophages. *Biochemical Pharmacology* 65:2065–2071.

Litwinienko, G. and Ingold, K.U. 2003. Abnormal solvent effects on hydrogen atom abstractions. 1. The reactions of phenols with 2,2-diphenyl-1-picrylhydrazyl (dpph•) in alcohols. *Journal of Organic Chemistry* 68:3433–3438.

Litwinienko, G. and Ingold, K.U. 2007. Solvent effects on the rates and mechanisms of reaction of phenols with free radicals. *Accounts of Chemical Research* 40:222–230.

Lizárraga, D., Vinardell, M.P., Noé, V. et al. 2011. A lyophilized red grape pomace containing proanthocyanidin-rich dietary fiber induces genetic and metabolic alterations in colon mucosa of female C57BL/6J mice. *Journal of Nutrition* 141:1597–1604.

Lodi, F., Jiménez, R., Moreno, L. et al. 2009. Glucuronidated and sulfated metabolites of the flavonoid quercetin prevent endothelial dysfunction but lack direct vasorelaxant effects in rat aorta. *Atherosclerosis* 204:34–39.

Loke, W.M., Proudfoot, J.M., Stewart, S., McKinley, A.J., Needs, P.W., Kroon, P.A., Hodgson, J.M., and Croft, K.D. 2008. Metabolic transformation has a profound effect on anti-inflammatory activity of flavonoids such as quercetin: Lack of association between antioxidant and lipoxygenase inhibitory activity. *Biochemical Pharmacology* 75:1045–1053.

Long, L.H., Clement, M.V., and Halliwell, B. 2000. Artifacts in cell culture: Rapid generation of hydrogen peroxide on addition of (−)-epigallocatechin, (−)-epigallocatechin gallate, (+)-catechin, and quercetin to commonly used cell culture media. *Biochemical and Biophysical Research Communications* 273:50–53.

Lozano, C., Juliá, L., Jiménez, A., Touriño, S., Centelles, J.-J., Cascante, M., and Torres, J.L. 2006. Electron transfer capacity of catechin derivatives and influence on the cell cycle and apoptosis in HT29 cells. *FEBS Journal* 273:2475–2486.

Lu, L., Chen, S.S., Zhang, J.Q., Ramires, F.J., and Sun, Y. 2004. Activation of nuclear factor-kappa B and its proinflammatory mediator cascade in the infarcted rat heart. *Biochemical and Biophysical Research Communications* 321:879–885.

Mackenzie, G.G., Delfino, J.M., Keen, C.L., Fraga, C.G., and Oteiza, P.I. 2009. Dimeric procyanidins are inhibitors of NF-κB-DNA binding. *Biochemical Pharmacology* 78:1252–1262.

Manach, C., Mazur, A., and Scalbert, A. 2005. Polyphenols and prevention of cardiovascular diseases. *Current Opinion in Lipidology* 16:77–84.

Martín, M.A., Ramos, S., Granado-Serrano, A.B., Rodríguez-Ramiro, I., Trujillo, M., Bravo, L., and Goya, L. 2010. Hydroxytyrosol induces antioxidant/detoxificant enzymes and Nrf2 translocation via extracellular regulated kinases and phosphatidylinositol-3-kinase/protein kinase B pathways in HepG2 cells. *Molecular Nutrition and Food Research* 54:956–966.

Mateos-Martín, M.L., Pérez-Jiménez, J., Fuguet, E., and Torres, J.L. 2012. Non-extractable proanthocyanidins from grapes are a source of bioavailable (epi)catechin and derived metabolites in rats. *British Journal of Nutrition* 108:290–297.

Matito, C., Agell, N., Sánchez-Tena, S., Torres, J.L., and Cascante, M. 2011. Protective effect of structurally diverse grape procyanidin fractions against UV-induced cell damage and death. *Journal of Agricultural and Food Chemistry* 59:4489–4495.

Matsuki, T., Watanabe, K., Fujimoto, J., Takada, T., and Tanaka, R. 2004. Use of 16S rRNA gene-targeted group-specific primers for real-time PCR analysis of predominant bacteria in human feces. *Applied and Environmental Microbiology* 70:7220–7228.

Medzhitov, R. 2008. Origin and physiological roles of inflammation. *Nature* 454:428–435.

Metodiewa, D., Jaiswal, A.K., Cenas, N., Dickancaite, E., and Segura-Aguilar, J. 1999. Quercetin may act as a cytotoxic prooxidant after its metabolic activation to semiquinone and quinoidal product. *Free Radical Biology and Medicine* 26:107–116.

Meydani, M. and Hasan, S.T. 2010. Dietary polyphenols and obesity. *Nutrients* 2:737–751.

Mittler, R., Vanderauwera, S., Suzuki, N., Miller, G., Tognetti, V.B., Vandepoele, K., Gollery, M., Shulaev, V., and Van Breusegem, F. 2011. ROS signaling: The new wave? *Trends in Plant Science* 16:300–309.

Miura, Y.H., Tomita, I., Watanabe, T., Hirayama, T., and Fukui, S. 1998. Active oxygen generation by flavonoids. *Biological and Pharmaceutical Bulletin* 21:93–96.

Monagas, M., Urpí-Sardá, M., Sánchez-Patán, F., Llorach, R., Garrido, I., Gómez-Cordovés, C., Andrés-Lacueva, C., and Bartolomé, B. 2010. Insights into the metabolism and microbial biotransformation of dietary flavan-3-ols and the bioactivity of their metabolites. *Food and Function* 1:233–253.

Morinobu, A., Biao, W., Tanaka, S., Horiuchi, M., Jun, L., Tsuji, G., Sakai, Y., Kurosaka, M., and Kumagai, S. 2008. (–)-Epigallocatechin-3-gallate suppresses osteoclast differentiation and ameliorates experimental arthritis in mice. *Arthritis and Rheumatism* 58:2012–2018.

Mouradov, A. and Spangenberg, G. 2014. Flavonoids: A metabolic network mediating plants adaptation to their real estate. *Frontiers in Plant Science* 5:1–16.

Murakami, A., Song, M., Katsumata, S.-I., Uehara, M., Suzuki, K., and Ohigashi, H. 2007. Citrus nobiletin suppresses bone loss in ovariectomized ddY mice and collagen-induced arthritis in DBA/1J mice: Possible involvement of receptor activator of NF-kappaB ligand (RANKL)-induced osteoclastogenesis regulation. *Biofactors* 30:179–192.

Murphy, K.J., Chronopoulos, A.K., Singh, I., Francis, M.A., Moriarty, H., Pike, M.J., Turner, A.H., Mann, N.J., and Sinclair, A.J. 2003. Dietary flavanols and procyanidin oligomers from cocoa (*Theobroma cacao*) inhibit platelet function. *American Journal of Clinical Nutrition* 77:1466–1473.

Noverr, M.C. and Huffnagel, G.B. 2004. Does the microbiota regulate immune responses outside the gut? *Trends in Microbiology* 12:562–568.

Nunes, C., Almeida, L., and Laranjinha, J. 2005. Synergistic inhibition of respiration in brain mitochondria by nitric oxide and dihydroxyphenylacetic acid (DOPAC) Implications for Parkinson's disease. *Neurochemistry International* 47:173–182.

Okuda, T. 2005. Systematics and health effects of chemically distinct tannins in medicinal plants. *Phytochemistry* 66:2012–2031.

Parkar, S.G., Stevenson, D.E., and Skinner, M.A. 2008. The potential influence of fruit polyphenols on colonic microflora and human gut health. *International Journal of Food Microbiology* 124:295–298.

Pérez-Jiménez, J., Hubert, J., Hooper, L., Cassidy, A., Manach, C., Williamson, G., and Scalbert, A. 2010. Urinary metabolites as biomarkers of polyphenol intake in humans: A systematic review. *American Journal of Clinical Nutrition* 92:801–809.

Qin, J., Li, R., Raes, J. et al. 2010. A human gut microbial gene catalogue established by metagenomic sequencing. *Nature* 464:59–65.

Queipo-Ortuño, M.I., Boto-Ordóñez, M., Murri, M., Gómez-Zumaquero, J.M., Clemente-Postigo, M., Estruch, R., Díaz, F.C., Andrés-Lacueva, C., and Tinahones, F.J. 2012. Influence of red wine polyphenols and ethanol on the gut microbiota ecology and biochemical biomarkers. *American Journal of Clinical Nutrition* 95:1323–1334.

Ramos, S. 2008. Cancer chemoprevention and chemotherapy: Dietary polyphenols and signalling pathways. *Molecular Nutrition and Food Research* 52:507–526.

Redman, L.M. and Ravussin, E. 2011. Caloric restriction in humans: impact on physiological, psychological, and behavioral outcomes. *Antioxidants and Redox Signaling* 14:275–287.

Rhee, S.G. and Woo, H.A. 2010. Multiple functions of peroxiredoxins: Peroxidases, sensors and regulators of the intracellular messenger H_2O_2, and protein chaperones. *Antioxidants and Redox Signaling* 15:781–794.

Rice-Evans, C.A., Miller, N.J., and Paganga, G. 1996. Structure-antioxidant activity relationships of flavonoids and phenolic acids. *Free Radical Biology and Medicine* 20:933–956.

Ristow, M. and Schmeisser, S. 2011. Extending life span by increasing oxidative stress. *Free Radical Biology and Medicine* 51:327–336.

Ristow, M. and Zarse, K. 2010. How increased oxidative stress promotes longevity and metabolic health: The concept of mitochondrial hormesis (mitohormesis). *Experimental Gerontology* 45:410–418.

Rivera, L., Morón, R., Sánchez, M., Zarzuelo, A., and Galisteo, M. 2008. Quercetin ameliorates metabolic syndrome and improves the inflammatory status in obese Zucker rats. *Obesity* 16:2081–2087.

Rivera, L., Morón, R., Zarzuelo, A., and Galisteo, M. 2009. Long-term resveratrol administration reduces metabolic disturbances and lowers blood pressure in obese Zucker rats. *Biochemical Pharmacology* 77:1053–1063.

Rodríguez-Mateos, A., Vauzour, D., Krueger, Shanmuganayagam, C.G.D., Reed, J., Calani, L., Mena, P., Rio, D.D., and Crozier, A. 2014. Bioavailability, bioactivity and impact on health of dietary flavonoids and related compounds: An update. *Archives of Toxicology* 88:1803–1853.

Romier, B., Van De Walle, J., During, A., Larondelle, Y., and Schneider, Y.J. 2008. Modulation of signalling nuclear factor-κB activation pathway by polyphenols in human intestinal Caco-2 cells. *British Journal of Nutrition* 100:542–551.

Roy, D., Weisz, J., and Liehr, J.G. 1990. The O-methylation of 4-hydroxyestradiol is inhibited by 2-hydroxyestradiol: Implications for estrogen-induced carcinogenesis. *Carcinogenesis* 11:459–462.

Rusznyák, S.P. and Szent-Györgyi, A. 1936. Vitamin P: Flavonols as vitamins. *Nature* 138:27.

Sang, S., Hou, Z., Lambert, J.D., and Yang, C.S. 2005. Redox properties of tea polyphenols and related biological activities. *Antioxidants and Redox Signaling* 7:1704–1714.

Sanz, Y. 2010. Effects of a gluten-free diet on gut microbiota and immune function in healthy adult humans. *Gut Microbes* 1:135–137.

Scalbert, A., Manach, C., Morand, C., Rémésy, C., and Jiménez, L. 2005. Dietary polyphenols and the prevention of diseases. *Critical Reviews in Food Science and Nutrition* 45:287–306.

Scalbert, A. and Williamson, G. 2000. Dietary intake and bioavailability of polyphenols. *Journal of Nutrition* 130:2073S–2085S.

Schulz, T.J., Zarse, K., Voigt, A., Urban, N., Birringer, M., and Ristow, M. 2007. Glucose restriction extends *Caenorhabditis elegans* life span by inducing mitochondrial respiration and increasing oxidative stress. *Cell Metabolism* 6:280–293.

Seibel, J., Molzberger, A.F., Hertrampf, T., Laudenbach-Leschowski, U., Degen, G.H., and Diel, P. 2008. In utero and postnatal exposure to a phytoestrogen-enriched diet increases parameters of acute inflammation in a rat model of TNBS-induced colitis. *Archives of Toxicology* 82:941–950.

Sekine, C., Sugihara, T., Miyake, S., Hirai, H., Yoshida, M., Miyasaka, N., and Kohsaka, H. 2008. Successful treatment of animal models of rheumatoid arthritis with small-molecule cyclin-dependent kinase inhibitors. *Journal of Immunology* 180:1954–1961.

Sekirov, I., Russell, S.L., Caetano, L., Antunes, M., and Finlay, B. 2010. Gut microbiota in health and disease. *Physiological Reviews* 90:859–904.

Selma, M.V., Espín, J.C., and Tomás-Barberán, F.A. 2009. Interaction between phenolics and gut microbiota: Role in human health. *Journal of Agricultural and Food Chemistry* 57:6485–6501.

Serra, A., Bladé, C., Arola, L., Macià, A., and Motilva, M.J. 2013. Flavanol metabolites distribute in visceral adipose depots after a long-term intake of grape seed proanthocyanidin extract in rats. *British Journal of Nutrition* 110:1411–1420.

Serra, A., Macià, A., Romero, M.-P., Reguant, J., Ortega, N., and Motilva, M.-J. 2012. Metabolic pathways of the colonic metabolism of flavonoids (flavonols, flavones and flavanones) and phenolic acids. *Food Chemistry* 130:383–393.

Shapiro, H., Singer, P., Halpern, Z., and Bruck, R. 2007. Polyphenols in the treatment of inflammatory bowel disease and acute pancreatitis. *Gut* 56:426–435.

Sies, H. 1991. *Oxidative Stress: Oxidants and Antioxidants*. New York: Academic Press.

Singh, B.N., Shankar, S., and Srivastava, R.K. 2011. Green tea catechin, epigallocatechin-3-gallate (EGCG): Mechanisms, perspectives and clinical applications. *Biochemical Pharmacology* 82:1807–1821.

Sjödin, M., Irebo, T., Utas, J.E., Lind, J., Merényi, G., Åkermark, B., and Hammarström, L. 2006. Kinetic effects of hydrogen bonds on proton-coupled electron transfer from phenols. *Journal of the American Chemical Society* 128:13076–13083.

Son, J., Pang, B., McFaline, J.L., Taghizadeh, K., and Dedon, P.C. 2008. Surveying the damage: The challenges of developing nucleic acid biomarkers of inflammation. *Molecular Biosystems* 4:902–908.

Soulat, T., Philippe, C., Sollier, C.B.D., Brezillon, C., Berge, N., Teissedre, P.-L., Callebert, J., Rabot, S., and Drouet, L. 2006. Wine constituents inhibit thrombosis but not atherogenesis in C57BL/6 apolipoprotein E-deficient mice. *British Journal of Nutrition* 96:290–298.

Spencer, J.P.E., Chaudry, F., Pannala, A.S., Kaila Srai, S., Debnam, E., and Rice-Evans, C. 2000. Decomposition of cocoa procyanidins in the gastric milieu. *Biochemical and Biophysical Research Communications* 272:236–241.

Spencer, J.P.E., Rice-Evans, C., and Williams, R.J. 2003. Modulation of pro-survival Akt/protein kinase B and ERK1/2 signaling cascades by quercetin and its in vivo metabolites underlie their action on neuronal viability. *Journal of Biological Chemistry* 278:34783–34793.

Spencer, J.P.E., Schroeter, H., Rechner, A.R., and Rice-Evans, C. 2001. Bioavailability of flavan-3-ols and procyanidins: Gastrointestinal tract influences and their relevance to bioactive forms *in vivo*. *Antioxidants and Redox Signaling* 3:1023–1039.

Spencer, J.P.E., Vafeiadou, K., Williams, R.J., and Vauzour, D. 2012. Neuroinflammation: Modulation by flavonoids and mechanisms of action. *Molecular Aspects of Medicine* 33:83–97.

Steffen, Y., Schewe, T., and Sies, H. 2007. (–)-Epicatechin elevates nitric oxide in endothelial cells via inhibition of NADPH oxidase. *Biochemical and Biophysical Research Communications* 359:828–833.

Taubert, D., Roesen, R., and Schomig, E. 2007. Effect of cocoa and tea intake on blood pressure: A meta-analysis. *Archives of Internal Medicine* 167:626–634.

Terao, J., Kawai, Y., and Murcita, K. 2008. Vegetable flavonoids and cardiovascular disease. *Asia Pacific Journal of Clinical Nutrition* 17:291–293.

Tipton, D.A., Carter, T.B., and Dabbous, M.K. 2014. Inhibition of interleukin 1β-stimulated interleukin-6 production by cranberry components in human gingival epithelial cells: Effects on nuclear factor κB and activator protein 1 activation pathways. *Journal of Periodontal Research* 49:437–447.

Torres, J.L., Lozano, C., Juliá, L., Sánchez-Baeza, F.J., Anglada, J.M., Centelles, J.J., and Cascante, M. 2002. Cysteinyl-flavan-3-ol conjugates from grape procyanidins. Antioxidant and antiproliferative properties. *Bioorganic and Medicinal Chemistry* 10:2497–2509.

Touriño, S., Pérez-Jiménez, J., Mateos-Martín, M.L., Fuguet, E., Vinardell, M.P., Cascante, M., and Torres, J.L. 2011. Metabolites in contact with the rat digestive tract after ingestion of a phenolic-rich dietary fiber matrix. *Journal of Agricultural and Food Chemistry* 59:5955–5963.

Tu, Y.-C., Lian, T.-W., Yen, J.-H., Chen, Z.-T., and Wu, M.J. 2007. Antiatherogenic effects of kaempferol and rhamnocitrin. *Journal of Agricultural and Food Chemistry* 55:9969–9976.

Turrens, J.F., Lariccia, J., and Gopal Nair, M. 1997. Resveratrol has no effect on lipoprotein profile and does not prevent peroxidation of serum lipids in normal rats. *Free Radical Research* 27:557–562.

Tzounis, X., Vulevic, J., Kuhnle, G.C., George, T., Leonczak, J., Gibson, G.R., Kwik-Uribe, C., and Spencer, J.P.E. 2008. Flavanol monomer-induced changes to the human faecal microflora. *British Journal of Nutrition* 99:782–792.

Van Dyke, T.E. and Kornman, K.S. 2008. Inflammation and factors that may regulate inflammatory response. *Journal of Periodontology* 79:1503–1507.

Vauzour, D., Vafeiadou, K., Rice-Evans, C., Williams, R.J., and Spencer, J.P.E., 2007. Activation of pro-survival Akt and ERK1/2 signalling pathways underlie the anti-apoptotic effects of flavanones in cortical neurons. *Journal of Neurochemistry* 103:1355–1367.

Viña, J., Borrás, C., Abdelaziz, K.M., García-Vallés, R., and Gómez-Cabrera, M.C. 2013. The free radical theory of aging revisited: The cell signaling disruption theory of aging. *Antioxidants and Redox Signaling* 19:779–787.

Walker, E.H., Pacold, M.E., Perisic, O., Stephens, L., Hawkins, P.T., Wymann, M.P., and Williams, R.L. 2000. Structural determinants of phosphoinositide 3-kinase inhibition by wortmannin, LY294002, quercetin, myricetin, and staurosporine. *Molecular Cell* 6:909–919.

Wang, J., Zhang, Q., Jin, S., He, D., Zhao, S., and Liu, S. 2008. Genistein modulate immune responses in collagen-induced rheumatoid arthritis model. *Maturitas* 59:405–412.

Weimer, S., Priebs, J., Kuhlow, D. et al. 2014. D-Glucosamine supplementation extends life span of nematodes and of ageing mice. *Nature Communications* 5:3563.

Williamson, G., Barron, D., Shimoi, K., and Terao, J. 2005. In vitro biological properties of flavonoid conjugates found in vivo. *Free Radical Research* 39:457–469.

Williamson, G. and Clifford, M.N. 2010. Colonic metabolites of berry polyphenols: The missing link to biological activity? *British Journal of Nutrition* 104:S48–S66.

Wilson, T., Knight, T.J., Beitz, D.C., Lewis, D.S., and Engen, R.I. 1996. Resveratrol promotes atherosclerosis in hypercholesterolemic rabbits. *Life Sciences* 59:PL15–PL21.

Wright, J.S., Johnson, E.R., and DiLabio, G.A. 2001. Predicting the activity of phenolic antioxidants: Theoretical method, analysis of substituent effects, and application to major families of antioxidants. *Journal of the American Chemical Society* 123:1173–1183.

Xia, M., Hou, M., Zhu, H. et al. 2005. Anthocyanins induce cholesterol efflux from mouse peritoneal macrophages: The role of the peroxisome proliferator-activated receptor γ-liver X receptor α-ABCA1 pathway. *Journal of Biological Chemistry* 280:36792–36801.

Yang, C.S., Lee, M.J., and Chen, L. 1999. Human salivary tea catechin levels and catechin esterase activities: Implication in human cancer prevention studies. *Cancer Epidemiology, Biomarkers and Prevention* 8:83–89.

Yang, G.-Y., Liao, J., Li, C., Chung, J., Yurkow, E.J., Ho, C.-T., and Yang, C.S. 2000. Effect of black and green tea polyphenols on c-Jun phosphorylation and H_2O_2 production in transformed and non-transformed human bronchial cell lines: Possible mechanisms of cell growth inhibition and apoptosis induction. *Carcinogenesis* 21:2035–2039.

Yang, H.L., Chen, S.C., Senthil, K.J., Ku, Y.N., Chao, P.D.L., Tsai, S.Y., Hou, Y.C., and Hseu, Y.C. 2012. Antioxidant and anti-inflammatory potential of hesperetin metabolites obtained from hesperetin-administered rat serum: An ex vivo approach. *Journal of Agricultural and Food Chemistry* 60:522–532.

Yoshida, Y., Shioi, T., and Izumi, T. 2007. Resveratrol ameliorates experimental autoimmune myocarditis. *Circulation Journal* 71:397–404.

Yuan, J.-P., Wang, J.-H., and Liu, X. 2007. Metabolism of dietary soy isoflavones to equol by human intestinal microflora: Implications for health. *Molecular Nutrition and Food Research* 51:765–781.

Zhu, B., Wang, X., and Li, L. 2010. Human gut microbiome: The second genome of human body. *Protein and Cell* 1:718–725.

Zucker, W.V. 1983. Tannins: Does structure determine function? An ecological perspective. *American Naturalist* 121:335–365.

3 Solid–Liquid Extraction of Polyphenols at Low Pressure

Eduardo Caballero-Valdés,
Araceli Olivares-Miralles, Carmen Soto-Maldonado,
and María Elvira Zúñiga-Hansen

CONTENTS

3.1 INTRODUCTION

Solubilization in organic solvents under low pressures (near atmospheric) is one of the most common processes for the recovery of compounds from vegetable matrices. The process considers different operating variables and depends on multiple factors (Teixeira et al., 2014). This chapter focuses mainly on the effect of operational conditions (type of solvent, temperature, pH, and raw material/solvent volume ratio) on polyphenols extraction efficiency and the activity of the obtained extracts. Special attention is given to the application of enzymes to enhance the recovery of polyphenols. Finally, examples of scaling up the process are presented, and some suppliers of commercial equipment are mentioned.

Extraction is the separation of one or more components by feedstock contact with an immiscible liquid (Couper et al., 2012). In the case of solid–liquid

TABLE 3.1
Classification Techniques for Solid–Liquid Extraction in Conventional and Nonconventional Methods

Extraction Method	Technique	Characteristics
Conventional	Solvent extraction (SE)	Compounds are separated based on their solubilities (immiscible liquids)
	Soxhlet	Compounds of lipidic nature
	Distillation	Liquids with different volatilities
Nonconventional	Pressurized liquid extraction (PLE)	Pressurized SE using water and organic solvents at high temperatures and pressures
	Microwave-assisted extraction (MAE)	Increases molecular movement and energy transfer (electromagnetic radiation) to solvent and solid
	Supercritical fluids extraction (SFE)	Better solubility and mass transfer than a solvent gas
	Turbo extraction	High-speed stirring induces hydrodynamic cavitation
	Electric extraction	Electric shocks are applied
	Ultrasound-assisted extraction (UAE)	High-frequency sound waves enhance the release of a solute from the solid
	Enzyme-assisted extraction	Enzymes help release intracellular polyphenols and liberate polyphenols bound to other molecules

Sources: Adapted from Azuola, R. and Vargas, P., *Tecnología en Marcha*, 20(4), 30, 2007; Gil-Chávez, G. et al., *Compr. Rev. Food Sci. Food Saf.*, 12, 5, 2013.

extraction, the mass transfer of one or more soluble components of an insoluble solid to the liquid phase is accomplished by this process. Solid–liquid extraction methods under low pressures can be divided into conventional and nonconventional (Table 3.1). The methods commonly used for the extraction are the conventional liquid–liquid or solid–liquid extraction, and the advanced methods (nonconventional) include pressurized liquid extraction (PLE), subcritical and supercritical extractions, microwave-assisted extraction (MAE), and ultrasound-assisted extraction (UAE) among others.

Even though there are many studies in literature regarding polyphenol content and antioxidant activity of fruits and vegetables, much less information about (1) extraction methods and (2) operating variables to optimize the extraction process is available. The extraction method strongly affects the extraction yield and activity. Solvent extraction is the most frequently used procedure to obtain extracts rich in phenolic compounds from fruits and vegetables. The selection of one or more solvents that are efficient for the extraction of polyphenols depends not only on the different physical and chemical characteristics of the plant material subjected to extraction but also on the specific phenolic compounds to be isolated.

Currently, there is great interest in solvent extraction of bioactive compounds, specifically polyphenols, and new techniques have been developed, such as ultrasound-assisted extraction, microwave-assisted extraction (Nayak et al., 2015). These methods have been developed to increase extraction yields and reduce extraction times. Despite these new developments, extraction with food-grade solvents, like aqueous ethanol mixtures, is the most commonly used at a commercial scale (Roopchand et al., 2013; Wood et al., 2013). Another alternative to conventional solvent extraction like hydroalcoholic mixtures, are the aqueous β-cyclodextrin solution (Alexandru et al., 2014). The principles of conventional solvent and Soxhlet extraction processes are based on the solubility of the solute in the solvent, which depends on temperature and solvent type (Azuola and Vargas, 2007; Prior et al., 2005). Techniques of nonconventional extraction methods, with the exception of the supercritical fluid extraction (SFE), are used to improve the extraction yields either by increasing the solvent–solute interaction or by cavitation (Azuola and Vargas, 2007).

3.2 PHENOLIC COMPOUNDS EXTRACTION: EFFECT OF OPERATIONAL CONDITIONS

Extraction with solvents in liquid phase is the most widely used method to separate soluble compounds by diffusion from a solid matrix. Specific extraction conditions are required to optimize the extraction of defined polyphenols, given their difference in chemical structure and the different composition and characteristics of the vegetable matrix. Phenolic compounds have been extracted from fresh, lyophilized (Sanal et al., 2005; Wijngaard et al., 2012), and dry samples (Conde et al., 2009) using a variety of solvents and sometimes by sequential use of solvents with increasing polarity (Escribano and Santos-Buelga, 2003).

Phenolic compounds are secondary metabolites produced by plants to comply with various biological functions (Naczk and Shahidi, 2004). A large variety of phenols are found in vegetables, such as benzoquinones, hydroxybenzoic acids, acetophenones, phenylacetic acids, hydroxycinnamic acids, phenylpropanoids, naphthoquinones, xanthones anthraquinones, flavonoids, isoflavones, lignans, neolignans, bioflavonoids, lignins, and condensed tannins with beneficial effects due to their antioxidant activity (Balasundram et al., 2006; Krishnaiah et al., 2011; Naczk and Shahidi, 2004, 2006). These phenolic compounds are not distributed uniformly in the vegetable matrix. Soluble phenols are encapsulated in plant vacuoles (Naczk and Shahidi, 2004, 2006) while insoluble polyphenols are present in the cell wall. The latter contribute to mechanical strength, plant growth, and morphogenesis. Some insoluble polyphenols that are linked to polysaccharides play a central role in cell–cell adhesion, and they serve as a site for the formation of lignin and contribute to the thermal stability of the cell. In grapes, the seeds contain about 63% of all phenolics (mostly tannins and carotenoids), the skin contains 34%, and the remaining 3% is found in the flesh (Paladino and Zuritz, 2012). Skin polyphenols (mostly anthocyanins) are found in the cell wall, within vacuoles, and associated with nuclear material. Extracts from plant materials are always a mixture of diverse soluble phenolic and non-phenolic compounds. The latter can be sugars, waxes, terpenes, chlorophyll and fats; hence, for commercial preparations, a further purification stage will be usually required. The degree of polymerization of grape polyphenols depends on where they are located; polyphenols found in the skin are mainly polymeric while monomeric polyphenols predominate in the seed (Gökmen et al., 2009).

The recovery of polyphenols from solid plant materials comprises two main stages: first polyphenols inside the solid matrix are dissolved, and then they diffuse through the pores of the solid matrix to the bulk of the solvent (Linares et al., 2011; Spigno et al., 2007; Tabart et al., 2007). The rate of diffusion depends on the molecular weight of the solute, permeability of the solid matrix, pore size, size of the solid particles, and polyphenols solubility in the given solvent (Cerpa-Calderon and Kennedy, 2008). Important operating variables that influence the efficiency of the extraction process are as follows: type of solvent, pH, contact time, temperature, number of extraction steps, and volume of solvent used at each step. Extraction temperature and contact time are critical since temperature strongly affects the degree of solubilization and degradation rate of polyphenols (Larrauri et al., 1997); optimal extraction conditions should be a compromise between both the effects.

3.2.1 Types of Solvents Used for the Extraction Process

The polarity of the polyphenol to be extracted defines the type of solvent. Since plant materials contain several polyphenols with different polarities, a sole solvent (or solvent mixture) will not provide optimal recovery for all polyphenols. The pH effect is especially complex in the case of anthocyanin extraction (Rodis et al., 2002). For example, it has been found that in grape seeds extraction, the efficiency and extract composition in total flavanols, procyanidins, and monomeric flavanols vary considerably depending on the type of solvent used (Jayaprakasha et al., 2001).

Soluble phenolic compounds are usually extracted using water, methanol, ethanol, ethyl acetate, acetone, and combinations of these solvents (Naczk and Shahidi, 2004), which can be mixed with a small amount of acid to increase the extraction yield. Glycosides are phenolic compounds bound to sugars, which are highly soluble in water. Mixtures of organic solvents with water are good options to extract plant materials rich in glycosides (Conde et al., 2009; Laroze et al., 2010) (see Figure 3.1). In contrast, less polar aglycones tend to be more soluble in nonaqueous solvents. Methanol at 70% or 80% has been very efficient in the extraction of hydroxycinnamic derivatives, flavones, flavonols, and catechins from some fruit and leguminous, especially grape seeds and marc (Bonilla et al., 1999; Escribano and Santos-Buelga, 2003). Flavones and flavonols are not generally extracted intact but as a form of their aglycones (Escribano and Santos-Buelga, 2003). The efficiency of proanthocyanidins extraction varies with the solvent and its polymerization degree. A variable proportion of proanthocyanidins is resistant to standard extraction, particularly those in old or oxidized tissues (García-Alonso et al., 2004).

Finally, Prommuak et al. (2008) and Casas et al. (2010) observed that the effect of the solvent/temperature pair in polyphenols extraction is statistically significant especially for water–alcohol mixtures under 70% of alcohol. Since at higher alcohol concentrations this effect is not noticeable, they recommended extracting at low

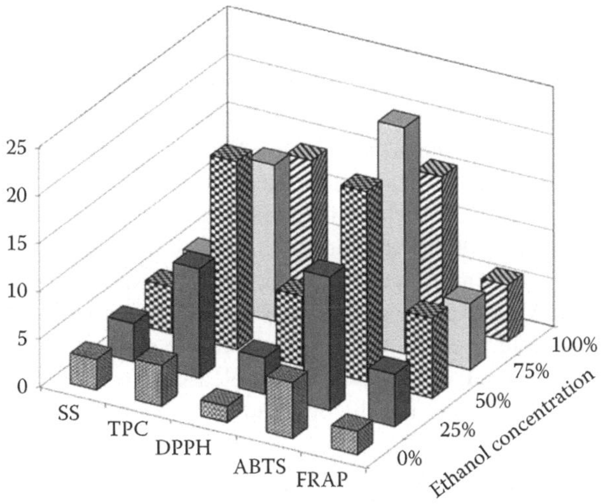

FIGURE 3.1 Effect of solvent types on the recovery of phenolic compounds from raspberry pomace. SS (g/g d.w.), TPC (mg gallic acid/g d.s.), DPPH (mmol TE/100 g d.s.), ABTS (mmol TE/100 g d.s.), and FRAP (mmol AAE/100 g d.s.). (Adapted from Laroze, L. et al., *Electron. J. Biotechnol.*, 13(6), 6, 2010.)

temperatures in these cases. Alcohol mixtures are commonly used nowadays to extract polyphenols; methanol and ethanol concentrations in the range 50%–85% produced the best results (Corrales et al., 2009; Durhing et al., 2007; Karacabey et al., 2010; Li et al., 2006; Pompeu et al., 2009). The use of different solvents allows for the selective extraction of bioactive compounds: anthocyanins are water soluble (Borowska et al., 2005; Luque-Rodriguez et al., 2007) but are also highly soluble in ethanol (Corrales et al., 2009). On the other hand, methanol has a higher affinity for catechin, epicatechin, benzoic acid monomers, flavan-3-ols, and flavanols (Dopico-Garcia, 2007; Naczk and Shahidi, 2004; Tabart et al., 2007).

3.2.2 Extraction Temperature

In a conventional solid/liquid extraction, temperature affects both the physical and chemical solvent process, as well as the characteristics of the extracts. Most of the time, extraction at high temperatures improves the extraction efficiency due to the obtained permeabilization of cell walls, increased solubility and diffusion coefficients of polyphenols, reduced viscosity of the solvent, reduced surface tension as well as enhanced hydrolysis and break-down reactions. Several authors have shown that the temperature rise is directly related to the amount of bioactive compounds recovered in the extract mainly due to increased solubility. This rise increases the movement of solutes in the active sites of the cellular matrix and reduces the strength of the hydrogen bonds, decreasing the energy required to break the matrix–solute interactions (Casas et al., 2010; Prommuak et al., 2008). In addition to affecting the solute/solvent interaction, temperature also modifies cell membranes. Temperatures above 65°C break part of the cell structure and enhance diffusion of substances from the intracellular matrix to the environment (Harbourne et al., 2009; Li et al., 2006). At high temperatures the dielectric constant of the solvent decreases, favoring the dissolution and extraction of low-polarity compounds (Budrat and Shotipruk, 2009; Hassas-Roudsari et al., 2009). In addition to this, high temperatures increase the diffusion coefficient (Corrales et al., 2009; Hassas-Roudsari et al., 2009; Luque-Rodríguez et al., 2007; Silva et al., 2007) and reduce the surface tension between the solvent and the solid matrix (Pompeu et al., 2009), reducing the contact time and the volume of solvent. Moreover, in many cases, extracts obtained at high temperatures are rich in simple phenolic compounds (Marete et al., 2009; Pompeu et al., 2009; Ruenroengklin et al., 2008), which are released after the hydrolysis of tannins (Harbourne et al., 2009) and the breakdown of phenolic–protein complexes (Al-Farsi and Yong, 2008).

However, increasing temperature may negatively affect the activity of the extracts obtained due to degradation or volatilization of bioactive compounds (Larrauri et al., 1997). These can also react with other components in the plants, preventing their extraction in some cases. Moreover, some combinations of time and temperature would affect the stability of the antioxidant capacity. Anthocyanins may lose some of their activity at temperatures above 70°C, depending on the contact time (Luque-Rodriguez et al., 2007). Dopico-García et al. (2007) indicated that high temperatures mainly affect the stability of flavonoids while organic acids are not significantly affected. Flavonols require much energy to be released (Silva et al., 2007); therefore, high temperatures improve their extraction. Azizah et al. (1999) studied the effect of temperature between 30°C and 90°C on the antioxidant activity of extracts of cocoa products, as well as Lecoupeau et al. (2008). Stable extracts were obtained at 50°C; higher temperature extraction resulted in decreased antioxidant activity. Moreover, in the extraction of anthocyanins from berries, extraction yields are reduced at temperatures above 40°C, finding an optimum condition between 30°C and 35°C (Cacace and Mazza, 2003; Riihinen et al., 2008). The authors report an increase of solubility in anthocyanin and diffusion coefficients with the temperature, but temperatures between 40°C and 74°C cause degradation of these phenolic compounds. Larraurri et al. (1997) found that grape pulp drying at temperatures between 100°C and 140°C affects the subsequent extraction of phenolic compounds. The total polyphenol content was reduced from 33% to 19%, the condensed tannins from 17% to 11%, and the antioxidant capacity from 50% to 28%, respectively. Drying at 60°C did not significantly affect the extraction yield value.

Some enzymes found in the vegetable matrix are responsible for the natural oxidation of polyphenols. These enzymes, such as polyphenol oxidase (PPO, EC 1.14.18.1) and polyphenol peroxidase (POD, EC 1.11.1.7), play a fundamental role in the antioxidant activity of the extracts. In the vegetable matrix, enzymes and polyphenols are located in separate compartments; their contact only occurs when the cell configuration is altered (Zhao, 1999). Often, these enzymes have a range of inactivation between 60°C and 80°C (Anil et al., 2008; Fortea et al., 2009; Marete et al., 2009; Terefe et al., 2010). Nevertheless, Shiferaw et al. (2010) reported heat-resistant PPOs in strawberry varieties, which remained active at 100°C.

In summary, extraction temperatures should be high enough to increase mass transfer and solubility of the solutes but low enough to minimize degradation of phenolic compounds (Herrero et al., 2012). Improving extraction efficiencies will require longer contact times, ranging from a few minutes to several days. However, very long periods can cause degradation and oxidation of the target molecules (Naczk and Shahidi, 2004).

3.2.3 pH

The stability of the cell membrane is strongly affected by the pH and solvent (Naczk and Shahidi, 2004; Türker and Erdogdu, 2006). In addition to this, acidic conditions may enhance the solubility of phenolic compounds and modify their structure; both effects increase the antioxidant activity of the extracts (Baublis et al., 2000). Lemanska et al. (2001) investigated the effect of pH on the chemical structure of a series of hydroxyflavones, observing that pH affects the deprotonation of the hydroxyl portion, resulting in an increase of free radical–scavenging activity. The authors indicate that electron donation is the dominant antioxidant mechanism after deprotonation of hydroxyflavones.

3.2.4 Effect of the Raw Material/Solvent Volume Ratio

Another important variable in recovering polyphenols and bioactive compounds is the ratio (R) of raw material to solvent volume. Extraction conditions, such as temperature, solvent type, and number of extraction stages have been extensively studied while the solid/solvent ratio has been much less studied (Cacace and Mazza, 2003; Hojnik et al., 2008; Luthria, 2008; Naczk and Shahidi, 2006; Nawaz et al., 2006; Pompeu et al., 2009). Phenolic compounds interact with various plant components, including carbohydrates and proteins, leading to the formation of insoluble complexes. The solubility of phenolic compounds is also affected by the polarity of the solvent used. Therefore, the amount and type of solvent used in the extraction of the raw material from vegetal origin strongly influence the concentration and composition of the extracts obtained. Hence, previous exhaustive laboratory studies with the specific vegetal material must be carried out to select the type and amount of solvent used. For example, varying R from 1/5 to 1/10, the extraction yield and total phenols recovery increase using commercial canola plants and acetone (70%) (Naczk and Shahidi, 2006). Hence, improved polyphenol extraction yields can be achieved with higher solvent/solid ratios, as long as the solvent is not saturated (Moure et al., 2000; Takeoka and Dao, 2003); however, polyphenol concentrations in the resulting extract are usually lower. After extraction with high solvent/solid ratios, it is advisable to concentrate the extract so as to achieve higher antioxidant activities; this concentration process should be carried out under mild conditions to avoid activity loss.

To understand the influence of R, either in the laboratory or in industrial processes, it is necessary to know how the compounds of interest solubilize and diffuse out of the vegetable matrices. Industrial solid–liquid extraction requires mass transfer models to analyze the extraction from an engineering standpoint. The extraction of polyphenols is a multiphase and multicomponent mass transfer operation where a clear steady state is difficult to define. This process involves the solubilization of more than one solute from the matrix (Sánchez Guerrero et al., 2008), and their concentrations vary constantly. For modeling purposes, it is commonly assumed that the extraction process occurs under the following conditions (Naczk and Shahidi, 2006; Sánchez Guerrero et al., 2008):

1. The solid material is an assembly of particles of the same size.
2. The diffusion of the solute and other compounds is performed in parallel, and there is no interaction between them.
3. The diffusivity of the solute to be extracted is constant.
4. It is a porous solid.
5. The concentration of polyphenols is time varying.
6. The extracted polyphenols are considered equivalent to a particular compound, usually gallic acid.

Two consecutive steady-state mass-transfer steps can model the solvent extraction of a solute from a solid phase. At first, the dissolution is fast, where the solid particle boundary layer resistance defines the transfer rate. Then, the extraction rate is much slower where the diffusivity within the particle controls the process (Hojnik et al., 2008).

3.2.5 Successive Extraction Steps

Multistage successive extraction steps (up to six) improve the overall extraction efficiency. The highest recovery of target compounds is obtained in one or two stages, indicating that a large amount of bioactive compounds can be initially removed. More than three stages rarely produce a significant improvement (Naczk and Shahidi, 2006).

3.2.6 Effect of Enzyme Application in Antioxidant Recovery

The conditioning of the raw material can have a strong effect on process performance. Commercial enzymes, due to their hydrolytic activity, help to degrade the cell wall structure of the vegetable materials. This effect favors the release of desired compounds, increasing the extraction yield. Hydrolytic enzymes have been widely applied in food industry processes, such as conversion of starch in high-fructose syrups, baking, clarification of fruit juices, extraction of vegetable oils as well as in the making of beer, wine, and cheese (Hamada, 1997). Enzymes are highly efficient and specific biological catalysts, which offer many advantages over chemical catalysts. They come from renewable and biodegradable sources; they increase the specificity of the reaction and the product purity, as well as reduce the environmental impact (Cherry and Fidantsef, 2003). However, the application of enzymes in industrial processes under aggressive conditions such as high temperatures, organic solvents, and extreme pH, could be limited. Currently, obtaining low-cost enzymes with better stability and higher versatility is possible. The low or no cost of agroindustrial waste could support the feasibility of their application to obtain antioxidants and bioactives (Cruz et al., 2004; Kiassos et al., 2009).

During the processing of fruit juices and vegetables, a combination of pectinases, cellulases, and hemicellulases is used to facilitate maceration, liquefaction, and clarification, thus, reducing production costs and improving yields (Bhat, 2000). In the extraction of oilseeds, cellulolytic, hemicellulolytic, and pectinolytic enzymes increase the release of vegetable oil (Cerda et al., 2013; Collao et al., 2006; Concha et al., 2004; Rosenthal et al., 1996; Shabati et al., 1998; Shankar et al., 1997; Soto et al., 2004, 2007). In addition, oil extraction assisted by enzymes improves the nutritional quality of the product (Zuñiga et al., 2003).

Few studies have analyzed the application of enzymes in processes that use organic solvents or water to extract phenolic compounds from plant matrices. In these studies, the enzymatic pretreatment is performed only to facilitate the subsequent extraction with solvent. The enzyme-aided extraction of antioxidants from grape marc was reported by Meyer et al. (1998), observing an increase in the extraction of phenolic compounds when the commercial enzyme Grindamyl (originated from *Aspergillus niger*) was used, obtaining 3072 mg GAE/L in comparison to 2677 mg GAE/L with nonenzymatic extraction. Other authors, who have studied the extraction with different solvents from defatted grape seed (Bonilla et al., 1999; Tobar et al., 2004), wild black-berry (Céspedes et al., 2010) and residual flour borage (Soto et al., 2004, 2008), have reported an increase in the extraction of phenolic compounds and their antioxidant capacity by using an enzymatic treatment prior to the oil extraction process. In the case reported by Tobar et al. (2004), total recovery of phenolic compounds with enzymatic treatment increased from two to four times compared with a nonenzymatic process. However, no increment in the antioxidant activity was observed, probably due to the degradation of some phenolic compounds caused by the same enzymatic treatment. Landbo and Meyer (2001) used enzymes to recover antioxidants from black currant pomace. They analyzed the effect of the type of enzyme, hydrolysis time, presence or absence of seeds in the pomace, particle size in the pomace, temperature, and type of solvent used for the extraction. The study applied a factorial design 2^6 and hydrolysis time of 0.5 and 8 h, using several pectinases (Macer8 FJ, Macer8 R, Pectinex BE) and an acid protease (Novozym 89), and included two types of waste (pomace with and without seeds). Hydrolysis was carried out at 50°C with an enzyme/solute ratio (E/S) of 10% and particle size <250 μm. The study of Landbo and Meyer (2001) showed that the enzyme treatment improved the polyphenol extraction, although the extraction yield of anthocyanins was variable due to degradation. This may be attributable to the presence of β-glycosidase, β-galactosidase, or α-L-arabinosidase side activities present in the pectinolytic commercial enzyme used. These can hydrolyze glycosylated anthocyanins to produce unstable aglycones. They concluded that pectinase was more effective than cellulase. A reduction of pomace particle size significantly increases the antioxidant activity as a result of an increment in polyphenol extractability due to a better enzymatic degradation of polysaccharides. The smaller the particle size, the best of antioxidant activity of extracts obtained with enzymes is achieved.

To improve the nutritional quality of rye bread, Andreasen et al. (1999) studied the release of hydroxybenzoic and hydroxycinnamic acids from rye grains with four different commercial enzyme preparations that degrade plant cell walls (Macer8 FJ, Grindamyl Pectinase, and Rapidase BE super Viscozyme). With all enzymatic

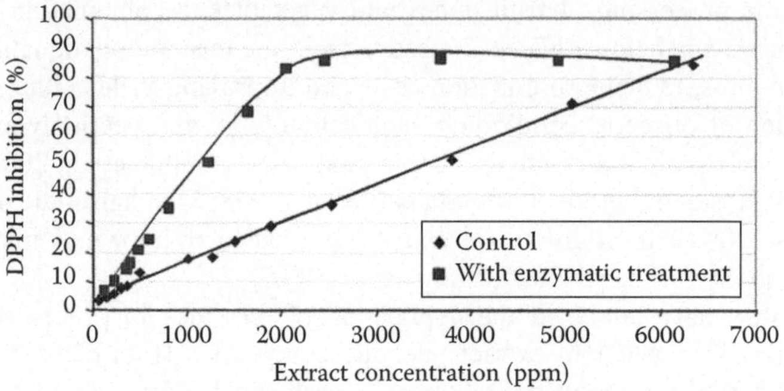

FIGURE 3.2 DPPH radical scavenging of two methanolic antioxidant extracts obtained from defatted borage seed flour with and without previous enzymatic treatment.

treatments, the extraction of hydroxycinnamic acids significantly increased. In the enzymatically hydrolyzed rye, the concentration of phenolic acids was higher than that obtained with NaOH hydrolysis, even though the total hydrolysis was higher in that process. Pardo et al. (1999) evaluated the effect of four enzymes (Endozym, Biopectinase, 100G, and Rapidase Ultrazym Excolor) on color quality, fractionation of pigments, as well as the content of anthocyanins and total phenols in Muscat red wines. All enzyme treatments increased the amount of extracted phenols and anthocyanins. Weinberg et al. (1999) observed an increase in phenolic content in the silage of rosemary and sage assisted by enzymes (cellulases, hemicellulases, and pectinases). Moreover, several studies report an increase in phenolic content and improved antioxidant activity when enzyme treatment is applied to oil recovery from residual flours from rose hip, evening primrose, and borage (Collao et al., 2006; Concha et al., 2004; Soto et al., 2008) (Figure 3.2). Laroze et al. (2010) studied the extraction of antioxidants from raspberry solid waste assisted by cellulase, hemicellulase, and pectinase. All enzyme treatments increased the soluble solids in the extracts between two and three times as compared with the nonenzymatic process; Grindamyl and Maxoliva offered the best recovery of polyphenols.

3.3 LABORATORY TECHNOLOGIES TO OBTAIN POLYPHENOL COMPOUNDS

At a laboratory scale, the most commonly used extraction technologies are the conventional methods, maceration and Soxhlet extraction. In maceration, the raw material is in contact with a solvent in batch mode, that is, they are put together in a vessel with or without stirring during a given time that can last several days. Most of the time, the liquid extract containing dissolved antioxidants is recovered by filtration. Subsequently, this extract may be clarified, concentrated, and dried to stabilize it. The maceration process is easy to implement at an industrial level, requiring only simple equipment, such as a maceration tank, a filter, a concentrator, and a dryer (Figure 3.3).

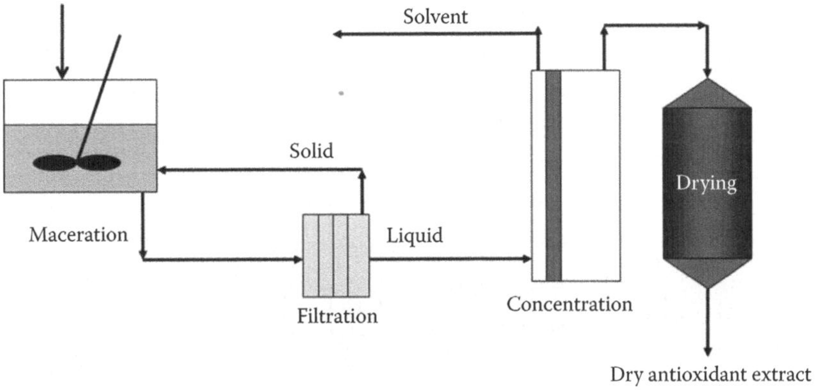

FIGURE 3.3 Schematic diagram of antioxidant extract obtained using the maceration process.

Soxhlet extraction is a semicontinuous process originally used for fat recovery at laboratory scale, even though it can also be applied at industrial scale. In this process, the raw material is in contact with the type of solvent in an immersion–percolation vessel, where the solute of interest is being recovered according to its solubility. Main operational variables are type of solvent and extraction time. The boiling point of the solvent defines the extraction temperature (Chanda and Kaneria, 2012; Hossain et al., 2013; Murugan and Parimelazhagan, 2014; Sultana et al., 2009).

Other less conventional techniques for the recovery of extracts rich in phenolic compounds are microwave-assisted extraction, ultrasound-assisted extraction, pressurized liquid extraction, and supercritical fluid extraction (Table 4.1). Applications of PLE and SFE to polyphenols recovery are described in detail in Chapters 4 and 5, respectively.

High selectivities as well as low extraction times and environmental impacts are achieved with UAE and MAE (Alexandru et al., 2014; Chen et al., 2015; Pingret et al., 2012) (see Figure 3.4).

In MAE, microwave energy and conventional solvent extraction are combined. The ability to quickly heat the sample–solvent mixture is the main advantage of this mode, reducing the extraction times and the amount of solvent needed (Sparr-Eskilsson and Björklund, 2000). The efficiency of the extraction process is limited by the stability of the solvent and target polyphenols at the temperature reached in the microwave (Dong et al., 2014; Escribano and Santos-Buelga, 2003).

In UAE, ultrasound energy and conventional solvent extraction are combined. High local pressure and heating are achieved through sound waves (with frequencies higher than 20 kHz). Mechanical vibrations cause cycles of expansion and compression of the media, leading to a subsequent cavitation. These changes promote cell membrane disruption, enhancing mass transfer from the solid matrix to the solvent (Moraes et al., 2013). Compared to maceration, UAE increases polyphenols extraction, and reduces processing time and energy consumption (Meullemiestre et al., 2016; Virot et al., 2010).

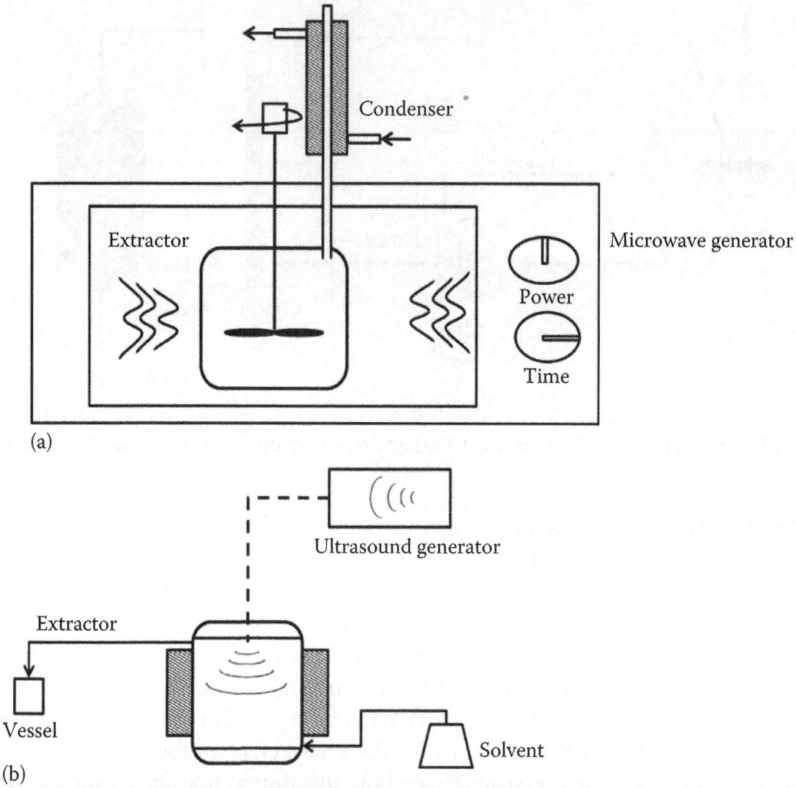

FIGURE 3.4 Diagram of equipment used for antioxidant extraction. MAE, microwave-assisted extraction (a); UAE, ultrasound-assisted extraction (b).

3.4 POLYPHENOL PRODUCTION AT THE INDUSTRIAL LEVEL

3.4.1 Polyphenol-Producing Companies

Understanding the conventional processes to obtain polyphenols from vegetable sources is the focus of this chapter. This knowledge should be used toward industrial applications to reach a high-quality product with health benefits for consumers. Therefore, it is important to know some companies that are involved in producing polyphenol extracts from natural sources at an industrial level. A list of some of these companies, their main products, and technologies are discussed next.

Ajinomoto Natural Specialties have more than 60 years of expertise in the extraction of botanicals, being one of the leaders among the polyphenol producers. The main products of this company are tannic acids (hydrolyzable polyphenols also known as gallotannins). Since 1947, they have developed a leading position in the polyphenols market. Their products are applied in several industries, such as brewery, foods, textiles, and pharma. This company also provides services to custom extractions, including process development, joint product development, industrial extraction and purification of ingredients for the food, nutraceutical, and cosmetics markets.

Bioingemar is working on the production of wine polyphenols. This company took advantage of the large amount of wine and grapes produced in Chile, and the concomitant generation of large amounts of waste, which are a good source of polyphenols. The process technology developed by Bioingemar produces a powder rich in polyphenols, which involves water and ethanol as solvents, low temperatures, and vacuum pressure. The process can have several steps as follows: hydroalcoholic extraction, concentration in a vacuum oven or in a concentrator (rotary vaporizer), and drying in a freeze drier or in a spray drier. This company also possesses a pilot plant containing an extrusion extractor, presses, a rotary vaporizer, a spray dryer, and a freeze dryer.

Natural Ingredient Bio-Resources (Changsha) Co. Ltd. produces over 300 kinds of tea extracts, 1000 standardized botanical herbal extracts, and natural ingredients for healthy foods. For example, this company produces polyphenol, L-theanine, theaflavin, caffeine, GTP, and bilberry anthocyanidin extracts; *Echinacea purpurea* herb-phenolic compounds, and grape seed proanthocyanidins.

Botanical Extracts and Specialties, a Nexira Health Division offers a wide range of high-quality natural botanical extracts, specialties, and powder with standardized active content. This company has extensive expertise in manufacturing technologies, including spray drying and microencapsulation, to make ingredients easier to use and perform better. In addition, *Ajinomoto Natural Specialties* offers customized contract manufacturing services to the food industry, which seems to be a good chance to develop new products and capture new customers who require innovation. This type of collaborative work for the development of new extracts rich in polyphenols is possible due to their infrastructure and equipment including spray dryers, co-spray dryers, encapsulators, microencapsulators, and equipment to carry out granulation and coating.

Iprona AG owns BerryPharma®, a trademark and a seal of quality for functional and nutritional polyphenol extracts for the food, nutritional, cosmetic, and pharmaceutical industries. They develop liquid and powdered polyphenol products from red berries, such as elderberry, blackcurrant, bilberry, chokeberry, and cranberry. Their products are validated and standardized to active analytical parameters. They offer customized products, and their technology does not use organic solvents. They receive analytical and scientific support from the German Institute for Nutrition in Karlsruhe, the universities in Giessen and Düsseldorf, the Charité in Berlin, the China Medical University in Taiwan, the Scottish Crop and Research Institute, and the Kuopio University in Finland.

There are laboratories that specialize in the quantification and identification of polyphenol compounds as well as carry out research to obtain them. This is the case of *Polyphenols Laboratories*, a company in Norway, which is based specifically on the production of anthocyanins and other polyphenols. Their customers range from the food, wine, and pharmaceutical industries, to medical, chemical, and biological laboratories. Currently, this company is able to provide 36 different products extracted from natural sources, including various plants, fruits, and berries. All their products (except mixtures) have an HPLC purity above 97% and are available from 10 mg samples. Most of their products are available at 50 and 100 mg as well. The main goal of this company is the production of specific kinds of polyphenols with a high purity degree, which could be used as standards to identify specific compounds in complex mixtures.

3.4.2 Industrial Technologies to Obtain Polyphenol Compounds

Most nonconventional technologies are only applied in research laboratories, since technical and financial restraints impede their application at a commercial scale. Therefore, industrial plants and processes that obtain polyphenol-rich extracts apply conventional technologies.

3.4.2.1 Conventional Industrial Plants

The work principle in a conventional plant is to put some feedstock (leaves, roots, flowers, seeds, or another vegetable source) into contact with water or another solvent (like ethanol) in an extractor. Then heat is applied to favor the dissolution of the desired component from the vegetable source. After the extraction stage, a pump will transfer the liquid into a concentrator to evaporate the water or solvent, which will be later condensed and collected in a receiver. The recovered solvent can be used again in the next batch. The component (mainly polyphenols) in the concentrator will become a cream or a liquid extract depending on the requirements of the client.

3.4.2.2 Microwave-Assisted Extraction Plants

This configuration is useful to obtain sweetener extracts, flower extracts, essential oils, and tea polyphenol extracts. Their main characteristic is to reduce the temperature gradient induced by heat conduction or radiation, which is replaced by a simultaneous inside and outside heating, increasing the extraction quality, and preserving the bioactive compounds of the feedstock. This technology has some advantages in comparison with the conventional extraction technology: lower extraction temperatures, lower extraction times, no thermal inertia (so it is easily controlled), higher efficiency, lesser usage of solvent, lower operating cost, and reduced working space. There are extractor capacities from 10 to 8000 L, microwave frequency of 2450 MHz, stirring speed between 10 and 120 rpm, at atmospheric pressure, and extraction temperatures between 25°C and 40°C.

REFERENCES

Al-Farsi, M. and Lee, C. 2008. Optimization of phenolics and dietary fibre extraction from date seeds. *Food Chemistry* 108(3):977–985.

Alexandru, L., Binello, A., Mantegna, S., Boffa, L., Chemat, F., and Cravotto, G. 2014. Efficient Green extraction of polyphenols from post-harvested agro-industry vegetal sources in Piedmont. *Compes Rendes Chimie* 17(3):212–217.

Andreasen, M.F., Christensen, L.P., Meyer, A.S., and Hansen, Å. 1999. Release of hydroxycinnamic and hydroxybenzoic acids in rye by commercial plant cell wall degrading enzyme preparations. *Journal of the Science of Food and Agriculture* 79(3):411–413.

Anil Kumar, V.M., Kishor Mohan, T.C., and Murugan, K. 2008. Purification and kinetic characterization of polyphenol oxidase from Barbados cherry (*Malpighia glabra* L.). *Food Chemistry* 110(2):328–333.

Azizah, A.H., Nik Ruslawati, N.M., and Swee Tee, T. 1999. Extraction and characterization of antioxidant from cocoa by-products. *Food Chemistry* 64(2):199–202.

Azuola, R. and Vargas, P. 2007. Extracción de sustancias asistida por ultrasonido (EUA). *Tecnología en Marcha* 20(4):30–40.

Balasundram, N., Sundram, K., and Samman, S. 2006. Phenolic compounds in plants and agri-industrial by-products: Antioxidant activity, occurrence, and potential uses. *Food Chemistry* 99(1):191–203.

Baublis, A., Decker, E.A., and Clydesdale, F.M. 2000. Antioxidant effect of aqueous extracts from wheat based ready-to-eat breakfast cereals—Importance of their interaction with biomembranes. *Food Chemistry* 68(1):1–6.

Bhat, M.K. 2000. Cellulases and related enzymes in biotechnology. *Biotechnology Advances* 18:355–383.

Bonilla, F., Mayen, M., Merida, J., and Medina, M. 1999. Extraction of phenolic compounds from red grape marc for use as food lipid antioxidants. *Food Chemistry* 66(2):209–215.

Borowska, E.J., Szajdek, A., and Borowski, J. 2005. Antioxidant properties of fruits, vegetables and their products. *Fruit Process* 1:38–43.

Budrat, P. and Shotipruk, A. 2009. Enhanced recovery of phenolic compounds from bitter melon (*Momordica charantia*) by subcritical water extraction. *Separation and Purification Technology* 66(1):125–129.

Cacace, J. and Mazza, G. 2003. Mass transfer process during extraction of phenolic compounds from milled berries. *Journal of Food Engineering* 59: 379–389.

Casas, L., Mantell, C., Rodríguez, E., Martínez, E., Roldán, A., De Ory, I., Caro, I., and Blandino, A. 2010. Extraction of resveratrol from the pomace of Palomino fino grapes by supercritical carbon dioxide. *Journal of Food Engineering* 96(2):304–308.

Cerda, A., Martínez, M.E., Soto, C., Poirrier, P., Perez-Correa, J.R., Vergara-Salinas, J.R., and Zúñiga, M.E. 2013. The enhancement of antioxidant compounds extracted from *Thymus vulgaris* using enzymes and the effect of extracting solvent. *Food Chemistry* 139(1–4):138–143.

Cerpa-Calderón, F. and Kennedy, J.A. 2008. Berry integrity and extraction of skin and seed proanthocyanidins during red wine fermentation. *Journal of Agricultural and Food Chemistry* 56(19):9006–9014.

Céspedes, C., Valdez-Morales, M., Avila, J., El-Hafidi, M., Alarcón, J., and Paredes-López, O. 2010. Phytochemical profile and the antioxidant activity of Chilean wild blackberry fruits, *Aristotelia chilensis* (Mol) Stuntz (Elaeocarpaceae). *Food Chemistry* 119(3):886–895.

Chanda, S.V. and Kaneria, M. 2012. Optimization of conditions for the extraction of antioxidants from leaves of *Syzygium cumini* L. using different solvents. *Food Analytical Methods* 5:332–338.

Chen, M., Zhao, Y., and Yu, S. 2015. Optimisation of ultrasonic-assisted extraction of phenolic compounds, antioxidants, and anthocyanins from sugar beet molasses. *Food Chemistry* 172:543–550.

Cherry, J.R. and Fidantsef, A. 2003. Directed evolution of industrial enzymes: An update. *Current Opinion in Biotechnology* 14:438–443.

Collao, C.A., CurottO, E., and Zúñiga, M.E. 2006. Tratamiento enzimático en la extracción de aceite y obtención de antioxidantes a partir de semilla de onagra, Oenothera biennis, por prensado en frío. *Grasas y Aceites* 58(1):10–14.

Concha, J., Soto, C., Chamy, R., and Zúñiga, M.E. 2004. Effect of enzymatic pretreatment on rose hip oil extraction: Hydrolysis and pressing conditions. *Journal of the American Oil Chemists' Society* 81(6):550–552.

Conde, E., Cara, C., Moure, A., Ruiz, E., Castro, E., and Domínguez, H. 2009. Antioxidant activity of the phenolic compounds released by hydrothermal treatments of olive tree pruning. *Food Chemistry* 114(3):806–812.

Corrales, M., Fernández, A., Butz, P., and Tauscher, B. 2009. Extraction of anthocyanins from grape skins assisted by high hydrostatic pressure. *Journal of Food Engineering* 90(4):415–421.

Couper, J.R., Penney, W.R., Fair, J.R., and Walas, S.M. 2012. *Chemical Process Equipment*, 3rd ed., Chapter 14, pp. 487–528. Boston, MA: Butterworth-Heinemann.

Cruz, J.M., Domínguez, H., and Parajó, J.C. 2004. Assessment of the production of antioxidants from winemaking waste solids. *Journal of Agricultural and Food Chemistry* 52(18):5612–5620.

Dong, Z., Gu, F., Xu, F., and Wang, Q. 2014. Comparison of four kinds of extraction techniques and kinetics of microwave-assisted extraction of vanillin from *Vanilla planifolia* Andrews. *Food Chemistry* 149:54–61.

Dopico-García, M., Valentão, P., Guerra, L., Andrade, P., and Seabra, R. 2007. Experimental design for extraction and quantification of phenolic compounds and organic acids in white "Vinho Verde" grapes. *Analytica Chimica Acta* 583(1):15–22.

Durhing, N., Catchpole, O., Grey, J., Webby, R., Mitchell, K., Foo, Y., and Perry, N. 2007. Extraction of phenolics and essential oil from dried sage (*Salvia officinalis*) using ethanol–water mixtures. *Food Chemistry* 101(4):1417–1424.

Escribano-Bailon, M.T. and Santos-Buelga, C. 2003. Polyphenols extraction from foods. In *Methods in Polyphenol Analysis*, eds. C. Santos-Buelga and G. Williamson, pp. 1–16. Cambridge, UK: Royal Society of Chemistry.

Fortea, M.I., López-Miranda, S., Serrano-Martínez, A., Carreño, J., and Núñez-Delicado, E. 2009. Kinetic characterisation and thermal inactivation study of polyphenol oxidase and peroxidase from Table grape (Crimson Seedless). *Food Chemistry* 113(4):1008–1014.

García-Alonso, M., de Pascual-Teresa, S., Santos-Buelga, C., and Rivas-Gonzalo, J. 2004. Evaluation of the antioxidant properties of fruits. *Food Chemistry* 84:13–18.

Gil-Chávez, G., Villa, J., Ayala-Zavala, J.F., Heredia, B., Sepúlveda, D., Yahia, E., González-Aguilar, G. 2013. Technologies for extraction and production of bioactive compounds to be used as nutraceuticals and food ingredients: An overview. *Comprehensive Reviews in Food Science and Food Safety* 12:5–23.

Gökmen, V., Serpen, A., and Fogliano, V. 2009. Direct measurement of the total antioxidant capacity of foods: The "QUENCHER" approach. *Trends in Food Science and Technology* 20(6–7):278–288.

Hamada, J.S. 1997. Large-scale high-performance liquid chromatography of enzymes for food applications. *Journal of Chromatography A* 760:81–87.

Harbourne, N., Jacquier, J., and O'Riordan, J. 2009. Optimization of the aqueous extraction conditions of phenols from meadowsweet (*Filipendula ulmaria* L.) for incorporation into beverages. *Food Chemistry* 116(3):722–727.

Hassas-Roudsari, M., Chang, P., Pegg, R., and Tyler, R. 2009. Antioxidant capacity of bioactives extracted from canola meal by subcritical water, ethanolic and hot water extraction. *Food Chemistry* 114(2):717–726.

Herrero, M., Plaza, M., Cifuentes, A., and Ibáñez, E. 2012. Extraction techniques for the determination of phenolic compounds in food. In *Comprehensive Sampling and Sample Preparation*, ed. J. Pawliszyn, pp. 159–180. Oxford, UK: Academic Press.

Hojnik, M., Mojca, S., and Knez, Z. 2008. Extraction of lutein from Marigold flower petals—Experimental kinetics and modeling. *Food and Science Technology* 41:2008–2016.

Hossain, A., AL-Mijizy, Z.H., Al-Rashdi, K.K., Weli, A.M., and Al-Riyami, Q. 2013. Effect of temperature and extraction process on antioxidant activity of various leaves crude extracts of *Thymus vulgaris* Mohammad. *Journal of Coastal Life Medicine* 1(2):130–134.

Jayaprakasha, G.K., Singh, R.P., and Sakariah, K.K. 2001. Antioxidant activity of grape seed (*Vitis vinifera*) extracts on peroxidation models in vitro. *Food Chemistry* 73(3):285–290.

Karacabey, E. and Mazza, G. 2010. Optimisation of antioxidant activity of grape cane extracts using response surface methodology. *Food Chemistry* 119(1):343–348.

Kiassos, E., Mylonaki, D., Makris, P., and Kefalas, P. 2009. Implementation of response surface methodology to optimise extraction of onion (*Allium cepa*) solid waste phenolics. *Innovative Food Science and Emerging Technologies* 10(2):246–252.

Krishnaiah, D., Sarbatly, R., and Nithyanandam, R. 2011. A review of the antioxidant potential of medicinal plant species. *Food and Bioproducts Processing* 89(3):217–233.

Landbo, A.K. and Meyer, A.S. 2001. Enzyme-assisted extraction of antioxidative phenols from black currant juice press residue (*Ribes nigrum*). *Journal of Agricultural and Food Chemistry* 49:3169–3177.

Laroze, L., Soto, C., and Zúñiga, M.E. 2010. Phenolic antioxidants extraction from raspberry waste assisted by-enzymes. *Electronic Journal of Biotechnology* 13(6):1–11.

Larrauri, J., Rupérez, P., and Saura-Calixto, F. 1997. Antioxidant dietary fiber product: A new concept and a potential food ingredient. *Journal Agricultural and Food Chemistry* 45:1390–1393.

Lecoupeau, J.P. and Vercauteren, J. 2008. Method for obtaining cocoa bean polyphenol extracts, resulting extracts and uses thereof. U.S. Patent 7,368,144.

Lemańska, K., Szymusiak, H., Tyrakowska, B., Zielinski, S., Soffers, A., and Rietjens, I. 2001. The influence of pH on antioxidant properties and the mechanism of antioxidant action of hydroxyflavones. *Free Radical Biology and Medicine* 31(7):869–881.

Li, B., Smith, M., and Hossain, M. 2006. Extraction of phenolics from citrus peels: I. Solvent extraction method. *Separation and Purification Technology* 48(2):182–188.

Linares, I.B., Arráez-Román, D., Herrero, M., Ibáñez, E., Segura-Carretero, A., and Fernández-Gutiérrez, A. 2011. Comparison of different extraction procedures for the comprehensive characterization of bioactive phenolic compounds in *Rosmarinus officinalis* by reversed-phase high-performance liquid chromatography with diode array detection coupled to electrospray time-of-flight mass spectrometry. *Journal of Chromatography A* 1218(42):7682–7690.

Luque-Rodríguez, J., Luque de Castro, M., and Pérez-Juan, P. 2007. Dynamic superheated liquid extraction of anthocyanins and other phenolics from red grape skins of wine-making residues. *Bioresource Technology* 98(14):2705–2713.

Luthria, D. 2008. Influence of experimental conditions on the extraction of phenolic compounds from parsley (*Petroselinum crispum*) flakes using a pressurized liquid extractor. *Food Chemistry* 107:745–752.

Marete, E., Jacquier, J., and Riordan, D. 2009. Effects of extraction temperature on the phenolic and parthenolide contents, and colour of aqueous feverfew (*Tanacetum parthenium*) extracts. *Food Chemistry* 117(2):226–231.

Meullemiestre, A., Petitcolas, E., Maache-Rezzoug, Z., Chemat, F., and Rezzoug, S.A. 2016. Impact of ultrasound on solid–liquid extraction of phenolic compounds from maritime pine sawdust waste: Kinetics, optimization and large scale experiments. *Ultrasonics Sonochemistry* 28:320–329.

Meyer, A.S., Jepsen, S.M., and Sorensen, N.S. 1998. Enzymatic release of antioxidants for human low-density lipoprotein from grape pomace. *Journal of Agricultural and Food Chemistry* 46:2439–2446.

Moraes, M., Zabot, G., Prado, J., and Meireles, M.A. 2013. Obtaining antioxidants from botanic matrices applying novel extraction techniques. *Food and Public Health* 3(4):195–214.

Moure, A., Franco, D., Sineiro, J., Domínguez, H., Núñez, M.J., and Lema, J.M. 2000. Evaluation of extracts from Gevuina avellana hulls as antioxidants. *Journal of Agricultural and Food Chemistry* 48(9):3890–3897.

Murugan, R. and Parimelazhagan, T. 2014. Comparative evaluation of different extraction methods for antioxidant and anti-inflammatory properties from Osbeckia parvifolia Arn: An in vitro approach. *Journal of King Saud University: Science*. 26:267–275.

Naczk, M. and Shahidi, F. 2004. Extraction and analysis of phenolics in food. *Journal of Chromatography A* 1054(1–2):95–111.

Naczk, M. and Shahidi, F. 2006. Phenolics in cereals, fruits and vegetables: Occurrence, extraction and analysis. *Journal of Pharmaceutical and Biochemical Analysis* 41:1523–1542.

Nawaz, H., Shi, J., Mittal, G., and Kakuda, Y. 2006. Extraction of polyphenols from grape seeds and concentration by ultrafiltration. *Separation and Purification Technology* 48(2):176–181.

Nayak, B., Dahmoune, F., Moussi, K., Remini, H., Dairi, S., Aoun, O., and Khodir, M. 2015. Comparison of microwave, ultrasound and accelerated-assisted solvent extraction for recovery of polyphenols from *Citrus sinensis* peels. *Food Chemistry* 187:507–516.

Paladino, S. and Zuritz, C. 2012. Antioxidant grape seed extract (*Vitis vinifera* L.) concentration, drying and comparison with commercial antioxidants. *Revista de la Facultad de Ciencias Agrarias: Universidad Nacional de Cuyo* 44(2):131–143.

Pardo, F., Salinas, M.R., Alonso, G.L., Navarro, G., and Huerta, M.D. 1999.Effect of diverse enzyme preparations on the extraction and evolution of phenolic compounds in red wines. *Food Chemistry* 67:135–142.

Pingret, D., Fabiano-Tixier, A.S., Bourvellec, C.L., Renard, C.M.G.C., and Chemat, F. 2012. Lab and pilot-scale ultrasound-assisted water extraction of polyphenols from apple pomace. *Journal of Food Engineering* 11(1):73–81.

Pompeu, D., Silva, E., and Rogez, H. 2009. Optimization of the solvent extraction of phenolic antioxidants from fruits of *Euterpe oleracea* using Response Surface Methodology. *Bioresource Technology* 100(23):6076–6082.

Prior, R.L., Wu, X., and Schaich, K. 2005. Standardized methods for the determination of antioxidant capacity and phenolics in foods and dietary supplements. *Journal of Agricultural and Food Chemistry* 53(10):4290–4302.

Prommuak, C., De-Eknamkul, W., and Shotipruk, A. 2008. Extraction of flavonoids and carotenoids from Thai silk waste and antioxidant activity of extracts. *Separation and Purification Technology* 62(2):444–448.

Riihinen, K., Jaakola, L., Kärenlampi, S., and Hohtola, A. 2008. Organ-specific distribution of phenolics compounds in bilberry (*Vaccinium myrtillus*) and "northblue" blueberry (*Vaccinium corymbosum* × *V. angustifolium*). *Food Chemistry* 110:156–160.

Rodis, P., Karathanos, V., and Mantzavinou, A. 2002. Partitioning of olive oil antioxidants between oil and water phases. *Journal of Agricultural and Food Chemistry* 50(3):596–601.

Roopchand, D., Krueger, C., Moskal, K., Fridlender, B., Lila, M., and Raskin, I. 2013. Food-compatible method for the efficient extraction and stabilization of cranberry pomace polyphenols. *Food Chemistry* 141(4):3664–3669.

Rosenthal, A., Pyle, D.L., and Niranjan, K. 1996. Aqueous and enzymatic processes for edible oil extraction. *Enzyme and Microbial Technology* 19(6):402–420.

Ruenroengklin, N., Zhong, J., Duan, X., Yang, B., Li, J., and Jiang, Y. 2008. Effects of various temperatures and pH values on the extraction yield of phenolics from litchi fruit pericarp tissue and the antioxidant activity of the extracted anthocyanins. *International Journal of Molecular Science* 9:1333–1341.

Sanal, I.S., Bayraktar, E., Mehmetoglu, Ü., and Çalimli, A. 2005. Determination of optimum conditions for SC-(CO_2 + ethanol) extraction of β-carotene from apricot pomace using response surface methodology. *Journal of Supercritical Fluids* 34:331–338.

Sánchez Guerrero, M., Sineiro Torres, J., and Nuñez, M.J. 2008. Extraction of polyphenols from white distilled grape pomace: Optimization and modelling. *Bioresource Technology* 99(5):1311–1318.

Shabtai, Y., Gotshal, N., Ramot, O., and Wisniak, J. 1998. Enzymatic pretreatment of jojoba seeds to facilitate oil extraction. *Indian Journal of Chemical Technology* 5:124–130.

Shankar, D., Agrawal, Y.C., Sarkar, B.C., and Singh, B.P.N. 1997. Enzymatic hydrolysis in conjunction with conventional pretreatments to soybean for enhanced oil availability and recovery. *Journal of the American Oil Chemists' Society* 74:1543–1547.

Shiferaw Terefe, N., Hong Yang, Y., Knoerzer, K., Buckow, R.,and Versteeg, C. 2010. High pressure and thermal inactivation kinetics of polyphenol oxidase and peroxidase in strawberry puree. *Innovative Food Science and Emerging Technologies* 11(1):52–60.

Silva, E.M., Rogez, H., and Larondelle, Y. 2007. Optimization of extraction of phenolics from Inga edulis leaves using response surface methodology. *Separation and Purification Technology* 55:381–387.

Soto, C.G., Chamy, R., and Zúñiga, M.E. 2004. Effect of enzymatic application on borage (*Borago officinalis*) oil extraction by cold pressing. *Journal of Chemical Engineering of Japan* 37(2):326–331.

Soto, C.G., Chamy, R., and Zúñiga, M.E. 2007. Enzymatic hydrolysis and pressing conditions effect on borage oil extraction by cold pressing. *Food Chemistry* 102:834–840.

Soto, C., Concha, J., and Zúñiga, M.E. 2008. Antioxidant content of oil and defatted meal obtained from borage seeds by an enzymatic-aided cold pressing process. *Process Biochemistry* 43(6): 696–699.

Spigno, G., Tramelli, L., and De Faveri, M. 2007. Effects of extraction time, temperature and solvent on concentration and antioxidant activity of grape marc phenolics. *Journal of Food Engineering* 81(1):200–208.

Sparr-Eskilsson, C., Björklund, E. 2000. Analytical-scale microwave-assisted extraction. *Journal of Chromatography A* 902, 227–250.

Sultana, B., Anwar, F., and Ashraf, M. 2009. Effect of extraction solvent/technique on the antioxidant activity of selected medicinal plant extracts. *Molecules* 14:2167–2180.

Tabart, J., Kevers, C., Sipel, A., Pincemail, J., Defraigne, J., and Dommes, J. 2007. Optimisation of extraction of phenolics and antioxidants from black currant leaves and buds and of stability during storage. *Food Chemistry* 105(3):1268–1275.

Takeoka, G.R. and Dao, L.T. 2003. Antioxidant constituents of almond (*Prunus dulcis* (Mill.) D.A. Webb) hulls. *Journal of Agricultural Food Chemistry* 51(2):496–501.

Terefe, N.S., Yang, Y.H., Knoerzer, K., Buckow, R., Versteeg, C. 2010. High-pressure and thermal inactivation kinetics of polyphenol oxidase and peroxidase in strawberry puree. *Innovative Food Science Emerging Technologies* 11(1):52–60.

Teixeira, A., Baenas, N., Dominguez-Perles, R., Barros, A., Rosa, E., Moreno, D., and García-Viguera, C. 2014. Natural bioactive compounds from winery by-products as health promoters: A review. *International Journal Molecular Sciences* 15:15638–15678.

Tobar, P., Moure, A., Soto, C., Chamy, R., and Zúñiga, M.E. 2005. Winery solid residue revalorization into oil and antioxidant with nutraceutical properties by an enzyme assisted process. *Journal Water Science and Technology* 51(1):47–52.

Türker, N. and Erdogdu, F. 2006. Effects of pH and temperature of extraction medium on effective diffusion coefficient of anthocynanin pigments of black carrot (*Daucus carota* var. L.). *Journal of Food Engineering* 76(4):579–583.

Virot, M., Tomao, V., Le Bourvellec, C., Renard, C., and Chemat, F. 2010. Towards the industrial production of antioxidants from food processing by-products with ultrasound-assisted extraction. *Ultrasonics Sonochemistry* 17:1066–1074.

Weinberg, Z.G., Akiri, B., Potoyevski, E., and Kanner, J. 1999. Enhancement of polyphenol recovery from rosemary (*Rosmarinus officinalis*) and sage (*Salvia officinalis*) by enzyme-assisted ensiling (ENLAC). *Journal of Agricultural and Food Chemistry* 47(7):2959–2962.

Wijngaard, H., Hossain, M., Rai, D., and Brunton, N. 2012. Techniques to extract bioactive compounds from food by-products of plant origin. *Food Research International* 46:505–513.

Wood, R., Kroon, P., and Green, R. 2013. Polyphenol extraction process. U.S. Patent 8,470,380.

Zhao, Y. 1999. *Berry Fruit: Value-Added Products for Health Promotion*. Boca Raton, FL: CRC Press.

Zuñiga, M.E., Soto, C., Mora, A., Chamy, R., and Lema, J.M. 2003. Enzymic pre-treatment of *Guevina avellana* mol oil extraction by pressing. *Process Biochemistry* 39:51–57.

4 Extraction of Polyphenols by Pressurized Liquids

José Rodrigo Vergara-Salinas,
José Cuevas-Valenzuela,
and José Ricardo Pérez-Correa

CONTENTS

4.1 INTRODUCTION

In general, the extraction of polyphenols from plants has been based on the use of organic solvents, at boiling point temperatures or lower, and atmospheric pressure. Organic solvents, such as methanol, ethanol, and acetone, and their combinations with different proportions of water are commonly used (Dai and Mumper, 2010; Ignat et al., 2011; Shi et al., 2005). The conventional extraction methods imply heating, boiling, or refluxing the solvent for long periods of time to extract natural phenolic compounds. These conventional extraction methods could present several disadvantages, including the loss of polyphenols due to ionization, hydrolysis, and oxidation; these methods also use large amounts of questionable solvents, which are time consuming and expensive, in addition to the subsequent solvent-recovering processes (Deyama et al., 1987; Herrero et al., 2006; Li et al., 2005; Ohnishi et al., 1994;

Ollanketo et al., 2002; Paganga et al., 1999; Sakuma et al., 1982; Yan et al., 1999; Yu et al., 2001). To obtain polyphenols as natural food ingredients, a green, efficient, and economic process of extraction is required. According to Chemat et al. (2012), green extraction is based on the discovery and design of extraction processes that will reduce energy consumption, allow for the use of alternative solvents and renewable natural products, as well as ensure a safe and high-quality extract/product. They also present the *six principles of green extraction*, a guide for the industry and researchers to establish sustainability standards in both the innovation and development of the extraction process and all aspects associated with the solid–liquid extraction:

Principle 1: Innovation by selection of varieties and use of renewable plant resources
Principle 2: Use of alternative solvents, principally water or agro-solvents
Principle 3: Reduction of energy consumption by energy recovery and using innovative technologies
Principle 4: Production of coproducts instead of waste to include the bio- and agro-refining industry
Principle 5: Reduction of unit operations, favoring safe, robust, and controlled processes
Principle 6: Aim for a nondenatured and biodegradable extract without contaminants

Alternative extraction methods, which considerably reduce the solvent volume used and accelerate the extraction process, include microwave-assisted extraction (MAE), ultrasound-assisted extraction (UAE), and compressed fluid extraction, such as pressurized liquid extraction (PLE) and supercritical fluid extraction (SFE) (Garcia-Salas et al., 2010). PLE, especially using pressurized hot water (PHW) as a solvent, is a promising emerging green technology. The PLE method involves the use of liquid solvents at temperatures above their atmospheric boiling point by applying high pressure. In these conditions, the solubility and mass transfer properties are considerably enhanced compared to conventional extraction methods (Cacace and Mazza, 2006b; Clifford, 2005; Garcia-Salas et al., 2010; Hawthorne et al., 2002; Herrero et al., 2006, 2013; King and Grabiel, 2007; Mendiola et al., 2007; Mukhopadhyay and Panja, 2010; Mustafa and Turner, 2011; Ong et al., 2006; Smith, 2002; Srinivas and King, 2010; Teo et al., 2010; Wiboonsirikul and Adachi, 2008). PLE has the advantages of being relatively inexpensive, and it minimizes or totally avoids the use of objectionable solvents and provides a facile means for supplying a wide source of concentrated phytochemicals for use in food formulation, dietary supplements, and phytopharmaceutical applications (King and Grabiel, 2007).

This chapter focuses on the use of PLE for the recovery of polyphenols from plant matrices, especially agroindustrial by-products, for their use as food ingredients. It delves into the enhancement of extractive properties of solvents at high temperatures and pressures, as well as the use of food-grade solvents and green solvents. Several studies focused on the optimization of the extraction parameters at laboratory scale are also presented. Finally, an approach to the design and scale-up of pressurized extractor vessels and information regarding commercial large-scale equipment are given.

4.2 EXTRACTION ENHANCEMENT BY RISING TEMPERATURE AND PRESSURE

The extraction of polyphenols and bioactive compounds, in general from vegetable matrices, can be described in a few steps. The first step involves the release of compounds from their site at a cellular level and their diffusion through the organic part of the vegetable matrix to reach the matrix–solvent interface. Then, the compound diffuses through the extraction phase in the pore network within the matrix. Finally, the compounds reach the extraction-phase zone affected by convection and migrate toward the solvent bulk (Kallithraka et al., 1995; Mustafa and Turner, 2011). These events are controlled by many related factors or phenomena (Figure 4.1). The first factor to be considered is the nature of the vegetable matrix, which includes the type, content, localization, and bonding of the target compound(s), particle size, geometry of the pore network, and water content. Another factor to be considered is the rate-limiting step in the extraction process: desorption, solubilization, and diffusion. In plant material extraction, solubilization and diffusion are commonly the rate-limiting steps, which strongly depend on solvent type and temperature.

The merit of PLE, and what differentiates it from conventional extraction methods, is the use of liquid solvents at temperatures above their atmospheric boiling point, with a significant enhancement of solubility and mass transfer properties (Mustafa and Turner, 2011). High temperatures in PLE, usually within the range of 50°C–200°C (Hossain et al., 2015; Pasrija and Anandharamakrishnan, 2015), enhance the extraction efficiency of target compounds (in this case, polyphenols) since several physicochemical properties of solvents are modified (Herrero et al., 2013; Shi et al., 2005). The dielectric constant (a rough measure of solvent polarity)

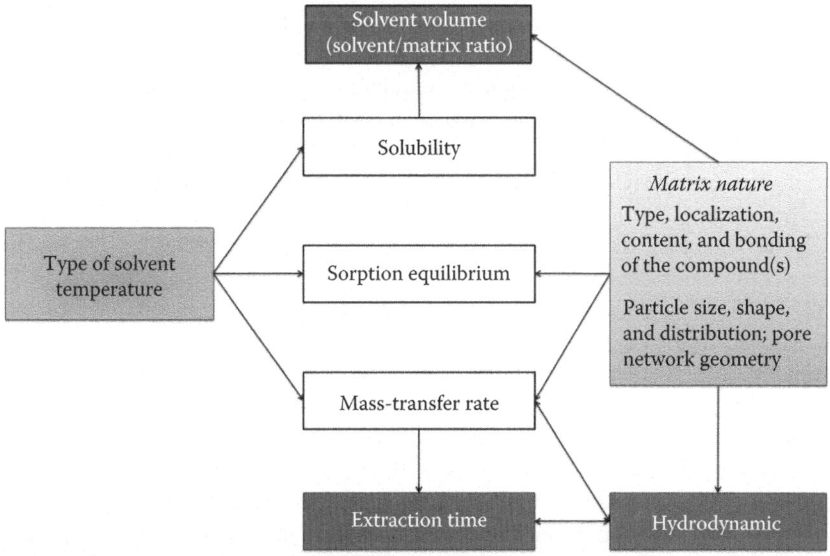

FIGURE 4.1 Extraction controlling factors and their interdependence.

is decreased, which allows for the solubilization of less polar compounds, viscosity and surface tension of the solvent are decreased, diffusion rates are increased, a better disruption of the strong solute–matrix bonds is reached, and hence, solubility and the mass-transfer rate of the solute are increased (Casazza et al., 2012; Hofmann et al., 2015; Liu et al., 2014; Mustafa and Turner, 2011; Ramos et al., 2002; Richter et al., 1996). High pressure (up to 3000 psi according to Pasrija and Anandharamakrishnan, 2015) is mainly used in PLE to maintain the solvent in its liquid state at temperatures above the boiling point, but it also might enhance solubilization of target compounds by permitting a better penetration of the solvent into the pores of the matrix (Mustafa and Turner, 2011; Ramos et al., 2002).

The effects of high temperatures and pressures on the solubility of polyphenols, mass transfer properties of the polyphenol–solvent system, and the formation of non-desired compounds are described in the following sections.

4.2.1 Solubility of Polyphenols at High Temperatures and Pressures

Solubility is a key index to evaluate the effects of both high temperatures and pressures on extraction efficiency; solubility data become an important input for the design and scale-up of PLE processes.

Solubility of polyphenols depends not only on temperature and pressure but also on the type of solvent used for extraction (fundamentally its polarity) and on the molecular structure and size of polyphenols, for example, degree of polymerization, amount and distribution of hydroxyl groups, and length of hydrocarbons (Kabir et al., 2015; Sarraf, 2015). Also, it can be strongly affected by the molecular interactions (i.e., hydrogen bonding) between polyphenols and other macromolecules present in the matrix (usually plant material), such as proteins and polysaccharides, which can lead to the formation of insoluble complexes (Magalhaes et al., 2007; Xu et al., 2006). Thus, the isolation of polyphenols at the levels of required purity for solubility measurements can be unmanageable, expensive, and time consuming, which largely explains the high prices of polyphenol standards available at commercial and recognized labs (e.g., Sigma-Aldrich, Cayman Chemical, Polyphenols Laboratory AS). Despite this issue, experimental methods for isolating polyphenols are widely covered in the literature (Callemien et al., 2005; Ignat et al., 2011; Khoddami et al., 2013; Mueller-Harvey, 2001; Naczk and Shahidi, 2004; Santos-Buelga and Williamson, 2003; Serrano et al., 2009) and can be applied at laboratory scale using standard and specialized equipment.

Once the polyphenols are either purchased from recognized laboratories or properly isolated following literature methods, their solubility in particular solvents or solvent mixtures can be experimentally measured at different temperatures and pressures through a shake-flask method coupled with high-performance liquid chromatography (HPLC) and/or ultraviolet–visible spectrophotometry (UV/VIS) analytical system. Several versions of this method can be found in the literature (data sources in Table 4.1), but they all fundamentally comprise (1) pouring of solvent and excess polyphenol powder into a flask (properly designed to avoid evaporation of the solvent and to maintain the required temperature and pressure), (2) agitation of the solvent–polyphenol mixture at the required temperature and pressure until

TABLE 4.1
Data Sources on Experimental Solubility of Several Polyphenols in Different Solvents at Specific Temperature Ranges

Polyphenol Class[a]	Polyphenol	Solvent	Temperature Range	Data Source[b]
Flavonoids	Apigenin	(i) Ethanol–water mixtures	(i) 273.15–323.2 K	(i) Xiao, Yan, and Zhang (2010)
		(ii) Dimethylsulfoxide (DMSO)	(ii)–(vii) 298.15 K	(ii)–(vii) Li, Robinson, and Birt (1997)
		(iii) Ethanol		
		(iv) Acetone		
		(v) *n*-Octanol		
		(vi) Propylene glycol		
		(vii) Water		
	(+)-Catechin	(i) Water	(i)–(ii) 277.6–331.1 K	(i)–(ii) Cuevas-Valenzuela et al. (2014)
		(ii) Ethanol–water mixtures		
	Chrysin	(i) Water	(i)–(iii) 293.15–323.15 K	(i)–(iii) Zhou et al. (2014)
		(ii) Ethanol	(iv) 298.15 K	(iv) Rastogi and Jana (2016)
		(iii) Ethanol–water mixtures	(v) 323.15 K	(v) Chebil et al. (2007)
		(iv) Phosphate buffer solution		
		(v) Acetonitrile		
	Daidzein	(i) Water	(i) 298.15 K	(i) Zhao et al. (2011)
	Daidzin	(i) Tetrahydrofuran	(i)–(xi) 283.2–323.2 K	(i)–(xi) Fan et al. (2015)
		(ii) Acetone		
		(iii) Methanol		
		(iv) 1-Butanol		
		(v) Acetonitrile		
		(vi) 2-Propanol		
		(vii) Ethyl acetate		

(Continued)

TABLE 4.1 (*Continued*)

Data Sources on Experimental Solubility of Several Polyphenols in Different Solvents at Specific Temperature Ranges

Polyphenol Class[a]	Polyphenol	Solvent	Temperature Range	Data Source[b]
		(viii) Trichloromethane		
		(ix) Cyclohexane		
		(x) Water		
		(xi) Ethanol–water mixtures		
	(−)-Epicatechin	(i) Ethanol–water mixtures	(i) 293–303 K	(i) Sevillano et al. (2013)
	(−)-Epicatechin gallate	(i) Ethanol–water mixtures	(i) 293–303 K	(i) Sevillano et al. (2013)
	(−)-Epigallocatechin	(i) Ethanol–water mixtures	(i) 293–303 K	(i) Sevillano et al. (2013)
	(−)-Epigallocatechin gallate	(i) Ethanol–water mixtures	(i) 293–303 K	(i) Sevillano et al. (2013)
	Genistein	(i) Water	(i)–(vi) 280–333 K	(i)–(vi) Wu et al. (2010)
		(ii) Ethanol		
		(iii) Methanol		
		(iv) 2-Propanol		
		(v) 1-Butanol		
		(vi) Ethyl acetate		
	Hesperetin	(i) Methanol–water mixtures	(i)–(iii) 298.2 K	(i)–(iii) Ferreira et al. (2013)
		(ii) Ethanol–water mixtures	(iv.a), (v.a), (vi.a), (vii), and (viii.a) 298.15–313.15 K	(iv.a), (v.a), (vi.a), (vii), and (viii.a) Ferreira and Pinho (2012)
		(iii) Acetone–water mixtures	(iv.b), (v.b), (vi.b), (ix), and (x) 288.2–323.2 K	(iv.b), (v.b), (vi.b), (ix), and (x) Liu and Chen (2008)
		(iv) Acetone	(viii.b) 323.15–343.15 K	(viii.b) Chebil et al. (2007)
		(v) Ethanol		
		(vi) Methanol		
		(vii) Ethyl acetate		
		(viii) Acetonitrile		
		(ix) 1-Butanol		
		(x) Water		

(*Continued*)

TABLE 4.1 (*Continued*)
Data Sources on Experimental Solubility of Several Polyphenols in Different Solvents at Specific Temperature Ranges

Polyphenol Class[a]	Polyphenol	Solvent	Temperature Range	Data Source[b]
	Isoquercitrin	(i) Tert-amyl alcohol	(i)–(iii) 323.15–343.15 K	(i)–(iii) Chebil et al. (2007)
		(ii) Acetone		
		(iii) Acetonitrile		
	Luteolin	(i) Ethanol–water mixtures	(i) 273.15–323.15 K	(i) Peng and Yan (2009)
	Naringenin	(i) Water	(i) 310.15 K	(i) Tozuka et al. (2010)
		(ii) Acetonitrile	(ii) 323.15–343.15 K	(ii) Chebil et al. (2007)
	Naringin	(i) Water	(i) 318.15–356.15 K	(i) Pulley (1936)
	Puerarin	(i) Methanol–acetic acid mixtures	(i) 293.15–333.15 K	(i) Wei and Zhang (2013)
	Quercetin	(i) Phosphate buffer solution	(i.a) 298.15 K	(ii.b) Rastogi and Jana (2016)
		(ii) Water	(i.b)–(ii) 310.15 K	(i)–(ii.a) Cadena et al. (2013)
		(iii) Ethanol–water mixtures	(iii)–(iv) 298.2–333.8 K	(iii)–(iv) Razmara et al. (2010)
		(iv) Methanol–water mixtures	(v)–(vi) 323.15 K	(v)–(vii) Chebil et al. (2007)
		(v) Tert-amyl alcohol	(vii) 323.15–343.15 K	
		(vi) Acetone		
		(vii) Acetonitrile		
	Rutin	(i) Phosphate buffer solution	(i) 298.15 K	(i) Rastogi and Jana (2016)
		(ii) Ethanol–water mixtures	(ii) 273.15–323.15 K	(ii) Peng et al. (2009)
		(iii) Tert-amyl alcohol	(iii)–(v) 323.15 K	(iii)–(v) Chebil et al. (2007)
		(iv) Acetone		
		(v) Acetonitrile		
Phenolic acids	Caffeic acid	(i) Phosphate buffer solution	(i) 298.15 K	(i) Rastogi and Jana (2016)
		(ii) Ethanol–water mixtures	(ii) 278.2–298.2 K	(iii) Zhang et al. (2012)
		(iii) Water	(iii) 288–323 K	(ii) Mota et al. (2008)

(*Continued*)

TABLE 4.1 (*Continued*)
Data Sources on Experimental Solubility of Several Polyphenols in Different Solvents at Specific Temperature Ranges

Polyphenol Class[a]	Polyphenol	Solvent	Temperature Range	Data Source[b]
	o-Coumaric acid	(i) Water	(i) 288–323 K	(i) Queimada et al. (2009)
	Ellagic acid	(i) Water (ii) Phosphate buffer solution (iii) Methanol	(i.a) 288–323 K (i.b)–(iii) 310.15 K	(i.a) Queimada et al. (2009) (i.b)–(iii) Bala et al. (2006)
	Ferulic acid	(i) Water (ii) Chlorides solutions	(i.a) 288–323 K (i.b) and (ii) 293.15–318.15 K	(i.a) Mota et al. (2008) (i.b) and (ii) Noubigh et al. (2007)
	Gallic acid	(i) Phosphate buffer solution (ii) Water (iii) Methanol (iv) Ethanol (v) Ethyl acetate (vi) Aqueous sodium sulfate solutions (vii) Chlorides solutions	(i) 298.15 K (ii.a)–(v) 298.2–333.2 K (ii.b) 288–323 K (ii.c) and (vi) 293.15–318.15 K (ii.d) and (vii) 293.15–318.15 K (ii.e) 273.15–363.15 K	(i) Rastogi and Jana (2016) (ii.a)–(v) Daneshfar et al. (2008) (ii.b) Mota et al. (2008) (ii.c) and (vi) Noubigh et al. (2008) (ii.d) and (vii) Noubigh et al. (2007) (ii.e) Lu and Lu (2007)
	Protocatechuic acid	(i) Methanol (ii) Ethanol (iii) Ethyl acetate (iv) Methyl acetate (v) Water (vi) Aqueous sodium sulfate solutions (vii) Chlorides solutions	(i)–(iv) 293.15–318.15 K (v.a) 288–323 K (v.b) and (vi) 293.15–318.15 K (v.c) and (vii) 293.15–318.15 K	(i)–(iv) Noubigh et al. (2015) (v.a) Queimada et al. (2009) (v.b) and (vi) Noubigh et al. (2008) (v.c) and (vii) Noubigh et al. (2007)

(*Continued*)

TABLE 4.1 (*Continued*)
Data Sources on Experimental Solubility of Several Polyphenols in Different Solvents at Specific Temperature Ranges

Polyphenol Class[a]	Polyphenol	Solvent	Temperature Range	Data Source[b]
	Syringic acid	(i) Water (ii) Aqueous sodium sulfate solutions (iii) Chlorides solutions	(i.a) 288–323 K (i.b) and (ii) 293.15–318.15 K (i.c) and (iii) 293.15–318.15 K	(i.a) Queimada et al. (2009) (i.b) and (ii) Noubigh et al. (2008) (i.c) and (iii) Noubigh et al. (2007)
	Trans-cinnamic acid	(i) Water	(i) 288–323 K	(i) Mota et al. (2008)
	Vanillic acid	(i) Water (ii) Chlorides solutions	(i)–(ii) 293.15–318.15 K	(i)–(ii) Noubigh et al. (2007)
Stilbenes	Resveratrol	(i) Phosphate buffer solution (ii) Water	(i.a) 298.15 K (i.b) and (ii) 310.15 K	(i.a) Rastogi and Jana (2016) (i.b) and (ii) Cadena et al. (2013)
	Trans-resveratrol	(i) Water (ii) Buffer solution (iii) Ethanol (iv) Methanol (v) 1-Propanol (vi) 2-Propanol (vii) 1-Butanol (viii) 1-Pentanol (ix) 1-Hexanol (x) Ethyl acetate (xi) Tetrahydrofuran (xii) Acetone (xiii) Ethanol–water mixtures (xiv) Acetone–water mixtures	(i.a) and (ii) 310.15 K (i.b) and (iii)–(xii) 278.2–318.2 K (xiii)–(xiv) 273.15–323.15 K	(i.a) and (ii) Zupančič et al. (2015) (i.b) and (iii)–(xii) Sun et al. (2008a) (xiii) and (xiv) Sun et al. (2008b)

(Continued)

TABLE 4.1 (*Continued*)
Data Sources on Experimental Solubility of Several Polyphenols in Different Solvents at Specific Temperature Ranges

Polyphenol Class[a]	Polyphenol	Solvent	Temperature Range	Data Source[b]
Other polyphenols	Tyrosol	(i) Water	(i) 288–323 K	(i) Queimada et al. (2009)
	Vanillin	(i) Water	(i.a), (ii), and (iii) 298.15–318.15 K	(i.a), (ii), and (iii) Shakeel et al. (2015b)
		(ii) Carbitol	(i.b) and (iv)–(xii) 298–318 K	
		(iii) Carbitol–water Mixtures	(i.c) and (xiii) 293.15–318.15 K	(i.b) and (iv)–(xii) Shakeel et al. (2015a)
		(iv) Ethanol	(i.d) and (xiv) 293.15–318.15 K	
		(v) Propylene glycol		(i.c) and (xiii) Noubigh et al. (2008)
		(vi) Transcutol		
		(vii) Polyethylene Glycol-400		(i.d) and (xiv) Noubigh et al. (2007)
		(viii) Ethylene glycol		
		(ix) Ethyl acetate		
		(x) 2-Propanol		
		(xi) 1-Butanol		
		(xii) 2-Butanol		
		(xiii) Aqueous sodium sulfate solutions		
		(xiv) Chlorides solutions		

[a] Polyphenol classes are defined using the Phenol Explorer database (phenol-explorer.eu).

[b] Shake-flask methods coupled with HPLC and/or UV/VIS systems were used to determine the solubility. Only Pulley (1936) and Sevillano et al. (2013) applied other methods (gravimetric and a high-throughput approach, respectively).

saturation is reached, (3) overnight settling of suspended solids at the same conditions of temperature and pressure, (4) taking and storing samples from the supernatant at the same conditions, and (5) measurement of concentration of the corresponding polyphenol using HPLC and/or UV/VIS systems. Other methods do exist, such as the high-throughput approaches (Alsenz and Kansy, 2007; Sevillano et al., 2013), but they are mainly used to determine the solubility of drugs in pharmaceutics.

Because polyphenols tend to degrade at higher temperatures (see Section 4.2.3 for more information), most of the solubility data available in the literature (references cited in Table 4.1) deal with atmospheric conditions, that is, pressure of 101,325 Pa and temperatures below the boiling point of the corresponding solvent or solvent mixture. Scarce data on the solubility of polyphenols in pressurized solvents at high temperatures and pressures can be found in the literature. For instance, Srinivas et al. (2010a,b) determined the solubility of gallic acid, protocatechuic acid, (+)-catechin, and quercetin in pressurized water at temperatures from 298.75 to 415.85 K using a dynamic flow apparatus coupled with an HPLC system designed to avoid thermal degradation of polyphenols. In general, three main findings can be derived from the experimental data considered here: (1) the solubility of polyphenols at atmospheric conditions or at high temperatures and pressures will increase as temperature increases; (2) the addition of salt hinders the solubilization of polyphenols, promoting a salting-out effect; and (3) polyphenols are better dissolved in solvents with intermediate polarity, such as alcohols, hydroalcoholic mixtures, and acetone. The latter finding is supported by the work of Galanakis et al. (2013), who analyzed the solubilization process of a set of polyphenols in different solvents through the calculation of the corresponding activity coefficients.

Scarcity of data on solubility of polyphenols at high temperatures and pressures shows how difficult it is to measure this experimentally. Thus, it is important to develop mathematical models that enable the correlation and extrapolation of experimental data at atmospheric conditions, inferring unknown solubilities at high temperatures and pressures. Various approaches have been addressed to achieve such a goal: empirical equations, excess solubility models, activity coefficient models, and associating equations of state. Empirical equations such as the simplified version of the Williamson equation (Apelblat and Manzurola 1999, Williamson 1944), also known as the modified Apelblat equation, and the λh (Buchowski and Khiat 1986, Buchowski et al. 1980), Heidmann (Heidman et al. 1985), Miller (Miller et al. 1998), Richards (Richards 1959), van't Hoff (a good description can be found in Esina and Korchuganova, 2015) and Yalkowsky-Roseman (Yalkowsky and Roseman 1981) models have provided good agreements with experimental solubility data of different classes of polyphenols, results that can be checked in various of the references of Table 4.1. Regarding the excess solubility models, those originally developed by Gude et al. (1996) and Jouyban et al. (2011) have been recently applied to correlate the solubility data of green tea catechins in ethanol–water mixtures at atmospheric conditions, giving a good fit (Sevillano et al., 2013) (in Table 4.1). In the case of activity coefficient models, UNIQUAC (Abrams and Prausnitz, 1975), UNIFAC (Fredenslund et al., 1975), modified UNIFAC (Larsen et al., 1987), modified UNIQUAC (Peres and Macedo, 1996), NRTL-SAC (Chen and Crafts, 2006; Chen and Song, 2004), and MPP (modified polyphenol) UNIFAC (Sevillano et al., 2014) have been successfully used to calculate the solubility of several polyphenols, mostly phenolic acids and flavonoids, at atmospheric pressure and temperatures below the normal boiling point of the corresponding solvent (Ferreira and Pinho, 2012; Galanakis et al., 2013; Queimada et al., 2009; Sevillano et al., 2013, 2014; Sun et al., 2008b; Zhang et al., 2012). Special attention should be given to the MPP-UNIFAC model (Sevillano et al., 2014), since it was specifically developed for polyphenols by regressing new interaction parameters from literature solubility data and by defining a new set of UNIFAC molecular groups and subgroups. Within the existing associating equations of state (EoS), the cubic-plus-association EoS (CPA) (Kontogeorgis et al., 1996) has been applied to adequately correlate the solubility of phenolic acids and vanillin in water at 101,325 Pa and temperatures below 373 K (Mota et al., 2008, 2010; Queimada et al., 2009).

Since they intrinsically take into account the effects of molecular structures and intermolecular interactions in their calculations, the CPA EoS and the UNIFAC-type activity coefficient models exhibit improved predictive capabilities relative to the empirical equations or the excess solubility approaches. Therefore, the CPA EoS or the UNIFAC-type models should be applied to extrapolate the existing solubility data of polyphenols and to generate new data at high temperatures and pressures. However, their suitability should be first carefully analyzed by considering their intrinsic limitations (Gmehling, 2009; Kontogeorgis, 2013) and the availability and reliability of the fusion properties, that is, enthalpy of fusion and melting point temperature of pure polyphenols (required for the solubility calculations). If not available, the fusion properties can be experimentally obtained by differential scanning calorimetry (DSC) or determined through group contribution methods widely described in the literature (Jain and Yalkowsky, 2006; Jain et al., 2004; Marrero and Gani, 2001; Qiang et al., 2009).

4.2.2 Mass Transfer Properties at High Temperatures and Pressures

In addition to solubility, several other physicochemical properties of liquids such as viscosity, surface tension, and disassociation constant (in the case of water) can be manipulated through the change in temperature to improve the extraction process (Hawthorne et al., 2002; King et al., 2010). The increase in temperature results in a decrease in viscosity and surface tension (Figure 4.2), enhancing solvent penetration and wetting of the matrix, as well as mass transfer properties (Mustafa and Turner, 2011; Teo et al., 2010). Additionally, increasing temperature and pressure results in the disruption of surface equilibrium, decreasing the activation energy required for the desorption process (Clifford, 2005; Hawthorne et al., 2002; Ong et al., 2006; Smith, 2006).

As in conventional processes, the mass transfer at high temperature and pressure is determined from diffusion and hydrodynamic data. The rate at which a component is transferred from one phase to another depends on the mass transfer coefficient and the deviation level of the system from the equilibrium; the transfer ends when the equilibrium is reached.

Mass transfer coefficients are essential for process design, scale-up, and optimization; regulating the speed at which equilibrium is reached also controls the time

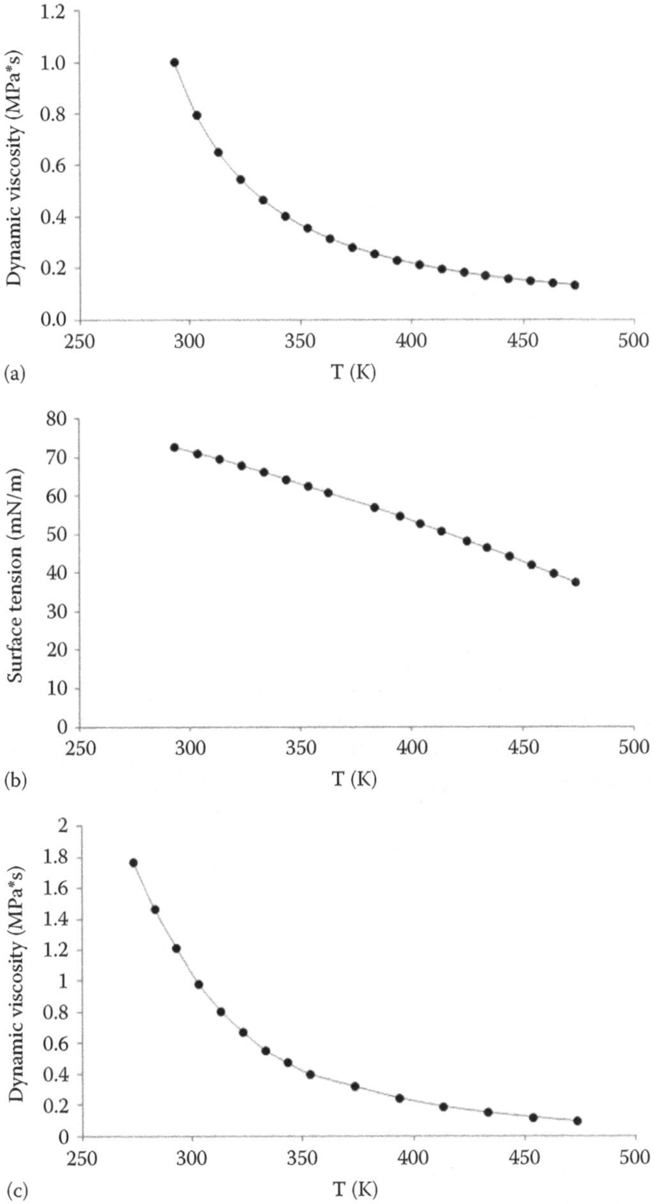

FIGURE 4.2 Change in viscosity and surface tension of water (a, b) and ethanol (c, d) between 293 and 473 K. (Data obtained from Dortmund Data Bank [DDB].) (From Gmehling, J. and Weidlich, U., *Chemie Ingenieur Technik*, 57, 447, 1985; Onken, U. et al., *Int. J. Thermophys.*, 10, 739, 1989.) *(Continued)*

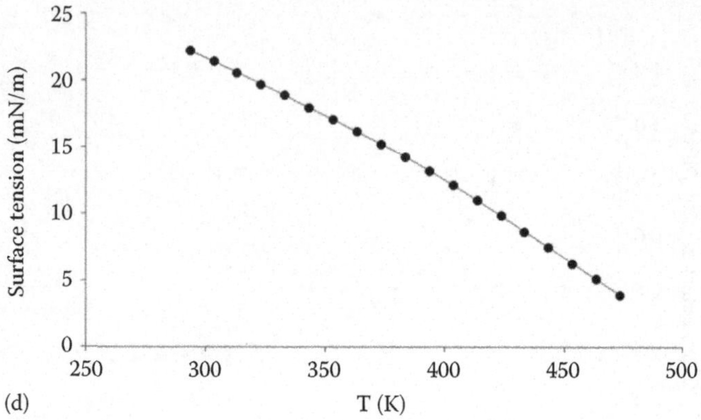

FIGURE 4.2 (*Continued*) Change in viscosity and surface tension of water (a, b) and ethanol (c, d) between 293 and 473 K. (Data obtained from Dortmund Data Bank [DDB].) (From Gmehling, J. and Weidlich, U., *Chemie Ingenieur Technik*, 57, 447, 1985; Onken, U. et al., *Int. J. Thermophys.*, 10, 739, 1989.)

required for the separation process and, therefore, the size and cost of equipment (Philip and Jessy Van, 2011; Treybal, 1968).

The mass transfer mechanism depends on the system dynamics. There are two mechanisms of mass transfer molecular diffusion and convection. In molecular diffusion, the mass is transferred through individual and random movement of molecules in fluids due to concentration differences. The molecular diffusion occurs under either static or dynamic fluids. In contrast, in convection, the mass is transferred due to global movement of the fluid. The movement is carried out under laminar or turbulent regimes. Turbulent flow is the movement of large groups of molecules and is influenced by the dynamic characteristics of flow, such as density and viscosity. In turbulent conditions, molecular diffusion is relatively unimportant, and transfer coefficients become more similar for all components (Philip and Jessy Van, 2011; Treybal, 1968).

The rate of diffusion of a single species in a single fluid can be described by Fick's law:

$$J_A = -cD_{AB}\frac{dx_A}{dz} \tag{4.1}$$

where

J_A is the rate of diffusion (mol m^{-2} s^{-1})
c is the concentration (mol m^{-3})
D_{AB} is the diffusion coefficient of the solute A in solvent B (m^2 s^{-1})
x_A is the mol fraction of A in B
z is the direction of diffusion

For dilute solutions of nonelectrolytes, the empirical correlation of Wilke and Chang (1955) is recommended for determining the diffusion coefficient value:

$$D_{AB}^{0} = \frac{\left(117.3\times10^{-18}\right)\left(\varphi\times M_{B}\right)^{0.5}\times T}{\mu\times V_{A}^{0.6}} \tag{4.2}$$

where

D_{AB}^{0} is diffusion coefficient or diffusivity of A in a dilute solution in solvent B (m^2 s^{-1})

M_B molecular weight of solvent (kg mol^{-1})

φ is the association parameter (multiple of nominal molecular weight of solvent to give effective value)

T is the temperature (K)

μ is the dynamic viscosity of solution (kg m^{-1} s^{-1})

V_A is the molal volume of solute at normal boiling point (m^3 $kmol^{-1}$)

The Wilke and Chang correlation clearly shows the dependence of diffusivity on temperature and viscosity. The diffusivity is directly proportional to temperature and inversely proportional to solution viscosity. This relationship can also be derived from the Stokes–Einstein equation, which predicts the dependence of the diffusivity on temperature and viscosity according to

$$\frac{D_{T_1}}{D_{T_2}} = \frac{T_1}{T_2}\times\frac{\mu_{T_2}}{\mu_{T_1}} \tag{4.3}$$

Therefore, according to Fick's law of diffusion, the mass-transfer rates are accelerated by increasing temperature, leading to improved extraction efficiencies and reduced extraction times.

When extraction of polyphenol from vegetable materials is controlled by internal diffusion, the process can be described by Fick's second law (Boussetta et al., 2012; Sineiro et al., 1996):

$$\frac{\partial C}{\partial t} = D_{eff}\frac{\partial^2 C}{\partial r^2} \tag{4.4}$$

where

C is the solute concentration (g L^{-1})

t is the time (s)

r is the radial position into the particle (m)

D_{eff} is the effective diffusion coefficient or effective diffusivity (m^2 s^{-1})

To solve the aforementioned equation, the geometry of the particle must be defined. For example, in the study of polyphenols' extraction from grape seeds, Boussetta et al. (2012) considered grape seeds as thin slabs of uniform thickness h, since polyphenols are located mainly in the seed coat. They found the following solution of diffusion equation for a limited volume of liquid and negligible external mass transfer:

$$Y(r,t) = \frac{C - C_e}{C_0 - C_e} = \sum_{n=1}^{\infty} \frac{2\alpha(1+\alpha)}{1+\alpha+q_n^2\alpha^2} \exp\left[\frac{-q_n^2 t}{\tau}\right] \tag{4.5}$$

where

C_e is the polyphenol concentration in the grape seed extract at equilibrium conditions (g m^{-3})

C_0 is the initial polyphenol concentration inside the grape seeds (g m^{-3})

α is the ratio of the volume of extraction water to the volume of solution inside the grape seeds

$\tau = (h^2/D_{eff})$ is the effective diffusion time (s)

q_n's are the nonzero positive roots of the equation $\tan q_n = \alpha q_n$

And, for long diffusion periods and negligible external resistance, Fick's second law equation can be further simplified to only the first term of the series, which can be rewritten in logarithmic form as

$$\ln Y = \ln\left(\frac{2\alpha(1+\alpha)}{1+\alpha+q_1^2\alpha^2}\right) - \frac{q_1^2}{\tau}t \tag{4.6}$$

The plot of ln Y versus time should give a straight line with a slope of q_1^2/τ.

Boussetta et al. (2012) reported effective diffusion times between 13 and 139 min for different extraction treatments in water–ethanol mixture at 50°C. Similarly, in the study of polyphenols mass transfer during the leaching of apple slices immersed in mannitol aqueous solutions at 20°C, Kebe et al. (2014) found different diffusivity values in two apple varieties: 0.2–0.1 × 10^{-9} and 0.2 × 10^{-9} m^2 s^{-1}, for Golden Delicious and Granny Smith, respectively. In the study on polyphenol extraction from sunflower press cake with 96% (v/v) at 50°C, Sineiro et al. (1996) found effective diffusivities between 1.568 × 10^{-12} and 2.21 × 10^{-13} m^2 s^{-1}.

In parallel, it is possible to infer that the larger the area over which diffusion can take place, the higher the rate of mass transfer. However, in the reduction of matrix particle size, certain limitations should be considered, such as (1) degradation of labile compounds due to the operation conditions of the size reduction process, (2) channeling of solvent in compacted bed of particles, (3) increasing extraction rate of undesired compounds and hence decreasing extraction selectivity and extract purity, and (4) filter blocking in subsequent solid–liquid separation (Philip and Jessy Van, 2011; Pronyk and Mazza, 2009).

4.2.3 Thermal Degradation and Formation of Nondesired Compounds

In addition to the enhancement of solubility and mass transfer properties, two main phenomena must be considered in PLE of polyphenols from vegetable matrices: thermal degradation and formation of neo-antioxidant and toxic compounds (Martins et al., 2000; Yilmaz and Toledo, 2005). These two phenomena are highly dependent on temperature and time of extraction (Duan et al., 2009; Plaza et al., 2010b; Rodriguez-Meizoso et al., 2006; Srinivas et al., 2011).

Polyphenols are compounds sensitive to high temperatures, especially above 100°C (Nakamura et al., 1998; Palma et al., 2001; Srinivas et al., 2011); therefore, the extraction yield will be determined by their solubility and mass-transfer rate, as well as their degradation rate. Several studies have demonstrated the negative effect of using high temperatures and long extraction times on stability and, therefore, on the extraction yield of polyphenols. Vergara-Salinas et al. (2012) found that by increasing temperature up to 150°C and 200°C and extraction times up to 15 and 30 min, the recovery of hydroxycinnamic acids, flavones, and flavonols/flavanones decreased. Likewise, Ju and Howard (2005) reported that increasing the temperature from 100°C to 110°C produced an increase in the extraction of anthocyanins, although extraction temperatures above 110°C resulted in lower recovery of individual and total anthocyanins.

Furthermore, many new compounds can also be found in the extracts obtained at high temperatures, particularly products from Maillard, thermoxidation, and caramelization reactions (He et al., 2012; Plaza et al., 2010a). Although Maillard reaction products show high antioxidant activity, they can also be toxic, mutagenic, and carcinogenic (Husøy et al., 2008; Martins et al., 2000; Yilmaz and Toledo, 2005). Therefore, care should be taken when maximizing the extraction yield of polyphenols from plant materials by PLE. Plaza et al. (2013) evaluated the effect of temperature and time on the flavonols recovery and formation of Maillard reaction products in the PHW extraction (PHWE) of apple by-products. By using the response surface methodology, they determined the condition that maximizes the flavonols extraction and minimizes the formation of Maillard reaction products, 120°C–125°C and 3 min.

4.3 SELECTING THE RIGHT TYPE OF SOLVENT

The first step in the development of extraction processes is the selection of a suitable type of solvent for a specific purpose. Many aspects must be analyzed during the solvent selection including safety, productivity (economics), and sustainability. This analysis should be more exhaustive for food and nutraceutical applications.

From a process safety point of view, PLE allows for the use of a wide range of solvents except those that possess an autoignition point equal to or lower than the extraction temperature, which is usually between 100°C and 200°C (e.g., carbon disulfide, diethyl ether, and 1,4-dioxane) (Green, 2008). Also, corrosive substances such as strong bases and acids should be especially avoided. Corrosiveness increases as temperature and pressure increase, causing damage to the extraction system.

From a process economy focus, aspects such as solubility of the desired analyte(s) and its diffusivity in the solvent, as well as the characteristics of the sample, must be reviewed in the choice of the right type of solvent (Mustafa and Turner, 2011; Pronyk and Mazza, 2009; Rizvi, 2010). Selecting the right type of solvent affects the amount of polyphenols extracted and the extraction rate (Dai and Mumper, 2010; Xu and Chang, 2007). The solvent polarity is a very important parameter. The rule of the thumb in solvent choice is "like dissolves like," which indicates the use of polar solvents for polar analytes and likewise nonpolar analytes are dissolved in nonpolar solvents (Mustafa and Turner, 2011). Furthermore, the polarities of phenolic compounds vary significantly because their structures range from simple (e.g., phenolic acids and flavonoids) to highly polymerized compounds (e.g., tannins). Moreover, polyphenols may have multiple hydroxyl groups that can be conjugated to sugars, acids, or alkyl groups. Thus, it is difficult to develop a single method for the optimum extraction of all phenolic compounds (Garcia-Salas et al., 2010).

Solvents, such as methanol, ethanol, acetone, ethyl acetate, n-hexane, isooctane, and their combinations have been used for the extraction of polyphenols from plant materials, often with different proportions of water (Alothman et al., 2009; Dai and Mumper, 2010). The best results have been obtained using alcohol-based polar solvents, especially for low-molecular-weight polyphenols. For these simple polyphenols, the best extraction yields have been obtained using methanol or acidified methanol (Ignat et al., 2011; Kapasakalidis et al., 2006). However, in the food industry, ethanol is preferred due to the methanol toxicity (Ignat et al., 2011). Moreover, adding water to ethanol can increase its solvating ability and mass-transfer rate. Water is usually important to aid breaking matrix and matrix–analyte (hydrogen) bonding. Thus, using solvent mixtures may considerably improve the extraction efficiency. For example, in a binary mixture, one solvent may increase the solubility of a target compound while the other solvent would increase its desorption (Mustafa and Turner, 2011). However, a high water content increases the release of other compounds present in the vegetable matrix such as sugars, which generate a less pure extract with a lower concentration of polyphenols (Spigno et al., 2007; Tomsone et al., 2012). On the other hand, high-molecular-weight polyphenols, such as proanthocyanidins, are preferably extracted with water/acetone mixtures (Dai and Mumper, 2010; Guyot et al., 2001; Labarbe et al., 1999).

Particularly, in PLE applications, the solvents commonly used have been methanol, methanol/water (56%–90% of methanol), methanol/water/acid (80/19/1), ethanol, ethanol/water (40%–70% of ethanol), ethanol/water/acid (70/25/5), 1-propanol/water (3/1), and acetone/water (70/30) (Herrero et al., 2013). A special case is the use of PHW to obtain polyphenols. PHW extraction is a green process that provides high selectivities and short extraction times to recover polyphenols from plant materials (Vergara-Salinas et al., 2015). It takes advantage of the properties (polarity, viscosity, surface tension) of liquid water at high temperatures (between 100°C and 374°C) to enhance the dissolution of target compounds from solid matrices (Clifford, 2005; Hawthorne et al., 2002; Herrero et al., 2013; Ong et al., 2006; Smith, 2006; Teo et al., 2010; Vergara-Salinas et al., 2015). However, in PHWE and PLE in general, care must be taken with the degradation of target compounds and possible generation of undesired substances due to high temperatures (Ju and Howard, 2005; Vergara-Salinas et al., 2012, 2013). The addition of small amounts of organic solvents such as ethanol and surfactants could strongly improve the solvating capacity of PHW, allowing for a reduction in the extraction temperature (Ong et al., 2006).

Finally, regarding food and environmental safety, the most suitable condition for producing food-relevant polyphenols is the application of food-grade solvents and emerging green solvents. A food-grade substance is the one that can be ingested up to a specified amount without causing deterioration or damage to health. Food-grade compounds include those "generally recognized as safe" (GRAS) by the U.S. Food and Drug Administration (FDA). They are listed as approved under the Code of Federal

Regulations (CFR), specifically 21 CFR (Bartnick et al., 2006). Extracting solvents used in food manufacturing have been labeled as food additive, solvent, defoaming agent, component of secondary food and color additives, minor constituent, or incidental additive, depending on the application. Incidental additives are "additives that are present in a food at insignificant levels and do not have a technical or functional effect in that food." They include substances that are added to a food during processing but removed from the food before it is packaged and are exempt from label declaration in the final food product. However, their use must be in accordance with the good manufacturing practices (GMP) (Wan and Wakelyn, 1997). In terms of human consumption, the solvents classified as Class 3 by the FDA (Table 4.2) may be regarded as less toxic and of lower risk to human health. Class 3 includes solvents such as acetone, ethanol, ethyl acetate, 1-propanol, 2-propanol, and propyl acetate, which are acceptable in small residual percentages, according to GMP. It is considered that amounts of these residual solvents of 50 mg per day or less would be acceptable without justification. Other solvents, particularly those classified as Class 2 by the FDA, such as acetonitrile, chloroform, hexane, methanol, toluene, ethylmethylketone, and dichloromethane, can be used under specific conditions and present limitations concerning pharmaceutical and food products because of their inherent toxicity (Meireles, 2008).

Although Class 3 solvents are recommended by the FDA for extractive purposes, many of them are not environment friendly, not very suitable for the extraction of health-promoting compounds, and they also are expensive or nonselective (Cacace and Mazza, 2006a; Luque De Castro et al., 1999). Emerging green solvents appear as an alternative to common food-grade solvents. The concept of green solvents, in addition to their food-grade quality, also includes the reduction of the environmental impact of the use of solvent in chemical processes (Mustafa and Turner, 2011). Green solvents play an important role in the replacement of petrochemical solvents (Chemat et al., 2012). These solvents are renewable resources produced from biomasses such

TABLE 4.2
Class 3 Solvents That Should Be Limited by GMP or Other Quality-Based Requirements

Acetic acid	Heptane
Acetone	Isobutyl acetate
Anisole	Isopropyl acetate
1-Butanol	Methyl acetate
2-Butanol	3-Methyl-1-butanol
Butyl acetate	Methylethyl ketone
tert-Butylmethyl ether	Methylisobutyl ketone
Dimethyl sulfoxide	2-Methyl-1-propanol
Ethanol	Pentane
Ethyl acetate	1-Pentanol
Ethyl ether	1-Propanol
Ethyl formate	2-Propanol
Formic acid	Propyl acetate

as wood and starch, which have a high solvent power, are nontoxic and biodegradable. The bioethanol is the most common green solvent, which is obtained by fermentation of sugar-rich materials. It is available at a low price, in high purity, and it is completely biodegradable. Green solvents also include terpenes (e.g., α-pinene and *d*-limonene), fatty acid methyl esters (extraction of nonpolar substances), ionic liquids, and glycerol. The characteristics (high solvent power, high chemical and thermal stability, nonflammability, and nonvolatility) of ionic liquids make them potentially applicable in PLE. Pressurized ionic liquids were successfully used for the recovery of rutin and quercetin from *Flos Sophorae Immaturus* herb; optimum conditions included the use of C4mim-Cl (1 mol L^{-1}) at 120°C and 5 min (Wu et al., 2012). Glycerol is a by-product from the trans-esterification of vegetable oils, which is very common in the cosmetic industry as a solvent for maceration of herbs and spices. Apostolakis et al. (2014) found that a mixture of 9.3% (w/v) of glycerol in water at 80°C provided extraction yield of polyphenols higher than that attained with 60% (v/v) ethanol in water, at pH 2 and 24°C. Moreover, the polyphenolic profile of water/glycerol mixture was significantly different than that of water/ethanol mixture, evidencing differences in their selectivities.

The extractability of solvents and green solvents can be improved by adding another miscible solvent and/or by adding substances such as surfactants or acids (Mustafa and Turner, 2011).

4.4 SELECTING THE OPTIMAL EXTRACTION CONDITIONS: LABORATORY/PILOT-SCALE STUDIES

The determination of the optimal PLE extraction conditions (mode of operation, temperature, and pressure) and the final selection of the extracting solvent (i.e., proportion of different compounds, inclusion of additives) must be done following thorough studies at laboratory or pilot-scale process units. All these procedures are essential for an adequate extraction of phenolic compounds from different food matrices and represent an important step in the scale-up procedure (Garcia-Salas et al., 2010; Pronyk and Mazza, 2009). Since the molecular characteristics and physicochemical properties of the polyphenols to be extracted strongly affect the extraction process (as discussed in previous sections), the results obtained during these procedures could drastically vary from one type of polyphenol to another. They will be finally determined by a balance between polyphenols' solubility, extraction rate and selectivity of solvents, thermal degradation, and formation of toxic compounds (Duan et al., 2009; Martins et al., 2000; Plaza et al., 2010a,b; Rodriguez-Meizoso et al., 2006; Srinivas et al., 2011; Yilmaz and Toledo, 2005).

Within the extraction conditions, the mode of operation and its corresponding parameters must be carefully analyzed and optimized. Two modes are possible: static, where the solid matrix is kept in the extraction unit together with a given volume of solvent at fixed conditions of temperature and pressure, and dynamic, where the pressurized solvent passes continuously through the solid matrix kept in the extraction vessel. Since in dynamic mode the phase equilibria might be displaced increasing the efficiency of the extraction, this mode should be selected over the static (Herrero et al., 2013). However, dynamic PLE generates more diluted extracts than those obtained in static mode, increasing posttreatment costs (Herrero et al., 2013).

The contact time and cycles of extraction (replacement of the solvent to deplete the matrix) are the key parameters to be optimized in the static mode. The extraction time, which is strongly affected by the mass-transfer properties (discussed in Section 4.2.2), must allow for the saturation of the solvent with the target compounds and minimize the exposure of the compounds to high temperatures (Vergara-Salinas et al., 2015). Such saturations, given by the solubility of the compounds, have shown to be reached in a period ranging from 5 to 30 min during static PLE (Mukhopadhyay and Panja, 2010). Evaluating the effect of time (5, 15, and 30 min) in PHWE of polyphenols from deodorized thyme, Vergara-Salinas et al. (2012) found that at 50°C, the extraction yields of flavones and total polyphenols increased with extraction time, whereas those of flavonols and hydroxycinnamic acids remained practically the same. In turn, at 150°C, all yields were negatively affected by the exposure time. On the other hand, the cycles of extraction (number of sequential extractions) can be determined through a balance between the solvent cost and the extraction yield improvement. In the dynamic mode, the solvent flow rate and the mean residence time become the main parameters to be optimized (Teo et al., 2010). If the flow rate is large enough, the diffusion of the solute within the matrix (internal mass transfer) controls the extraction rate, and therefore, further increases in the flow rate would have little effect on the extraction rate (Pronyk and Mazza, 2009). If the flow rate is low enough, the removal of the solute from the surface of the matrix decreases, and then external mass transfer from the solid to the solvent controls the extraction. In this case, an increase in the flow rate would result in an increase of the extraction rate of the solute (Pronyk and Mazza, 2009). Regarding the mean residence time of the solvent (calculated as bed depth/superficial velocity), this should be long enough to allow the solvent penetration through the solid pores. In large-scale vessels, higher superficial velocities must be used compared to smaller-scale vessels in order to keep similar residence times (Pronyk and Mazza, 2009). Ho et al. (2008) studied the feasibility of extracting lignans from flaxseed meal using PHW in a fixed-bed extraction cell, evaluating the flow rate, bed depth, and solvent-to-solid ratio. They found that the extraction yield of lignans increased with increasing bed depth and was independent of the solvent-to-solid ratio, probably due to the longer residence time.

As widely discussed before, temperature and pressure are the most important parameters to be optimized in PLE. Herrero et al. (2011) made interesting findings regarding temperature: the extraction of polyphenols, such as hydroxytyrosol, was improved by increasing temperature up to 200°C, independent of the solvent used; flavonoids reached their highest recovery at medium extraction temperatures (100°C and 150°C); and maximum extraction yields for some flavones such as apigenin and diosmetin were obtained at 50°C. Vergara-Salinas et al. (2012) also studied the effects of temperature (50°C–200°C) on the static PHWE of polyphenols from deodorized thyme. The extraction yields of total polyphenols, hydroxycinnamic acids, flavones, and flavonols/flavanones increased with temperature up to 100°C. At 150°C and 200°C, the recovery of these phenolic compounds decreased due to thermal degradation. Only the recovery of dihydroxyphenyllactic acid (a hydroxyphenylpropanoic acid) increased constantly with temperature, reaching a maximum at 200°C, probably as a product of the thermal degradation of rosmarinic acid. Monrad et al. (2009) also found different optimal static extraction temperatures for different polyphenolic structures; procyanidin monomers and oligomers were preferentially extracted between 100°C and 140°C, depending on the solvent mixture used; and the extraction of procyanidin polymers was maximized at temperatures between 120°C and 140°C. Regarding the operating pressure in PLE, vapor pressure data and boiling temperatures of solvents must be considered to determine the optimal value of the operating pressure in PLE. As described in Section 4.2, this value should be high enough to maintain the solvent in its liquid state, that is, bigger than the vapor pressure corresponding to the operating temperature. Vapor pressure data of solvents can be obtained from literature or specialized databases, such as the Dortmund Data Bank (DDB, http://www.ddbst.com/) and DETHERM (http://www.dechema.de/en/detherm.html). If it is not available, the vapor pressure data can be calculated by using the Antoine equation and the Raoult's law, as shown in Table 4.3 for water, ethanol, and its combinations at temperatures between 353 and 473 K.

TABLE 4.3
Vapor Pressure of Liquid Water, Ethanol, and Water/Ethanol Mixtures at Temperatures between 353 and 473 K (80°C–200°C)

	Vapor Pressure (kPa)						
Temperature (K)	Water	Ethanol	Water/Ethanol x = 0.9	Water/Ethanol x = 0.8	Water/Ethanol x = 0.7	Water/Ethanol x = 0.6	Water/Ethanol x = 0.5
353	47.3	108.4	53.4	59.5	65.6	71.7	77.9
363	70.0	158.5	78.9	87.7	96.6	105.4	114.3
373	101.9	256.0	117.3	132.7	148.1	163.5	178.9
383	143.4	314.8	160.5	177.7	194.8	212.0	229.1
393	198.0	429.6	221.1	244.3	267.5	290.6	313.8
403	268.7	575.1	299.4	330.0	360.7	391.3	421.9
413	359.0	757.0	398.8	438.6	478.4	518.2	558.0
423	472.6	980.7	523.4	574.2	625.0	675.8	726.6
433	613.7	1252.3	677.6	741.5	805.3	869.2	933.0
443	787.0	1578.2	866.2	945.3	1024.4	1103.5	1182.6
453	997.5	1964.8	1094.2	1191.0	1287.7	1384.4	1481.1
463	1250.6	2418.6	1367.4	1484.2	1601.0	1717.8	1834.6
473	1552.0	2946.5	1691.4	1830.9	1970.3	2109.8	2249.2

Notes: Vapor pressures of water and ethanol were calculated according to the equation $P_v = 10^{A - B/(C+T)}$; where P_v is the vapor pressure (mmHg); T is the water temperature (°C); and A, B, and C are the Antoine constants. When water temperature is in the range of 80°C–99°C, A = 8.07131, B = 1730.63, C = 233.426. When the temperature is in the range of 100°C–200°C, A = 8.14019; B = 1810.94; and C = 244.485. When ethanol temperature is in the range of 80°C–200°C, A = 7.68117; B = 1332.04; and C = 199.2. Vapor pressure of water/ethanol mixtures were calculated according to Raoult's law $P_v = \Sigma P_i x_i$, where P_i is the vapor pressure of pure solvent and x_i is the corresponding mole fraction of that solvent. "i" is an indexing component that keeps track of each substance in the solution. (Antoine constants were obtained from NIST Standard Reference Database 69, June 2005 Release, NIST Chemistry WebBook.)

Finally, the definitive selection of the extracting solvent (i.e., proportion of different compounds, inclusion of additives) must be carried out. Given the solubility data of polyphenols (Section 4.2.1) and concerns profoundly described in Section 4.3, the most adequate solvents to extract polyphenols for food applications are ethanol, water, and the corresponding mixtures, which in fact are the commonly used solvents in several studies on PLE of plant polyphenols at the laboratory or pilot scale (Herrero et al., 2013). Therefore, only the proportion of ethanol and water in the mixtures must be optimized, together with solubility-enhancing additives. For instance, Herrero et al. (2011) studied the PLE of different polyphenols (hydroxytyrosol, caffeic acid, luteolin-7-glucoside, apigenin-7-glucoside, oleuropein, quercetin, apigenin, diosmetin) from olive leaves, using both water and ethanol, at four different temperatures (50°C, 100°C, 150°C, and 200°C), and using a static extraction laboratory equipment. The extraction time and pressure were maintained constant, 20 min and 10.34 MPa (1500 psi). A strong influence of the solvent polarity on the type of compound extracted was observed. Using pure water favors the extraction of more polar compounds, such as hydroxytyrosol and caffeic acid, whereas pure ethanol was more selective toward the extraction of less polar compounds: the flavones luteolin-7-glucoside, apigenin-7-glucoside, apigenin, and diosmetin and the flavonol quercetin. In most of the temperatures assessed (except 50°C), the highest recovery of total polyphenols (sum of individual polyphenols) was obtained using pure water, mainly due to the high extraction yield of hydroxytyrosol and oleuropein. However, many other studies show that using ethanol–water mixtures in the range of 40%–70% of ethanol results in higher extraction yields of polyphenols (Howard and Pandjaitan, 2008; Kim et al., 2010; Luthria, 2008; Monrad et al., 2009; Wijngaard and Brunton, 2009). Monrad et al. (2009) studied the PLE of procyanidin monomers, oligomers, and polymers from red grape pomace using several ethanol–water mixtures (0%, 10%, 30%, 50%, 70%, and 90%, v/v of ethanol) at different temperatures (40°C, 60°C, 80°C, 100°C, 120°C, and 140°C) in a static extraction laboratory equipment. No ideal solvent composition was found due to variation in structural complexity and polarity of the compounds; in general, total procyanidins were optimally extracted with 50% ethanol/water (v/v), especially polymers. Regarding the addition of solubility-enhancing additives, no standard compounds and compositions have been found. For example, resveratrol and other polyphenols from *Polygonum cuspidatum* roots were efficiently extracted in a water solution of β-cyclodextrin (1.5%) (Mantegna et al., 2012). Moreover, solutions of nonionic surfactants can be used as an alternative solvent system in PLE. Furthermore, using some surfactants could decrease the degradation of the target compounds. Also, the addition of protective antioxidants such as ascorbic acid can prevent the oxidation of the analytes in PLE. Another compound that can be used to improve PLE is carbon dioxide (CO_2). The addition of CO_2 reduces the pH of water, increasing its hydrolytic capacity and improving the release of compounds bound to cell structures (Mustafa and Turner, 2011).

Many other studies regarding the evaluation of PLE processes at laboratory scale can be found in several review articles (Herrero et al., 2006, 2013; Mendiola et al., 2007; Mukhopadhyay and Panja, 2010; Mustafa and Turner, 2011; Ong et al., 2006; Smith, 2002; Teo et al., 2010; Wiboonsirikul and Adachi, 2008), book sections

(Cacace and Mazza, 2006b; Clifford, 2005, Srinivas and King, 2010; Vergara-Salinas et al., 2015), and patents (Hawthorne et al., 2002; King and Grabiel, 2007).

4.5 DESIGN OF A LARGE-SCALE PLE PROCESS FOR POLYPHENOLS

Large-scale PLE processes should be designed by following a methodology commonly used in chemical engineering and thoroughly described in the specialized literature. Such a methodology includes (1) definition of the design objective, that is, requirements regarding the product to be obtained with the process; (2) setting the design basis, which fundamentally comprises the production rate, product characteristics and information about raw materials, fungibles, utility services, and others; (3) generation of possible design concepts, in this case, different process options developed from laboratory- and pilot-scale studies; (4) testing of design concepts, which includes the application of process modelling and simulation; (5) design selection by evaluating process economics; (6) detailed design and equipment selection, that is, scale-up and equipment sizing and characterization; and (7) procurement, construction, and process operation. Steps (3) to (7) are briefly described in the following text for a PLE process for polyphenols (Towler and Sinnott, 2007).

4.5.1 Generation of PLE Design Concepts and Their Testing

PLE processes are designed and constructed as batch or semicontinuous processes due to the intrinsic difficulties for continuously charging solid material into a high-pressure and high-temperature extraction vessel (Pronyk and Mazza, 2009). The extracting solvent can be poured together with the solid matrix into the vessel and remain static for a given amount of time to reach equilibrium, that is, static mode of operation, or can be fed through the solid bed at a certain flow rate, that is, dynamic mode of operation (as described previously). The basic configuration of a PLE process consists of solvent supply tanks, a high-pressure pump, one or more heat exchangers, the pressurized extraction vessel, a pressure-regulating system (fundamentally a back pressure valve), and a collection vessel for the extract (Vergara-Salinas et al., 2015). In addition, PLE processes may have different configurations, different levels of instrumentation and energy recovery, and gas purge systems, among other features. Different design concepts can be found in the literature, most of them have been developed at the laboratory and pilot scale (Afonso and Crespo, 2005; Monrad et al., 2012; Pronyk and Mazza, 2009). Design concepts can also be based on commercially available Accelerated Solvent Extractors (ASE) developed by Thermo Scientific Dionex (Salt Lake City, Utah), which have been extensively used to obtain polyphenols by PLE at the laboratory scale (Co et al., 2009; Dawidowicz et al., 2009; Herrero et al., 2011; Ko et al., 2011; Kumar et al., 2011; Vergara-Salinas et al., 2012, 2013).

A set of possible PLE design concepts for polyphenols must be defined by taking into account the production rate, raw material to be extracted (e.g., plant material), purity of the polyphenolic extract to be obtained, type of solvent, optimal operating conditions (pressure, temperature, extraction time), and other fundamental information obtained from laboratory- and/or pilot-scale studies (the aforementioned design basis). Phase equilibrium (i.e., solubility, see Section 4.2) and mass transfer data as well as other thermophysical properties for the polyphenol(s), solvent(s), and their mixtures are specially required for designing. Once defined, the possible PLE design concepts should be tested in terms of their capabilities to meet process requirements (i.e., design objective), particularly those related to budget constraints. This type of testing procedure can be carried out by using process modeling tools and simulators, such as Aspen Plus (Aspen Technology, Inc., Massachusetts), CHEMCAD (Chemstations, Inc., Texas), gProms (Process Systems Enterprise Ltd., London, United Kingdom), ICAS (CAPEC, Department of Chemical and Biochemical Engineering, Technical University of Denmark), ProSimPlus (ProSim, Toulouse, France), SuperPro Designer (Intelligen, Inc., New Jersey), and UniSim (Honeywell, New Jersey).

4.5.2 PLE Design Selection, Detailed Design, and Equipment Selection

Final selection of a PLE design concept for polyphenols must be done by means of an economic evaluation. Also, safety issues should be considered; a good revision of materials and process hazards and safety legislation can be found in Towler and Sinnott (2007) and references therein. The economic evaluation, which can be carried out by using the aforementioned simulators, fundamentally comprises estimating capital and operating costs of the process and the determination of some profitability metrics such as the return on investment and gross margin (Towler and Sinnott, 2007). Accuracy of the cost estimation strongly depends on the type of the design estimate; for example, an estimate of level 2 (project-planning estimate) gives an accuracy of approximately 30%, while an estimate of level 3 (preliminary engineering), less than 25% (Petrides, 2000). In this stage, both design estimates are suitable for evaluating the economics of PLE of polyphenols. After cost estimations and profitability metrics have been used to analyze and optimize all the possible design concepts, then the best design can be selected.

Once the most adequate design has been selected (i.e., the more economical and safest design), detailed design and equipment selection must be addressed. Detailed design comprises fundamentally sizing and specifying the PLE equipment based on the optimal extraction conditions previously determined at laboratory scale (i.e., type of solvent, temperature, pressure, and time) and considering all the data gathered during the stage of generation and testing of PLE design concepts (solubility and mass transfer data, process modeling and simulation). Sizing and specification of the pressure extraction vessel can be carried out through the scale-up of a laboratory- or pilot-scale PLE process, a procedure that follows three main steps (Kwon and Chung, 2015; Lagadec et al., 2000; Pronyk and Mazza, 2009; Teo et al., 2010):

1. Determination of an adequate scale-up factor, primarily considering the required production capacity
2. Selection of suitable scale-up parameters based on laboratory-scale studies and process modeling
3. Calculation of dimensions and definition of characteristics of the extraction vessel by using scale-up factors and parameters and knowledge on phase equilibria, solubility data, and mass transfer kinetics

A scale-up factor, defined as the ratio between large- and small-scale equipment sizes (throughputs or volumes), must be determined as a function of the required production capacity, which typically ranges from 10 to 10^4 ton year^{-1} for the food and pharmaceutical industries (Schmidt, 2009), the quantity and availability of the solid raw material to be used as a source of polyphenols (e.g., plant material, agroindustrial residues), the polyphenol concentration in the solid, and the amount of solvent for extraction (determined by the solvent–solid ratio). Mass and energy balances are essential to calculate adequate values for the scale-up factor, which can commonly reach a maximum of 500–1000 in extraction processes when the scale-up procedure is done from pilot to large/industrial-scale equipment (Moulijn et al., 2001).

Regarding scale-up parameters, fundamental inputs optimized at laboratory-scale studies such as temperature, pressure, and solvent-to-solid ratio are usually fixed at large scale in order to maintain efficiency, extraction rate, or extract concentration (Pronyk and Mazza, 2009). Selection of other suitable scale-up parameters depends on the PLE mode of operation, that is, static or dynamic. For instance, superficial velocity of the solvent (i.e., the velocity through the cross section of the extractor) is one of the key parameters in the operation of dynamic mode extractors. Because of its influence on the extraction rate, extract concentration, and, hence, the volume of solvent used, it is one of the parameters to remain fixed during the scale-up procedure (Pronyk and Mazza, 2009). For example, if the processing capacity is increased by 100 times (maintaining the same bulk density and porosity) by increasing the vessel diameter, this diameter should be increased 10 times due to the squared term in area equation. Because of the increase in the diameter, and in order to maintain the same superficial velocity, the flow should be increased 100 times (superficial velocity is defined as the solvent volumetric flow rate divided by the area of the extractor cross section). Since there is no increase in the length of the extractor, the mean residence time is kept constant (Pronyk and Mazza, 2009).

The height/diameter ratio is another parameter that can be set for scale-up of PLE equipment in either static or dynamic mode. Continuing the aforementioned example, if the processing capacity of the extractor is increased 100 times maintaining the height/diameter ratio, both height and diameter must be increased $\sqrt[3]{100}$ times. The mass of the material and solvent must increase 100 times in order to reach the same extract concentration, bulk density, and porosity. In order to keep the superficial velocity, the volumetric flow of solvent must be increased $(\sqrt[3]{100})^2$ times. Due to the increase in vessel height, the mean residence time of solvent increased, which may even increase the extraction yield (Ho et al., 2008).

For any scale-up parameter selected, the solvent must be properly dispersed in the extractor to avoid heterogeneous extraction, and problems associated with large-scale equipment, such as material agglomeration, increased pressure drop, flow channeling, and back mixing. To avoid these problems, it would be necessary to use material with higher particle size than that used in laboratory-scale studies. However, increasing particle size results in the decreasing mass-transfer rate. The size distribution and shape of the particles also influence the height/diameter ratio of the extractor (Bertucco and Vetter, 2001). As the costs for extractors depend not only on volume, but also mainly on the diameter, the proper selection of the extractor height/diameter ratio must be made with a preference for slim vessels, always considering the associated hydrodynamic problems. In the case of large particle sizes, the height/diameter ratio is usually high (e.g., 9:1). For matrices that tend to block the filter at the bottom of the vessel (small particle sizes and swelling material), low height/diameter ratios are used (e.g., 3:1). If the extractor is equipped with baskets, these must be equipped with multiple solvent distribution points (Bertucco and Vetter, 2001; Pronyk and Mazza, 2009). In PLE equipment, mainly at laboratory scale, height/diameter ratios generally range from 2:1 to 25:1 (Cacace and Mazza, 2006b; Ho et al., 2008; Richter et al., 1996).

The final step in the scale-up procedure entails the determination of the dimensions and characteristics of the pressurized extraction vessel. As mentioned before, volume, height, and diameter of the vessel can be calculated by applying phenomenological (mass transfer and phase equilibria) and hydrodynamic considerations and other mathematical equations fundamentally related to geometry and structural features (e.g., use of filters and baskets). In addition, the wall thickness of the vessel (usually vertical and cylindrical) can be calculated through the use of the following equation (ASME Boiler and Pressure Vessel Code—Section VIII: Rules for Construction of Pressure Vessel):

$$\sigma_\theta = \frac{p(r + 0.6t)}{tE} \tag{4.7}$$

where

- t is the thickness of the cylinder wall
- σ_θ is the hoop stress or stress in the circumferential direction
- p is the internal gauge pressure
- r is the inner radius of the cylinder
- E is the joint efficiency, whose value is 1 if the pressure vessel is made up of a single, machined block of solid stainless steel with no joints

If AISI 304 and 316 stainless steels, the most common grades of steel in chemical and pharmaceutical industry, are considered for the construction of the pressure vessel, the maximum values for $\sigma\theta$ will be 143 and 157 MPa, respectively (ASME Boiler and Pressure Vessel Code—Section VIII: Rules for Construction of Pressure Vessel). Using this information, it is possible to determine the wall thickness of pressurized vessels of different sizes (volume, height, and diameter) at different operating conditions (temperature, pressure) as exemplified in Table 4.4. The wall thickness increases with the increase in pressure, while it decreases with the increase of the height/diameter ratio. More information on pressure vessel design can be found in Bertucco and Vetter (2001), Farr and Jawad (2010), Megyesy and Buthod (2008), and Moss and Basic (2012).

Together with the size and characteristics of the pressure extraction vessel, a pressure relief system should be specified according to the ASME Boiler and Pressure Vessel Code (ASME BPV), Section VIII, Division 1, Parts UG-125 to UG-137 and Division 2 Part AR. This pressure relief system must be coupled to the pressure vessel to avoid an overpressure incident (when pressure exceeds the system design value), one of the most serious hazards in industrial processes. General considerations on this issue and details about the design, specification, and selection of adequate relief devices (actuated valves and bursting discs) are given in Towler and Sinnott (2007) and references therein.

TABLE 4.4
Wall Thickness of Pressure Vessels of Different Sizes and Materials of Construction (AISI 304 and 316 Stainless Steels) under Different Operating Conditions (Temperature and Pressure), Considering Water as a Solvent for the T–P Pairs According to the Antoine Equation (Table 4.3)

Temperature (K)	Pressure (kPa)	Volume (L)	Height/ Diameter Ratio	Wall Thickness AISI 304 (mm)	Wall Thickness AISI 316 (mm)
373	101.9	5	2	0.05	0.05
			5	0.04	0.04
			10	0.03	0.03
		50	2	0.11	0.10
			5	0.08	0.08
			10	0.07	0.06
		500	2	0.24	0.22
			5	0.18	0.16
			10	0.14	0.13
423	472.6	5	2	0.24	0.22
			5	0.18	0.16
			10	0.14	0.13
		50	2	0.52	0.48
			5	0.39	0.35
			10	0.31	0.28
		500	2	1.13	1.03
			5	0.83	0.76
			10	0.66	0.60
473	1552	5	2	0.80	0.73
			5	0.59	0.54
			10	0.47	0.43
		50	2	1.73	1.58
			5	1.28	1.16
			10	1.01	0.92
		500	2	3.73	3.40
			5	2.75	2.50
			10	2.18	1.99

Note: Information kindly provided by Caleuche Tecnologías (Valparaíso, Chile).

Once the pressure extraction vessel has been designed and specified, detailed design and selection of the main PLE auxiliary equipment must be carried out. In the case of the high-pressure pump and the back pressure regulator, which are proprietary equipment (i.e., designed and manufactured by specialist companies), detailed design is not required and only selection must be addressed, considering the advice and expertise of vendors and technical consultants. Selection and specification of the high-pressure pump must be mainly based on the required capacity (flow rate), system pressure, temperature range, and fluid characteristics, following general guidelines given by Davidson and Von Bertele (2000) and Karassik et al. (2007). Selection of the back pressure regulator, a valve that holds back the desired high pressure at the inlet (upstream, at the exit of the extraction vessel) by opening only when required, must be done, fundamentally taking into account the operating pressure and temperature, fluid rate, and fluid properties such as corrosiveness and solid content. Standards such as the ASME B16.34-2004 (given by the American Society of Mechanical Engineers) and relevant literature sources (Nesbitt, 2011; Smith and Zappe, 2004; Zappe, 1999) can be used for the valve selection.

The nonproprietary PLE auxiliary equipment, that is, the solvent supply tank, the collecting vessel for the polyphenolic extract, and the heat exchanger, must be designed in detail. The first two, which can be simply considered as atmospheric liquid storage tanks, should be sized and specified by following methodologies widely covered in the literature (Jawad and Farr, 1989; Myers, 1997) and international standards such as the British Standard BS 2654. The heat exchanger, located after the high-pressure pump and before the pressure extraction vessel to heat the solvent at the required temperature, can be dimensioned and characterized based principally on the fluid flow rates, temperatures, and heat transfer rate. The comprehensive work of Kakaç et al. (2012) can be used for the selection, rating, and thermal design of the required heat exchanger equipment. Fundamentals of heat transfer theory can be found in Incropera et al. (2013). Finally, consulting with recognized manufactures and suppliers is also advised for detailed design and selection of the required nonproprietary equipment.

4.5.3 Procurement, Construction, and Operation

When the stage of detailed design and equipment selection has been finalized, a last project cost estimation must be done before procurement (i.e., purchasing) of the PLE equipment. This estimation can be done following the guidelines described in Harrison et al. (2015) for a level 4 design estimate, the so-called detailed engineering estimate (accuracy less than 15%). After the capital to afford such a cost is approved, the equipment can be procured from different specialty firms. Particularly, the pressure vessel, the storage tanks, and the heat exchanger can be constructed, together with equipment fabricators and vendors, by using sketches and specification sheets obtained previously. Once the corresponding equipment is procured and/or constructed, the PLE process is ready to be installed and connected to other process units to operate in a plant for extraction and purification of polyphenols.

Engineering, procurement, and construction (EPC) companies, commonly known as contractors, with expertise in PLE or similar processes, can carry out the complete procedure described earlier. Contracting EPC companies can considerably reduce associated costs, primarily because those companies can place large orders on process items and get important discounts during equipment purchasing (Towler and Sinnott, 2007). Some EPC companies or similar alternatives include Applied Separations, Inc. (Pennsylvania), Bio-Process Group (Indiana), Chematur Engineering AB (Karlskoga, Sweden), Critical Processes Ltd. (North Yorkshire, United Kingdom), Flavourtech Pty. Ltd. (Griffith, Australia), and Thyssenkrupp

Industrial Solutions AG (Dortmund, Germany). Rigorous collaborative work with these types of companies can deliver a cost-effective and efficient large-scale PLE system for polyphenols.

4.6 CONCLUDING REMARKS

The PLE is a suitable alternative for replacing conventional solvent extraction of polyphenols from plant matrices. In the PLE method, several physicochemical properties (e.g., polarity, viscosity, and superficial tension) of liquid solvents can be manipulated in order to enhance the extraction process by increasing temperature and applying enough pressure to maintain their liquid state. The increase in temperature above the atmospheric boiling point of the solvent results in a significant increase in their solvating properties, as well as the solubility and diffusion rate of polyphenols. However, two main phenomena must be considered in PLE of polyphenols from vegetable matrices: thermal degradation of polyphenols and the formation of neo-antioxidant and toxic compounds. These two phenomena depend on several factors, mainly temperature and time of extraction as well as the type of polyphenol.

On the other hand, the recovery of polyphenols by PLE for food and nutraceutical applications requires an exhaustive analysis and selection of a suitable type of solvent considering many aspects such as safety, productivity (economics), and sustainability. Water, ethanol, and its mixtures and emerging green solvents such as glycerol appear to be the most suitable solvents for producing food-relevant polyphenols.

In conclusion, the extraction performance will be determined by a balance between polyphenols solubility, extraction rate and selectivity of solvents, thermal degradation, and formation of toxic compounds.

At an industrial level, PLE processes may have different configurations, different levels of instrumentation and energy recovery, and gas purge systems, among other features. The basic configuration of the extraction system consists of solvent supply tanks, a high-pressure pump, one or more heat exchangers, the pressurized extraction vessel, a pressure-regulating system (fundamentally a back pressure valve), and a collection vessel for the extract.

In general, large-scale PLE processes should be designed by following a methodology commonly used in chemical engineering and thoroughly described in the specialized literature. This methodology includes generation of possible design concepts; testing of design concepts, which includes the application of process modeling and simulation; design selection by evaluating process economics; detailed design and equipment selection (i.e., scale-up and equipment sizing and characterization); and finally, procurement, construction, and process operation.

4.7 FUTURE RESEARCH AND DEVELOPMENTS

Although they are extremely important to design and develop PLE processes, physicochemical properties of polyphenols at high temperatures and pressures, particularly solubility, are rarely available in the literature (scientific articles, books, industry papers, patents). Therefore, future research on this topic should be focused not only on obtaining experimental data at laboratory or bench scale (using state-of-the-art technologies) but also on enhancing existing molecular-based solubility models such as activity coefficient approaches (e.g., MPP-UNIFAC) and associating equations of state (e.g., CPA).

Together with solubility, mass transfer properties, such as diffusion and hydrodynamic data, become particularly important during the process of scale-up and optimization; they mostly determine the size and cost of the extraction equipment and operational costs. Therefore, a deep understanding of how those properties behave under particular conditions (i.e., type of raw material, solvent, temperature, pressure, type of polyphenols) and different dimensions (experimental, bench, and pilot scales) is required to develop cost-effective PLE processes for commercial applications in the food and related industries. Such an understanding will be reached only through several research studies and the development of PLE process prototypes, including the design and evaluation of new empirical models to adequately determine diffusivities of polyphenols in pressurized solvents.

Finally, new cost-effective strategies for the scale-up of PLE processes for polyphenols should be addressed in short term. Interesting options are dimensional analysis (Zlokarnik, 2006) and the use of mini-plant technologies (Bart and Pilz, 2011), which have shown good results in the chemical and bioprocess engineering field.

ACKNOWLEDGMENTS

The authors acknowledge the Grant Anillo en Ciencia y Tecnología ACT1105 and FONDECYT/Postdoctoral Project 3150622 from CONICYT–Chile for their financial support and Caleuche Tecnologías Ltda. for their technical support.

REFERENCES

Abrams, D.S. and Prausnitz, J.M. 1975. Statistical thermodynamics of liquid mixtures: A new expression for the excess Gibbs energy of partly or completely miscible systems. *AIChE Journal* 21:116–128.

Afonso, C.A.M. and Crespo, J.G. 2005. *Green Separation Processes*. Weinheim, Germany: Wiley-VCH Verlag GmbH & Co.

Alothman, M., Bhat, R., and Karim, A. 2009. Antioxidant capacity and phenolic content of selected tropical fruits from Malaysia, extracted with different solvents. *Food Chemistry* 115:785–788.

Alsenz, J. and Kansy, M. 2007. High throughput solubility measurement in drug discovery and development. *Advanced Drug Delivery Reviews* 59:546–67.

Apelblat, A. and Manzurola, E. 1999. Solubilities of o-acetylsalicylic, 4-aminosalicylic, 3, 5-dinitrosalicylic, and p-toluic acid, and magnesium-DL-aspartate in water from T = (278 to 348) K. *Journal of Chemical Thermodynamics* 31:85–91.

Apostolakis, A., Grigorakis, S., and Makris, D.P. 2014. Optimisation and comparative kinetics study of polyphenol extraction from olive leaves (*Olea europaea*) using heated water/glycerol mixtures. *Separation and Purification Technology* 128:89–95.

Bala, I., Bhardwaj, V., Hariharan, S., and Kumar, M.R. 2006. Analytical methods for assay of ellagic acid and its solubility studies. *Journal of Pharmaceutical and Biomedical Analysis* 40:206–210.

Bart, H.-J. and Pilz, S. 2011. *Industrial Scale Natural Products Extraction*. Weinheim, Germany: John Wiley & Sons.

Bartnick, D., Mohler, C., and Houlihan, M. 2006. Methods for the production of food grade extracts. U.S. Patent 20,060,088,627 A1.

Bertucco, A. and Vetter, G. 2001. *High Pressure Process Technology: Fundamentals and Applications*. Amsterdam, the Netherlands: Elsevier.

Boussetta, N., Vorobiev, E., Le, L.H., Cordin-Falcimaigne, A., and Lanoisellé, J.-L. 2012. Application of electrical treatments in alcoholic solvent for polyphenols extraction from grape seeds. *LWT: Food Science and Technology* 46:127–134.

Buchowski, H. and Khiat, A. 1986. Solubility of solids in liquids: One-parameter solubility equation. *Fluid Phase Equilibria* 25:273–278.

Buchowski, H., Ksiazczak, A., and Pietrzyk, S. 1980. Solvent activity along a saturation line and solubility of hydrogen-bonding solids. *Journal of Physical Chemistry* 84:975–979.

Cacace, J.E. and Mazza, G. 2006a. Pressurized low polarity water extraction of lignans from whole flaxseed. *Journal of Food Engineering* 77:1087–1095.

Cacace, J.E. and Mazza, G. 2006b. Pressurized low polarity water extraction of biologically active compounds from plant products. In *Functional Food Ingredients and Nutraceuticals*, ed. J. Shi, pp. 135–155. Boca Raton, FL: CRC Press.

Cadena, P.G., Pereira, M.A., Cordeiro, R.B., Cavalcanti, I.M., Neto, B.B., Carmo, C.M.D., Filho, J.L.L., Silva, V.L., and Santos-Magalhães, N.S. 2013. Nanoencapsulation of quercetin and resveratrol into elastic liposomes. *Biochimica et Biophysica Acta (BBA)—Biomembranes* 1828:309–316.

Callemien, D., Jerkovic, V., Rozenberg, R., and Collin, S. 2005. Hop as an interesting source of resveratrol for brewers: Optimization of the extraction and quantitative study by liquid chromatography/atmospheric pressure chemical ionization tandem mass spectrometry. *Journal of Agricultural and Food Chemistry* 53:424–429.

Casazza, A.A., Aliakbarian, B., Sannita, E., and Perego, P. 2012. High-pressure high-temperature extraction of phenolic compounds from grape skins. *International Journal of Food Science and Technology* 47:399–405.

Clifford, A.A. 2005. Separations using superheated water. In *Green Separation Processes*, eds. C.A.M. Afonso and J.G. Crespo. Weinheim, Germany: Wiley-VCH Verlag GmbH & Co. KGaA.

Co, M., Koskela, P., Eklund-Akergren, P., Srinivas, K., King, J.W., Sjoberg, P.J.R., and Turner, C. 2009. Pressurized liquid extraction of betulin and antioxidants from birch bark. *Green Chemistry* 11:668–674.

Cuevas-Valenzuela, J., González-Rojas, A., Wisniak, J., Apelblat, A., and Pérez-Correa, J.R. 2014. Solubility of (+)-catechin in water and water-ethanol mixtures within the temperature range 277.6–331.2 K: Fundamental data to design polyphenol extraction processes. *Fluid Phase Equilibria* 382:279–285.

Chebil, L., Humeau, C., Anthoni, J., Dehez, F., Engasser, J.-M., and Ghoul, M. 2007. Solubility of flavonoids in organic solvents. *Journal of Chemical and Engineering Data* 52:1552–1556.

Chemat, F., Vian, M.A., and Cravotto, G. 2012. Green extraction of natural products: Concept and principles. *International Journal of Molecular Sciences* 13:8615–8627.

Chen, C.-C. and Crafts, P.A. 2006. Correlation and prediction of drug molecule solubility in mixed solvent systems with the nonrandom two-liquid segment activity coefficient (NRTL-SAC) model. *Industrial and Engineering Chemistry Research* 45:4816–4824.

Chen, C.-C. and Song, Y. 2004. Solubility modeling with a nonrandom two-liquid segment activity coefficient model. *Industrial and Engineering Chemistry Research* 43:8354–8362.

Dai, J. and Mumper, R.J. 2010. Plant phenolics: Extraction, analysis and their antioxidant and anticancer properties. *Molecules* 15:7313–7352.

Daneshfar, A., Ghaziaskar, H.S., and Homayoun, N. 2008. Solubility of gallic acid in methanol, ethanol, water, and ethyl acetate. *Journal of Chemical and Engineering Data* 53:776–778.

Davidson, J. and Von Bertele, O. 2000. Process pumps selection: A systems approach. London, UK: Professional Engineering Pub.

Dawidowicz, A.L., Rado, E., and Wianowska, D. 2009. Static and dynamic superheated water extraction of essential oil components from *Thymus vulgaris* L. *Journal of Separation Science* 32:3034–3042.

Deyama, T., Ikawa, T., Kitagawa, S., and Nishibe, S. 1987. The constituents of *Eucommia ulmoides* Oliv. V. Isolation of dihydroxydehydrodiconiferyl alcohol isomers and phenolic compounds. *Pharmaceutical Bulletin* 35:1785–1789.

Duan, L., Wallace, S., Engelberth, A., Lovelady, J., Clausen, E., King, J., and Carrier, D. 2009. Extraction of co-products from biomass: Example of thermal degradation of silymarin compounds in subcritical water. *Applied Biochemistry and Biotechnology* 158:362–373.

Esina, Z. and Korchuganova, M. 2015. Modeling of the solubility of a solid in a liquid for organic systems. *Theoretical Foundations of Chemical Engineering* 49:313–321.

Fan, J.-P., Yang, D., Xu, X.-K., Guo, X.-J., and Zhang, X.-H. 2015. Solubility of daidzein in different organic solvents and (ethyl alcohol+ water) mixed solvents. *Journal of Chemical Thermodynamics* 88:85–89.

Farr, J.R. and Jawad, M.H. 2010. *Guidebook for the Design of ASME Section VIII Pressure Vessels*. New York: ASME.

Ferreira, O. and Pinho, S.P. 2012. Solubility of flavonoids in pure solvents. *Industrial and Engineering Chemistry Research* 51:6586–690.

Ferreira, O., SchröDer, B., and Pinho, S.P. 2013. Solubility of hesperetin in mixed solvents. *Journal of Chemical and Engineering Data* 58:2616–2621.

Fredenslund, A., Jones, R.L., and Prausnitz, J.M. 1975. Group-contribution estimation of activity coefficients in nonideal liquid mixtures. *AIChE Journal* 21:1086–1099.

Galanakis, C., Goulas, V., Tsakona, S., Manganaris, G., and Gekas, V. 2013. A knowledge base for the recovery of natural phenols with different solvents. *International Journal of Food Properties* 16:382–396.

Garcia-Salas, P., Morales-Soto, A., Segura-Carretero, A., and Fernández-Gutiérrez, A. 2010. Phenolic-compound-extraction systems for fruit and vegetable samples. *Molecules* 15:8813–8826.

Gmehling, J. 2009. Present status and potential of group contribution methods for process development. *Journal of Chemical Thermodynamics* 41:731–747.

Gmehling, J. and Weidlich, U. 1985. Die Dortmunder Datenbank—Basis für die Weiterentwicklung der UNIFAC-Methode. *Chemie Ingenieur Technik* 57:447–449.

Green, D.W. 2008. *Perry's Chemical Engineers' Handbook*. New York: McGraw-Hill.

Gude, M.T., Van Der Wielen, L.A., and Luyben, K.C.A. 1996. Phase behavior of α-amino acids in multicomponent aqueous alkanol solutions. *Fluid Phase Equilibria* 116:110–117.

Guyot, S., Marnet, N., and Drilleau, J.-F. 2001. Thiolysis-HPLC characterization of apple procyanidins covering a large range of polymerization states. *Journal of Agricultural and Food Chemistry* 49:14–20.

Harrison, R.G., Todd, P.W., Rudge, S.R., and Petrides, D.P. 2015. *Bioseparations Science and Engineering*. New York: Oxford University Press.

Hawthorne, S.B., Miller, D.J., Lagadec, A.J.-M., Hammond, P.J., and Clifford, A.A. 2002. Method of manipulating the chemical properties of water to improve the effectiveness of a desired process. Energy and Environmental Research Center Foundation, Grand Forks, ND. U.S. Patent 6,352,644 B1.

He, L., Zhang, X., Xu, H., Xu, C., Yuan, F., Knez, Z., Novak, Z., and Gao, Y. 2012. Subcritical water extraction of phenolic compounds from pomegranate (*Punica granatum* L.) seed residues and investigation into their antioxidant activities with HPLC–ABTS+ assay. *Food and Bioproducts Processing* 90:215–223.

Heidman, J., Tsonopoulos, C., Brady, C.J., and Wilson, G.M. 1985. High-temperature mutual solubilities of hydrocarbons and water. Part II: Ethylbenzene, ethylcyclohexane, and n-octane. *AIChE Journal* 31:376–384.

Herrero, M., Castro-Puyana, M., Mendiola, J.A., and Ibañez, E. 2013. Compressed fluids for the extraction of bioactive compounds. *Trends in Analytical Chemistry* 43:67–83.

Herrero, M., Cifuentes, A., and Ibañez, E. 2006. Sub- and supercritical fluid extraction of functional ingredients from different natural sources: Plants, food-by-products, algae and microalgae: A review. *Food Chemistry* 98:136–148.

Herrero, M., Temirzoda, T.N., Segura-Carretero, A., Quirantes, R., Plaza, M., and Ibañez, E. 2011. New possibilities for the valorization of olive oil by-products. *Journal of Chromatography A* 1218:7511–7520.

Ho, C.H.L., Cacace, J.E., and Mazza, G. 2008. Mass transfer during pressurized low polarity water extraction of lignans from flaxseed meal. *Journal of Food Engineering* 89:64–71.

Hofmann, T., Nebehaj E., and Albert, L. 2015. The high-performance liquid chromatography/multistage electrospray mass spectrometric investigation and extraction optimization of beech (Fagus sylvatica L.) bark polyphenols. *Journal of Chromatography A* 1393:96–105.

Hossain, M., Rawson, A., Aguiló-Aguayo, I., Brunton, N., and Rai, D. 2015. Recovery of steroidal alkaloids from potato peels using pressurized liquid extraction. *Molecules* 20:8560–8573.

Howard, L. and Pandjaitan, N. 2008. Pressurized liquid extraction of flavonoids from spinach. *Journal of Food Science* 73:C151–C157.

Husøy, T., Haugen, M., Murkovic, M., Jöbstl, D., Stølen, L.H., Bjellaas, T., Rønningborg, C., Glatt, H., and Alexander, J. 2008. Dietary exposure to 5-hydroxymethylfurfural from Norwegian food and correlations with urine metabolites of short-term exposure. *Food and Chemical Toxicology* 46:3697–3702.

Ignat, I., Volf, I., and Popa, V.I. 2011. A critical review of methods for characterisation of polyphenolic compounds in fruits and vegetables. *Food Chemistry* 126:1821–1835.

Incropera, F.P., Bergman, T.L., and Lavine, A.S. 2013. *Foundations of Heat Transfer.* Singapore: Wiley.

Jain, A. and Yalkowsky, S.H. 2006. Estimation of melting points of organic compounds-II. *Journal of Pharmaceutical Sciences* 95:2562–2618.

Jain, A., Yang, G., and Yalkowsky, S.H. 2004. Estimation of melting points of organic compounds. *Industrial and Engineering Chemistry Research* 43:7618–7621.

Jawad, M.H. and Farr, J.R. 1989. *Structural Analysis and Design of Process Equipment.* New York: Wiley.

Jouyban, A., Shayanfar, A., Panahi-Azar, V., Soleymani, J., Yousefi, B.H., Acree, W.E., and York, P. 2011. Solubility prediction of drugs in mixed solvents using partial solubility parameters. *Journal of Pharmaceutical Sciences* 100:4368–4382.

Ju, Z. and Howard, L.R. 2005. Subcritical water and sulfured water extraction of anthocyanins and other phenolics from dried red grape skin. *Journal of Food Science.* 70:S270–S276.

Kabir, F., Sultana, M.S., and Kurnianta, H. 2015. Polyphenolic contents and antioxidant activities of underutilized grape (*Vitis vinifera* L.) pomace extracts. *Preventive Nutrition and Food Science* 20:210.

Kakaç, S., Liu, H., and Pramuanjaroenkij, A. 2012. *Heat Exchangers: Selection, Rating, and Thermal Design*, 3rd ed. Boca Raton, FL: Taylor & Francis.

Kallithraka, S., Garcia-Viguera, C., Bridle, P., and Bakker, J. 1995. Survey of solvents for the extraction of grape seed phenolics. *Phytochemical Analysis* 6:265–267.

Kapasakalidis, P.G., Rastall, R.A., and Gordon, M.H. 2006. Extraction of polyphenols from processed black currant (*Ribes nigrum* L.) Residues. *Journal of Agricultural and Food Chemistry* 54:4016–4021.

Karassik, I., Messina, J., Cooper, P., and Heald, C. 2007. *Pump Handbook*. New York: McGraw-Hill.

Kebe, M., Renard, C., N'guessan Amani, G., and Maingonnat, J.-F. 2014. Kinetics of apple polyphenol diffusion in solutions with different osmotic strengths. *Journal of Agricultural and Food Chemistry* 62:9841–9847.

Khoddami, A., Wilkes, M.A., and Roberts, T.H. 2013. Techniques for analysis of plant phenolic compounds. *Molecules* 18:2328–2375.

Kim, M.-B., Park, J.-S., and Lim, S.-B. 2010. Antioxidant activity and cell toxicity of pressurised liquid extracts from 20 selected plant species in Jeju, Korea. *Food Chemistry* 122:546–552.

King, J.W. and Grabiel, R.D. 2007. Isolation of polyphenolic compounds from fruits or vegetables utilizing sub-critical water extraction. Represented by the Secretary of Agriculture, Washington, DC. U.S. Patent 7,208,181 B1.

King, J.W., Srinivas, K., Howard, L.R., and Monrad, J.K. 2010. Solubility of gallic acid, catechin, and protocatechuic acid in subcritical water from (298.75 to 415.85) K. *Journal of Chemical and Engineering Data* 55:3101–3108.

Ko, M.-J., Cheigh, C.-I., Cho, S.-W., and Chung, M.-S. 2011. Subcritical water extraction of flavonol quercetin from onion skin. *Journal of Food Engineering* 102:327–333.

Kontogeorgis, G.M. 2013. Association theories for complex thermodynamics. *Chemical Engineering Research and Design* 91:1840–1858.

Kontogeorgis, G.M., Voutsas, E.C., Yakoumis, I.V., and Tassios, D.P. 1996. An equation of state for associating fluids. *Industrial and Engineering Chemistry Research* 35:4310–4318.

Kumar, M.S.Y., Dutta, R., Prasad, D., and Misra, K. 2011. Subcritical water extraction of antioxidant compounds from Seabuckthorn (*Hippophae rhamnoides*) leaves for the comparative evaluation of antioxidant activity. *Food Chemistry* 127:1309–1316.

Kwon, H.-L. and Chung, M.-S. 2015. Pilot-scale subcritical solvent extraction of curcuminoids from *Curcuma long* L. *Food Chemistry* 185:58–64.

Labarbe, B., Cheynier, V., Brossaud, F., Souquet, J.-M., and Moutounet, M. 1999. Quantitative fractionation of grape proanthocyanidins according to their degree of polymerization. *Journal of Agricultural and Food Chemistry* 47:2719–2723.

Lagadec, A.J.M., Miller, D.J., Lilke, A.V., and Hawthorne, S.B. 2000. Pilot-scale subcritical water remediation of polycyclic aromatic hydrocarbon- and pesticide-contaminated soil. *Environmental Science and Technology* 34:1542–1548.

Larsen, B.L., Rasmussen, P., and Fredenslund, A. 1987. A modified UNIFAC group contribution model for prediction of phase equilibria and heats of mixing. *Industrial and Engineering Chemistry Research* 26:2274–2286.

Li, B., Robinson, D.H., and Birt, D.F. 1997. Evaluation of properties of apigenin and [G-3H] apigenin and analytic method development. *Journal of Pharmaceutical Sciences* 86:721–725.

Li, H., Chen, B., and Yao, S. 2005. Application of ultrasonic technique for extracting chlorogenic acid from *Eucommia ulmodies* Oliv. (*E. ulmodies*). *Ultrasonics Sonochemistry* 12:295–300.

Liu, L. and Chen, J. 2008. Solubility of hesperetin in various solvents from (288.2 to 323.2) K. *Journal of Chemical and Engineering Data* 53:1649–1650.

Liu, Z., Mei, L., Wang, Q., Shao, Y., and Tao, Y. 2014. Optimization of subcritical fluid extraction of seed oil from Nitraria tangutorum using response surface methodology. *LWT-Food Science and Technology* 56:168–174.

Lu, L.-L. and Lu, X.-Y. 2007. Solubilities of gallic acid and its esters in water. *Journal of Chemical and Engineering Data* 52:37–39.

Luque De Castro, M.D., Jiménez-Carmona, M.M., and Fernández-Pérez, V. 1999. Towards more rational techniques for the isolation of valuable essential oils from plants. *TrAC Trends in Analytical Chemistry* 18:708–716.

Luthria, D.L. 2008. Influence of experimental conditions on the extraction of phenolic compounds from parsley (*Petroselinum crispum*) flakes using a pressurized liquid extractor. *Food Chemistry* 107:745–752.

Magalhaes, P.J., Guido, L.F., Cruz, J.M., and Barros, A.A. 2007. Analysis of xanthohumol and isoxanthohumol in different hop products by liquid chromatography-diode array detection-electrospray ionization tandem mass spectrometry. *Journal of Chromatography A* 1150:295–301.

Mantegna, S., Binello, A., Boffa, L., Giorgis, M., Cena, C., and Cravotto, G. 2012. A one-pot ultrasound-assisted water extraction/cyclodextrin encapsulation of resveratrol from *Polygonum cuspidatum*. *Food Chemistry* 130:746–750.

Marrero, J. and Gani, R. 2001. Group-contribution based estimation of pure component properties. *Fluid Phase Equilibria* 183:183–208.

Martins, S.I.F.S., Jongen, W.M.F., and Van Boekel, M.J.A.S. 2000. A review of Maillard reaction in food and implications to kinetic modelling. *Trends in Food Science and Technology* 11:364–373.

Megyesy, E.F. and Buthod, P. 2008. *Pressure Vessel Handbook*. Oklahoma City, OK: Pv Pub Incorporated.

Meireles, M.A.A. 2008. *Extracting Bioactive Compounds for Food Products: Theory and Applications*. Boca Raton, FL: CRC Press.

Mendiola, J.A., Herrero, M., Cifuentes, A., and Ibanez, E. 2007. Use of compressed fluids for sample preparation: Food applications. *Journal of Chromatography A* 1152:234–246.

Miller, D.J., Hawthorne, S.B., Gizir, A.M., and Clifford, A.A. 1998. Solubility of polycyclic aromatic hydrocarbons in subcritical water from 298 K to 498 K. *Journal of Chemical and Engineering Data* 43:1043–1047.

Monrad, J.K., Howard, L.R., King, J.W., Srinivas, K., and Mauromoustakos, A. 2009. Subcritical solvent extraction of procyanidins from dried red grape pomace. *Journal of Agricultural and Food Chemistry* 58:4014–4021.

Monrad, J.K., Srinivas, K., Howard, L.R., and King, J.W. 2012. Design and optimization of a semicontinuous hot-cold extraction of polyphenols from grape pomace. *Journal of Agricultural and Food Chemistry* 60:5571–5582.

Moss, D.R. and Basic, M.M. 2012. *Pressure Vessel Design Manual*. Oxford, UK: Butterworth-Heinemann.

Mota, F.L., Queimada, A.J., Pinho, S.P., and Macedo, E.A. 2008. Aqueous solubility of some natural phenolic compounds. *Industrial and Engineering Chemistry Research* 47:5182–5189.

Mota, F.L., Queimada, A.J., Pinho, S.P., and Macedo, E.A. 2010. Water solubility of drug-like molecules with the cubic-plus-association equation of state. *Fluid Phase Equilibria* 298:75–82.

Moulijn, J.A., Makkee, M., and Van Diepen, A. 2001. *Chemical Process Technology*. New York: John Wiley & Sons.

Mueller-Harvey, I. 2001. Analysis of hydrolysable tannins. *Animal Feed Science and Technology* 91:3–20.

Mukhopadhyay, M. and Panja, P. 2010. Pressurised hot water as a novel extractant of natural products: A review. *Indian Chemical Engineer* 51:311–324.

Mustafa, A. and Turner, C. 2011. Pressurized liquid extraction as a green approach in food and herbal plants extraction: A review. *Analytica Chimica Acta* 703:8–18.

Myers, P.E. 1997. *Above Ground Storage Tanks*. New York: McGraw-Hill.

Naczk, M. and Shahidi, F. 2004. Extraction and analysis of phenolics in food. *Journal of Chromatography A* 1054:95–111.

Nakamura, Y., Ohto, Y., Murakami, A., and Ohigashi, H. 1998. Superoxide scavenging activity of rosmarinic acid from *Perilla frutescens* Britton var. acuta f. Viridis. *Journal of Agricultural and Food Chemistry* 46:4545–4550.

Nesbitt, B. 2011. *Handbook of Valves and Actuators: Valves Manual International*. San Diego, CA: Elsevier Science.

Noubigh, A., Abderrabba, M., and Provost, E. 2007. Temperature and salt addition effects on the solubility behaviour of some phenolic compounds in water. *Journal of Chemical Thermodynamics* 39:297–303.

Noubigh, A., Aydi, A., and Abderrabba, M. 2015. Experimental measurement and correlation of solubility data and thermodynamic properties of protocatechuic acid in four organic solvents. *Journal of Chemical and Engineering Data* 60:514–518.

Noubigh, A., Cherif, M., Provost, E., and Abderrabba, M. 2008. Solubility of gallic acid, vanillin, syringic acid, and protocatechuic acid in aqueous sulfate solutions from (293.15 to 318.15) K. *Journal of Chemical and Engineering Data* 53:1675–1678.

Ohnishi, M., Morishita, H., Iwahashi, H., Toda, S., Shirataki, Y., Kimura, M., and Kido, R. 1994. Inhibitory effects of chlorogenic acids on linoleic acid peroxidation and haemolysis. *Phytochemistry* 36:579–583.

Ollanketo, M., Peltoketo, A., Hartonen, K., Hiltunen, R., and Riekkola, M.-L. 2002. Extraction of sage (*Salvia officinalis* L.) by pressurized hot water and conventional methods: Antioxidant activity of the extracts. *European Food Research and Technology* 215:158–163.

Ong, E.S., Cheong, J.S.H., and Goh, D. 2006. Pressurized hot water extraction of bioactive or marker compounds in botanicals and medicinal plant materials. *Journal of Chromatography A* 1112:92–102.

Onken, U., Rarey-Nies, J., and Gmehling, J. 1989. The Dortmund Data Bank: A computerized system for retrieval, correlation, and prediction of thermodynamic properties of mixtures. *International Journal of Thermophysics* 10:739–747.

Paganga, G., Miller, N., and Rice-Evans, C.A. 1999. The polyphenolic content of fruit and vegetables and their antioxidant activities: What does a serving constitute? *Free Radical Research* 30:153–162.

Palma, M., Piñeiro, Z., and Barroso, C.G. 2001. Stability of phenolic compounds during extraction with superheated solvents. *Journal of Chromatography A* 921:169–174.

Pasrija, D. and Anandharamakrishnan, C. 2015. Techniques for extraction of green tea polyphenols: A review. *Food and Bioprocess Technology* 8:935–950.

Peng, B., Li, R., and Yan, W. 2009. Solubility of rutin in ethanol + water at (273.15 to 323.15) K. *Journal of Chemical and Engineering Data* 54:1378–1381.

Peng, B. and Yan, W. 2009. Solubility of luteolin in ethanol+ water mixed solvents at different temperatures. *Journal of Chemical and Engineering Data* 55:583–85.

Peres, A.M. and Macedo, E.A. 1996. Thermodynamic properties of sugars in aqueous solutions: Correlation and prediction using a modified UNIQUAC model. *Fluid Phase Equilibria* 123:71–95.

Petrides, D. 2000. Bioprocess design and economics. In *Bioseparations Science and Engineering*, ed. R.G. Harrison. New York: Oxford University Press.

Philip, J.L. and Jessy Van, W. 2011. Introduction to extraction in food processing. In *Enhancing Extraction Processes in the Food Industry*, eds. F. Chémat, N.I. Lebovka, and E. Vorobiev, pp. 1–24. Boca Raton, FL: CRC Press.

Plaza, M., Abrahamsson, V., and Turner, C. 2013. Extraction and neoformation of antioxidant compounds by pressurized hot water extraction from apple by-products. *Journal of Agricultural and Food Chemistry* 61:5500–5510.

Plaza, M., Amigo-Benavent, M., Del Castillo, M.D., Ibáñez, E., and Herrero, M. 2010a. Facts about the formation of new antioxidants in natural samples after subcritical water extraction. *Food Research International* 43:2341–2348.

Plaza, M., Amigo-Benavent, M., Del Castillo, M.D., Ibáñez, E., and Herrero, M. 2010b. Neoformation of antioxidants in glycation model systems treated under subcritical water extraction conditions. *Food Research International* 43:1123–1129.

Pronyk, C. and Mazza, G. 2009. Design and scale-up of pressurized fluid extractors for food and bioproducts. *Journal of Food Engineering* 95:215–226.

Pulley, G.N. 1936. Solubility of naringin in water. *Industrial and Engineering Chemistry Analytical Edition* 8:360–360.

Qiang, W., Peisheng, M., and Shifeng, N. 2009. Position group contribution method for estimation of melting point of organic compounds. *Chinese Journal of Chemical Engineering* 17:468–472.

Queimada, A.J., Mota, F.L., Pinho, S.P., and Macedo, E.A. 2009. Solubilities of biologically active phenolic compounds: Measurements and modeling. *Journal of Physical Chemistry B* 113:3469–3476.

Ramos, L., Kristenson, E.M., and Brinkman, U.A.Th. 2002. Current use of pressurised liquid extraction and subcritical water extraction in environmental analysis. *Journal of Chromatography A* 975:3–29.

Rastogi, H. and Jana, S. 2016. Evaluation of physicochemical properties and intestinal permeability of six dietary polyphenols in human intestinal colon adenocarcinoma Caco-2 cells. *European Journal of Drug Metabolism and Pharmacokinetics* 41:33–43.

Razmara, R.S., Daneshfar, A., and Sahraei, R. 2010. Solubility of quercetin in water+ methanol and water+ ethanol from (292.8 to 333.8) K. *Journal of Chemical and Engineering Data* 55:3934–3936.

Richards, F. 1959. A flexible growth function for empirical use. *Journal of Experimental Botany* 10:290–301.

Richter, B.E., Jones, B.A., Ezzell, J.L., Porter, N.L., Avdalovic, N., and Pohl, C. 1996. Accelerated solvent extraction: A technique for sample preparation. *Analytical Chemistry* 68:1033–1039.

Rizvi, S.S. 2010. *Separation, Extraction and Concentration Processes in the Food, Beverage and Nutraceutical Industries*. Oxford, UK: Woodhead Publishing.

Rodriguez-Meizoso, I., Marin, F.R., Herrero, M., Senorans, F.J., Reglero, G., Cifuentes, A., and Ibanez, E. 2006. Subcritical water extraction of nutraceuticals with antioxidant activity from oregano: Chemical and functional characterization. *Journal of Pharmaceutical and Biomedical Analysis* 41:1560–1565.

Sakuma, H., Matsushima, S., Munakata, S., and Sugawara, S. 1982. Pyrolysis of chlorogenic acid and rutin. *Agricultural and Biological Chemistry* 46:1311–1317.

Santos-Buelga, C. and Williamson, G. 2003. *Methods in Polyphenol Analysis*. Cambridge, UK: Royal Society of Chemistry.

Sarraf, C. 2015. Étude du potentiel nutraceutique de la culture du houblon au Québec. PhD thesis. Université Laval, Quebec City, Quebec, Canada.

Schmidt, L.D. 2009. *The Engineering of Chemical Reactions*. New York: Oxford University Press.

Serrano, J., Puupponen-Pimiä, R., Dauer, A., Aura, A.-M., and Saura-Calixto, F. 2009. Tannins: Current knowledge of food sources, intake, bioavailability and biological effects. *Molecular Nutrition and Food Research* 53:S310–S329.

Sevillano, D.M., Van Der Wielen, L.A., Hooshyar, N., and Ottens, M. 2014. MPP-UNIFAC: A predictive activity coefficient model for polyphenols. *Fluid Phase Equilibria* 384:82–88.

Sevillano, D.M., Van Der Wielen, L.A., Trifunovic, O., and Ottens, M. 2013. Model comparison for the prediction of the solubility of green tea catechins in ethanol/water mixtures. *Industrial and Engineering Chemistry Research* 52:6039–6048.

Shakeel, F., Haq, N., and Siddiqui, N.A. 2015a. Solubility and thermodynamic function of vanillin in ten different environmentally benign solvents. *Food Chemistry* 180:244–248.

Shakeel, F., Haq, N., Siddiqui, N.A., Alanazi, F.K., and Alsarra, I.A. 2015b. Solubility and thermodynamics of vanillin in Carbitol-water mixtures at different temperatures. *LWT: Food Science and Technology* 64:1278–1282.

Shi, J., Nawaz, H., Pohorly, J., Mittal, G., Kakuda, Y., and Jiang, Y.M. 2005. Extraction of polyphenolics from plant material for functional foods—Engineering and technology. *Food Reviews International* 21:139–166.

Sineiro, J., Domínguez, H., Núñez, M., and Lema, J. 1996. Ethanol extraction of polyphenols in an immersion extractor: Effect of pulsing flow. *Journal of the American Oil Chemists' Society* 73:1121–1125.

Smith, P. and Zappe, R.W. 2004. *Valve Selection Handbook: Engineering Fundamentals for Selecting the Right Valve Design for Every Industrial Flow Application.* Amsterdam, the Netherlands: Elsevier Science.

Smith, R.M. 2002. Extractions with superheated water. *Journal of Chromatography A* 975:31–46.

Smith, R.M. 2006. Superheated water: The ultimate green solvent for separation science. *Analytical and Bioanalytical Chemistry* 385:419–421.

Spigno, G., Tramelli, L., and Marco De Faveri, D. 2007. Effects of extraction time, temperature and solvent on concentration and antioxidant activity of grape marc phenolics. *Journal of Food Engineering* 81:200–208.

Srinivas, K. and King, J.W. 2010. Supercritical carbon dioxide and subcritical water: Complementary agents in the processing of functional foods. In *Functional Food Product Development*, eds. J. Smith and E. Charter, pp. 39–78. Chichester, UK: Wiley-Blackwell.

Srinivas, K., King, J.W., Howard, L.R., and Monrad, J.K. 2010a. Solubility and solution thermodynamic properties of quercetin and quercetin dihydrate in subcritical water. *Journal of Food Engineering* 100:208–218.

Srinivas, K., King, J.W., Howard, L.R., and Monrad, J.K. 2010b. Solubility of gallic acid, catechin, and protocatechuic acid in subcritical water from (298.75 to 415.85) K. *Journal of Chemical and Engineering Data* 55:3101–3108.

Srinivas, K., King, J.W., Howard, L.R., and Monrad, J.K. 2011. Binary diffusion coefficients of phenolic compounds in subcritical water using a chromatographic peak broadening technique. *Fluid Phase Equilibria* 301:234–243.

Sun, X., Peng, B., and Yan, W. 2008a. Measurement and correlation of solubility of trans-resveratrol in 11 solvents at T = (278.2, 288.2, 298.2, 308.2, and 318.2) K. *Journal of Chemical Thermodynamics* 40:735–738.

Sun, X., Shao, Y., and Yan, W. 2008b. Measurement and correlation of solubilities of trans-resveratrol in ethanol+ water and acetone+ water mixed solvents at different temperatures. *Journal of Chemical and Engineering Data* 53:2562–2566.

Teo, C.C., Tan, S.N., Yong, J.W.H., Hew, C.S., and Ong, E.S. 2010. Pressurized hot water extraction (PHWE). *Journal of Chromatography A* 1217:2484–2494.

Tomsone, L., Kruma, Z., and Galoburda, R. 2012. Comparison of different solvents and extraction methods for isolation of phenolic compounds from horseradish roots (*Armoracia rusticana*). *World Academy of Science, Engineering and Technology* 64: 903–908.

Towler, G. and Sinnott, R.K. 2007. *Chemical Engineering Design: Principles, Practice and Economics of Plant and Process Design.* Burlington, MA: Elsevier Science.

Tozuka, Y., Kishi, J., and Takeuchi, H. 2010. Anomalous dissolution property enhancement of naringenin from spray-dried particles with α-glucosylhesperidin. *Advanced Powder Technology* 21:305–309.

Treybal, R.E. 1968. *Mass-Transfer Operations*. New York: McGraw-Hill.

Vergara-Salinas, J.R., Bulnes, P., Zúñiga, M.C., Pérez-Jiménez, J., Torres, J.L., Mateos-Martín, M.L., Agosin, E., and Pérez-Correa, J.R. 2013. Effect of pressurized hot water extraction on antioxidants from grape pomace before and after enological fermentation. *Journal of Agricultural and Food Chemistry* 61:6929–6936.

Vergara-Salinas, J.R., Cuevas-Valenzuela, J., and Pérez-Correa, J.R. 2015. Pressurized hot water extraction of polyphenols from plant material. In *Biotechnology of Bioactive Compounds*, eds. V.K. Gupta, M.G. Tuohy, A. O'Donovan, and M. Lohani, pp. 63–101. Chichester, UK: John Wiley & Sons.

Vergara-Salinas, J.R., Pérez-Jiménez, J., Torres, J.L., Agosin, E., and Pérez-Correa, J.R. 2012. Effects of temperature and time on polyphenolic content and antioxidant activity in the pressurized hot water extraction of deodorized thyme (*Thymus vulgaris*). *Journal of Agricultural and Food Chemistry* 60:10920–10929.

Wan, P.J. and Wakelyn, P.J. 1997. *Technology and Solvents for Extracting Oilseeds and Nonpetroleum Oils*. Champaign, IL: American Oil Chemists Society.

Wei, D. and Zhang, X. 2013. Solubility of puerarin in the binary system of methanol and acetic acid solvent mixtures. *Fluid Phase Equilibria* 339:67–71.

Wiboonsirikul, J. and Adachi, S. 2008. Extraction of functional substances from agricultural products or by-products by subcritical water treatment. *Food Science and Technology Research* 14:319–319.

Wijngaard, H. and Brunton, N. 2009. The optimization of extraction of antioxidants from apple pomace by pressurized liquids. *Journal of Agricultural and Food Chemistry* 57:10625–10631.

Wilke, C. and Chang, P. 1955. Correlation of diffusion coefficients in dilute solutions. *AIChE Journal* 1:264–270.

Williamson, A. 1944. The exact calculation of heats of solution from solubility data. *Transactions of the Faraday Society* 40:421–436.

Wu, H., Chen, M., Fan, Y., Elsebaei, F., and Zhu, Y. 2012. Determination of rutin and quercetin in Chinese herbal medicine by ionic liquid-based pressurized liquid extraction–liquid chromatography–chemiluminescence detection. *Talanta* 88:222–229.

Wu, J.-G., Ge, J., Zhang, Y.-P., Yu, Y., and Zhang, X.-Y. 2010. Solubility of genistein in water, methanol, ethanol, propan-2-ol, 1-butanol, and ethyl acetate from (280 to 333) K. *Journal of Chemical and Engineering Data* 55:5286–5288.

Xiao, M., Yan, W., and Zhang, Z. 2010. Solubilities of apigenin in ethanol+ water at different temperatures. *Journal of Chemical and Engineering Data* 55:3346–3348.

Xu, B.J. and Chang, S. 2007. A comparative study on phenolic profiles and antioxidant activities of legumes as affected by extraction solvents. *Journal of Food Science* 72:S159–S166.

Xu, X., Xie, B., Pan, S., Yang, E., Wang, K., Cenkowski, S., Hydamake, A.W., and Rao, S. 2006. A new technology for extraction and purification of proanthocyanidins derived from sea buckthorn bark. *Journal of the Science of Food and Agriculture* 86:486–492.

Yalkowsky, S.H. and Roseman, Th.J. 1981. Solubilization of drugs by cosolvents, in *Techniques of Solubilization of Drugs*, ed. Yalkowsky, S.H. New York: Marcel Dekker. pp. 91–134.

Yan, X., Suzuki, M., Ohnishi-Kameyama, M., Sada, Y., Nakanishi, T., and Nagata, T. 1999. Extraction and identification of antioxidants in the roots of yacon (*Smallanthus sonchifolius*). *Journal of Agricultural and Food Chemistry* 47:4711–4713.

Yilmaz, Y. and Toledo, R. 2005. Antioxidant activity of water-soluble Maillard reaction products. *Food Chemistry* 93:273–278.

Yu, J., Vasanthan, T., and Temelli, F. 2001. Analysis of phenolic acids in barley by high-performance liquid chromatography. *Journal of Agricultural and Food Chemistry* 49:4352–4358.

Zappe, R.W. 1999. *Valve Selection Handbook: Engineering Fundamentals for Selecting Manual Valves, Check Valves, Pressure Relief Valves, and Rupture Discs*. Houston, TX: Gulf Publishing Company.

Zhang, L., Gong, X., Wang, Y., and Qu, H. 2012. Solubilities of protocatechuic aldehyde, caffeic acid, D-Galactose, and D-Raffinose pentahydrate in ethanol–water solutions. *Journal of Chemical and Engineering Data* 57:2018–2022.

Zhao, C., Wang, Y., Su, Y., Zhang, H., Ding, L., Yan, X., Zhao, D., Shao, N., Xiyun, Y., and Cheng, Y. 2011. Inclusion complexes of isoflavones with two commercially available dendrimers: Solubility, stability, structures, release behaviors, cytotoxicity, and antioxidant activities. *International Journal of Pharmaceutics* 421:301–309.

Zhou, L., Zhang, P., Yang, G., Lin, R., Wang, W., Liu, T., Zhang, L., and Zhang, J. 2014. Solubility of chrysin in ethanol and water mixtures. *Journal of Chemical and Engineering Data* 59:2215–2220.

Zlokarnik, M. 2006. *Scale-up in Chemical Engineering*. Weinheim, Germany: Wiley.

Zupančič, Š., Lavrič, Z., and Kristl, J. 2015. Stability and solubility of trans-resveratrol are strongly influenced by pH and temperature. *European Journal of Pharmaceutics and Biopharmaceutics* 93:196–204.

5 Supercritical Fluid Extraction of Polyphenols

José M. del Valle, Edgar Uquiche, and Loreto Valenzuela

CONTENTS

5.1 INTRODUCTION

SuperCritical (SC) Fluids (SCFs) can replace conventional solvents as separating agents to extract and purify phenolics and other natural antioxidants in vegetable matrices and other food materials. A SCF is a substance that is at state conditions above its corresponding critical temperature (T_c) and critical pressure (P_c). SCFs exhibit pronounced changes not only in density with state conditions but also in other relevant properties related to solvation power and mass-transfer rate such as the dielectric constant, viscosity, and self-diffusivity. Under SC conditions, molecules form clusters of high local density (the same as in the liquid state) that do not interact with each other to a great extent (the same as molecules in the gas state), which are respectively responsible for the solvation power and excellent transport properties (low viscosity, high self-diffusivity) of SCFs. The near-liquid density allows dissolution of target solutes from complex matrices, and the near-gas transport properties allow fast extractions and purifications. Moreover, in the absence of separate phases, there is no surface tension, which facilitates interfacial mass-transfer phenomena and infiltration of porous solid matrices (Akgerman et al., 1991). Undeniably, the physical properties of SCFs can be tuned to specific separation needs by changing the state conditions, especially near the critical point where an SCF is highly compressible. The recommended extraction conditions are a near-critical temperature [reduced temperature, T_r (= T/T_c) ≈ 1] and a supercritical pressure [reduced pressure, P_r (= P/P_c) ≥ 4] (Brunner and Peter, 1981).

Extraction processes for the food ingredient and nutraceutical industry require stable, inert, innocuous, and low-boiling solvents to facilitate their subsequent removal from treated substrates and to prevent deleterious chemical reactions of high-value functional compounds. CO_2 has become an ideal solvent in SCF Extraction (SCFE) for food applications because it is innocuous, nonflammable, and inexpensive (even when highly purified). Indeed, CO_2 has a "Generally Recognized As Safe" (GRAS) status. CO_2 has a near-environmental critical temperature of about 30°C (T_c = 304.1 K; Lemmon et al., 2010), which allows for the carrying out of extraction processes at relatively low temperature, thus avoiding thermal damage and loss of bioactivity to target compounds in vegetable matrices (Brunner, 1994). Furthermore, SC-CO_2 extractions are performed in the absence of air, thus avoiding oxidative damage to the target compounds. Because CO_2 is a gas under normal environmental conditions, it can be easily removed from the extract and the treated substrate during decompression that follows the extraction and the recovery of the extract. Finally, and very importantly, SC-CO_2 exhibits high-selectivity toward high-value compounds in vegetable substrates such as essential oils, triglycerides, and some minor lipids (Brogle, 1982).

SC-CO_2, however, has some limitations as a separating agent for extraction processes. CO_2 has a relatively high critical pressure (P_c = 7.38 MPa; Lemmon et al., 2010), which forces typical extraction processes to be carried out at >28 MPa. At a large scale, a pump or compressor should be selected to pressurize CO_2 up to those high levels, and an extraction vessel should be designed to withhold these pressures, which may cause investment cost to increase as compared to conventional, ambient-pressure processes. Furthermore, despite the claim of SC-CO_2 being an excellent solvent for extraction processes in the food industry, it is not a universal solvent. Just the opposite! The high selectivity of SC-CO_2, which allows for obtaining extracts with high concentrations of target compounds or the removal of toxic or otherwise undesirable components from a vegetable matrix without affecting its overall sensory value, is due to its limited and tunable solvation power. SC-CO_2 is a lipophilic solvent that solubilizes low-molecular-weight, nonpolar solutes such as monoterpene hydrocarbons and oxygenated monoterpenes in essential oils (the mutual miscibility of CO_2 and essential oils occurs at ≥15 MPa; del Valle et al., 2011). On the other hand, SC-CO_2 has difficulties solubilizing high-molecular-weight, polar solutes such as polyphenols.

The use of solvent mixtures instead of pure solvents eases selecting adequate extracting agents for specific SCFE applications such as the extraction of polyphenols (Joshi and Prausnitz, 1984). This may be the result of preferential intermolecular forces, for example, hydrogen bonds, between the solute and the modifier, a substance such as hexane or ethanol that has intermediate volatility between those of the solute and CO_2 and that increases the selectivity of the extraction process.

The use of "modified" instead of pure CO_2 as solvent, however, negates the most important advantage of SCFE in the food ingredient and nutraceutical industry, namely, the absence of objectionable traces of organic solvents in the extracts (Moyler, 1993). The food industry is regulated to use GRAS substances in compliance with Good Manufacturing Practice (GMP). The list of GRAS (liquid) solvents that can be used as modifiers includes butyl and ethyl acetate, acetone, and ethanol (Sanders, 1993). However, unlike in the case of ethanol, there are limits to acetone residues in spice oleoresins (30 ppm) and use of ethyl acetate in decaffeination of tea and coffee imposed by the U.S. legislation, and stricter regulations apply to other common solvents such as dichloromethane, hexane, and methanol (Sanders, 1993). Because of these antecedents, just water and ethanol can be considered as natural cosolvents for the extraction of food materials (Moyler, 1993).

The main purpose of this chapter is to describe the use of SCFs, particularly of SC-CO_2, as separating agents for the extraction and purification of phenolic compounds from biologic substrates, particularly plant material. About a decade ago, Díaz-Reinoso et al. (2006) published an authoritative review of this subject. Thus, our objective will be to update the work of Díaz-Reinoso et al. (2006) on phase equilibria, mass transfer, process design, and applications of SC-CO_2 technologies to the isolation and formulation of food polyphenols. Our chapter, however, will not include separate sections on sources (Chapter 1) and biologic effects (Chapter 2) of food polyphenols.

5.2 LITERATURE SURVEY

The subject of this chapter is covered in books on the application of extraction and related separation processes to foods. They include a chapter on SC-CO_2 extraction (Dunford et al., 2003) that does not cover the extraction of natural antioxidants, chapters on the extraction of natural antioxidants that cover SC-CO_2 extraction technologies succinctly (Oreopoulou, 2003) or more in depth (Conde et al., 2010), and a section containing detailed information on the SC-CO_2 extraction of natural antioxidants (Parajó et al., 2009). The subject of SC-CO_2 extraction of natural antioxidants is also included as separate chapters in books on the SC-CO_2 of food materials, including Chapter 8 of the book authored by Mukhopadyay (2000) and a chapter authored by Díaz-Reinoso et al. (2008).

Natural antioxidants in biologic substrates include both minor lipids (oil-soluble compounds that protect chemical energy storage compounds such as triglycerides and biologic membranes against oxidative damage such as carotenoids and tocochromanols) and water-soluble phenolics. Thus, it is necessary to exclude minor lipids from references alluding to the use of SC-CO_2 to process biologic substrates to obtain natural antioxidants as done by Díaz-Reinoso et al. (2006). This approach was also followed by Maróstica et al. (2010), but not by Shi et al. (2011) and Moraes et al. (2013). Maróstica et al. (2010) focused on the extraction and stabilization of natural phenolic compounds using SCFs. Shi at al. (2011) reviewed briefly the SC-CO_2

extraction of terpenoids, carotenoids, sterols, vitamin E, and alkaloids, in addition to phenolics. Moraes et al. (2013) reviewed advances in four novel techniques (SCFE, pressurized liquid extraction, microwave-assisted extraction, and ultrasound-assisted extraction) for the extraction of natural antioxidants such as polyphenols, terpenoids, and vitamin E. They pointed out that heavy, polar compounds such as flavonoids are poorly soluble in pure CO_2 and require a polar modifier such as ethanol to improve their solubility in SC-CO_2.

There are several wide encompassing reviews on the SC-CO_2 extraction of vegetable substrates that touch on the subject of antioxidant extraction. Herrero et al. (2006) addressed applications of SCFE in the period 2001–2003 to isolate natural products from plants, food by-products, algae, and microalgae, emphasizing antioxidant compounds because of their importance in human health and food preservation. They include, however, limited information on extraction of polyphenols such as their isolation from grape seeds using methanol-modified SC-CO_2 and point out that SCFE is faster and allows for the fractionation of phenolic compounds by changing the pressure and/or adding different types and amounts of modifiers. Herrero et al. (2010) reviewed developments of SCFE during the period of 2007–2009 in different fields, including food and natural product applications, but did not address in depth the extraction of polyphenols, the only exception being their recovery from cocoa kernels, rice wine lees, and pomegranate seeds. Pereira and Meireles (2010) reviewed the SC-CO_2 extraction of bioactive compounds from plant matrices, including descriptions of extraction and fractionation procedures, SC solvents, and main factors influencing solubility, extraction yields, and manufacturing costs. This chapter includes a section and a table devoted specifically to the extraction of polyphenols using pure and modified (i.e., ethanol, ethyl acetate, dichloromethane) SC-CO_2, including 16 references from the 2002–2008 period. Although Sovová and Stateva (2011) did not address specifically the extraction of polyphenols, they mentioned the extraction of polyphenols from apple fruit, peels, and pomace; of tannins, glycosides, and polyphenols from grape seed, pulp, and skin; of glycosylated flavonoids from passion fruit and leaves; and of polyphenols from peach pomace using pure or modified SC-CO_2. The review of Oman et al. (2013) includes a table that deals specifically with the extraction of plant phenolics. Souza-Machado et al. (2013) reviewed developments in SCFE between 2001 and 2010 emphasizing the use of the solubility parameters to optimize processing conditions for extraction and fractionation applications, including isolation of natural antioxidants from sustainable sources; they inform that the extraction of polyphenols with SC-CO_2 usually requires the use of methanol or ethanol as a modifier. DeMelo et al. (2014) reviewed the SCFE of naturally occurring molecules from vegetable matrices. Vegetable species that have been extensively studied include phenolic-rich eucalypt, grape, rosemary, and thyme, among others. SC-CO_2 extracts obtained from vegetable substrates are typical mixtures of triglycerides, fatty acids, fatty alcohols, terpenoids, phytosterols, tocopherols, tocotrienols, and phenolics. They indicate that depending on the vegetable species, phenolics obtained by SC-CO_2 can comprise several groups like coumarins, cinnamic acids, quinones, flavonoids, and lignans. Of 441 publications reviewed by de Melo et al. (2014), 38% include at least one experimental assay using modified SC-CO_2. Main polar modifiers applied include ethanol (53% of cases), methanol (21%), and water (5%).

There are other narrower-scope reviews on SC-CO_2 technologies that touch laterally on the extraction of phenolic compounds. For example, Sunarson and Ismadji (2009) reviewed the decontamination of solid matrices (e.g., extraction of herbicides and pesticides from plants and hazardous substances from animal tissues and foods) using SCFs. Ahluwalia et al. (2013) provided an overview on the applications of SCFE for the recovery of bioactive compounds from fruit and vegetable waste/by-products, citing briefly works suggesting extraction conditions for polyphenols from apple and peach pomace, grape pomace and peel, and guava seed. Finally, Capuzzo et al. (2013) address the SCFE of plant flavors and fragrances.

The remaining of the chapter will be organized as follows. First, we will describe phase equilibrium aspects affecting the solubility of food phenolics in SC-CO_2. Then, we will exemplify the SC-CO_2 technologies to isolate plant phenolics for rosemary diterpenes, grape phenolics, and tea phenolics as representative of processes that can be applied to alternative substrates such as herbs and spices; fruits and vegetables; seeds, beans, and grains; medicinal plants; three barks and leaves; and propolis. Section 5.4 will end with a summary of the multiunit processing schemes that can be applied with the same purpose. The multiunit processing schemes consider several separation processes using SC-CO_2 that are described in Section 5.5.

5.3 PHASE EQUILIBRIUM

Phase equilibrium information is critical in designing and assessing separation processes under supercritical conditions. Solute solubility in SC-CO_2 depends markedly on the operational conditions of pressure and temperature. This is dramatically illustrated in Figure 5.1 for the changes in solubility in SC-CO_2 of two simple phenolic compounds such as benzoic acid (a and b) and salicylic acid (c and d) as a function of system pressure, for different temperatures (a and c), and as a function of system temperature, for different pressures (b and d). In both cases, the molar fraction of the solute in a saturated SC-CO_2 phase (y_2) increases monotonously with CO_2 pressure (P) because, under supercritical conditions, density increases with pressure with the end result of a reduction in intermolecular distance between CO_2 molecules, thus contributing to enhanced solute–solvent interactions (i.e., solvation). The effect of system temperature (T) on solute solubility is more complex than that of CO_2 pressure. Indeed, there are two main effects on the solubility in SC-CO_2 when T increases isobarically (Brunner, 1994): the density and solvent power of CO_2 decrease, and the vapor pressure and volatility of the solute increase. These two effects have opposite consequences in solubility that decreases when solvent power decreases and increases when solute volatility increases. Depending on which of these two effects dominates, solubility decreases, remains constant, or increases as T increases isobarically. Density effects dominate near the critical point, where CO_2 is highly compressible and experiences large changes in density as a result from moderate changes in T. On the other hand, vapor pressure effects dominate away from the critical point, where CO_2 is less compressible and density changes moderately as a result from large changes in T. There is an intermediate pressure, the so-called crossover pressure (Foster et al., 1991), where the density and vapor pressure effects are similar, and the solubility remains constant when temperature changes. For benzoic

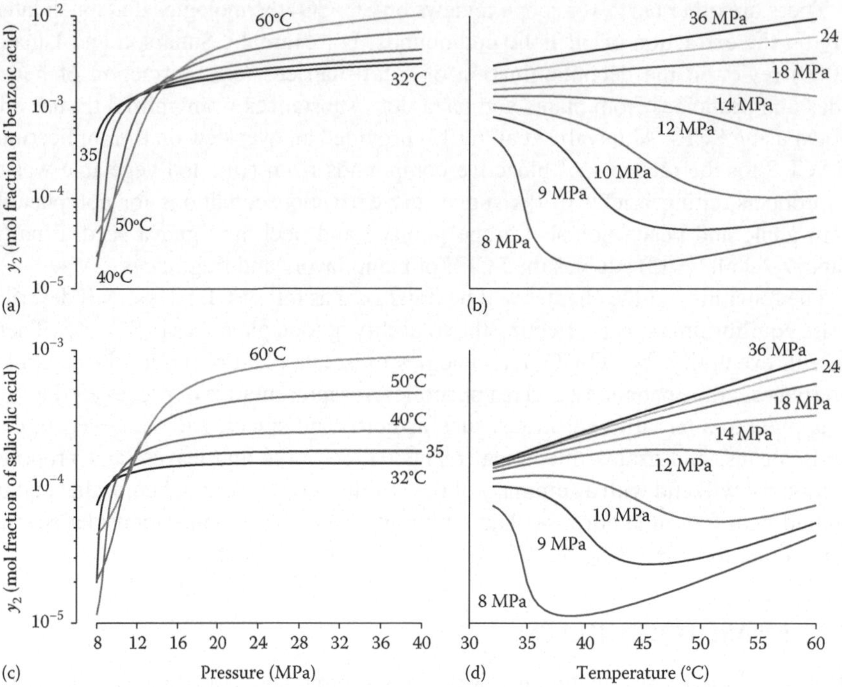

FIGURE 5.1 Solubility of (a, b) benzoic acid and (c, d) salicylic acid in supercritical CO_2 as a function of (a, c) system pressure (lines indicate solubility isotherms at noted temperatures) or (b, d) system temperature (lines indicate solubility isobars at noted pressures).

acid, the crossover pressure between 50°C and 60°C is ca. 16 MPa, and the crossover pressure decreases as the temperature decreases (Figure 5.1a). Thus, at ca. 10 MPa, the solubility of benzoic acid decreases as the temperature increases.

Phase equilibrium of different solutes at high pressure in systems containing CO_2 has been broadly reviewed. *Fluid Phase Equilibria* presents a series of articles reviewing systematically the experimental methods and investigated systems (Christov and Dohrn, 2002; Dohrn and Brunner, 1995; Dohrn et al., 2010; Fornari et al., 1990; Fonseca et al., 2011) for a period spanning three decades (1978–2008). Díaz-Reinoso et al. (2006) presented a thorough review on solubility of phenolic compounds in pure and ethanol-modified SC-CO_2. Subsequently, there were additional reviews covering the solubility of low-volatility solutes such as polyphenols in pure SC-CO_2 and CO_2 modified with polar cosolvents such as ethanol and water, including the handbook of Gupta and Shim (2007) and the review article of Škerget et al. (2011). Table 5.1 summarizes solubility data of pure phenolic compounds in pure or modified SC-CO_2, including the ones presented in Díaz-Reinoso et al. (2006).

Díaz-Reinoso et al. (2006) included information on the methods used to model solute solubility in SC-CO_2 that included (1) fundamental approaches based on equations of state with mixing and (2) combination rules to characterize solute–solvent interactions to more empirical approaches based on the simpler relationship between

TABLE 5.1
Solubility of Selected Phenolic Compounds in Pure or Modified Supercritical CO_2

Phenolic Compound	T (K)	P (MPa)	Modifier (%)	y_2 (–)	References
Aldehydes					
p-Anisaldehyde	323–373	5.5–13.7		$(0.2–14) \times 10^{-3}$	Mukhopadhyay and De (1995)
Protocatechuic aldehyde	313–333	10–50		$(0.01–4.57) \times 10^{-5}$	Murga et al. (2002)
Phenols					
Phenol	309–333	7.9–24.9		$(0.65–4.66) \times 10^{-2}$	Van Leer and Paulaitis (1980)
2,5-Xylenol	308	7.4–26.7		$(0.365–11.6) \times 10^{-3}$	Iwai et al. (1990)
3,4-Xylenol	308	8.2–26.2		$(2.34–9.10) \times 10^{-3}$	Mori et al. (1992)
Carvacrol	313–323	0.26–30.84		$(0.29–16.69) \times 10^{-2}$	Leeke et al. (2001)
Thymol	323–343	5.7–14		$(0.51–12.6) \times 10^{-3}$	Mukhopadhyay and De (1995)
Hydroquinone	308–318	8.4–31.3		$(0.204–1.495) \times 10^{-5}$	Coutsikos et al. (1995)
Carnosic acid	313–333	27.5–40	Ethanol 0.7%–10%	$(0.60–53.03) \times 10^{-6}$	Cháfer et al. (2005)
Phenolic acids					
Benzoic acid	318–338	12–28		$(0.32–9.83) \times 10^{-3}$	Kurnik et al. (1981)
Benzoic acid	308–343	10.1–36.4		$(0.12–12.8) \times 10^{-3}$	Schmitt and Reid (1986)
Benzoic acid	308	12–28		$(1.25–3.03) \times 10^{-3}$	Dobbs et al. (1987)
Benzoic acid	308	10–33	Acetone 3.5% mol	$(3.34–6.40) \times 10^{-3}$	Dobbs et al. (1987)
Benzoic acid	308	9–35	Methanol 3.5% mol	$(5.33–11.9) \times 10^{-3}$	Dobbs et al. (1987)
Benzoic acid	318	9–23		$(0.144–4.31) \times 10^{-3}$	Galia et al. (2002)
o-Toluic acid (*o*-methylbenzoic acid)	313–333	11–24.3		$(9.02–57.0) \times 10^{-4}$	Tsai and Tsai (1995)
m-Toluic acid (*m*-methylbenzoic acid)	313–333	11–24.3		$(1.84–53.4) \times 10^{-4}$	Tsai and Tsai (1995)
p-Toluic acid (*p*-methylbenzoic acid)	313–333	11–24.3		$(0.43–7.02) \times 10^{-4}$	Tsai and Tsai (1995)
Salicylic acid (*o*-hydroxybenzoic acid)	373	20.7–41.4		$(20.7–69.9) \times 10^{-4}$	Krukonis and Kurnik (1985)
Salicylic acid (*o*-hydroxybenzoic acid)	308–328	8.1–20.3		$(0.07–6.24) \times 10^{-4}$	Gurdial and Foster (1991)

(Continued)

TABLE 5.1 (*Continued*)
Solubility of Selected Phenolic Compounds in Pure or Modified Supercritical CO_2

Phenolic Compound	T (K)	P (MPa)	Modifier (%)	y_2 (–)	References
Phenolic acids					
Salicylic acid (o-hydroxybenzoic acid)	318–328	8.6–20.1	Acetone 3.5% mol	$(6.8–25.2) \times 10^{-4}$	Gurdial et al. (1993)
Salicylic acid (o-hydroxybenzoic acid)	318–328	9.6–20.1	Methanol 3.5% mol	$(12.5–51.2) \times 10^{-4}$	Gurdial et al. (1993)
Salicylic acid (o-hydroxybenzoic acid)	313–328	9.6–25		$(0.34–5.83) \times 10^{-4}$	Stassi et al. (2000)
Salicylic acid (o-hydroxybenzoic acid)	318	9.8–19.5		$(7.3–28.9) \times 10^{-4}$	Abbott et al. (2002)
Salicylic acid (o-hydroxybenzoic acid)	318–337	10.2–22.0		$(0.195–6.25) \times 10^{-4}$	Galia et al. (2002)
Salicylic acid (o-hydroxybenzoic acid)	308–318	8.7–15.8	Ethanol 0%–6.9% mol	$(0.97–126) \times 10^{-4}$	Ke et al. (1996)
Salicylic acid (o-hydroxybenzoic acid)	318–328	10.1–20.3		$(0.32–5.62) \times 10^{-4}$	Lucien and Foster (1996)
m-Hydroxybenzoic acid	373	20.7–41.4		$(1.12–7.54) \times 10^{-5}$	Krukonis and Kurnik (1985)
m-Hydroxybenzoic acid	318–328	9.1–20.1	Acetone 3.5% mol	$(1.50–5.47) \times 10^{-5}$	Gurdial et al. (1993)
m-Hydroxybenzoic acid	318–328	9.1–20.1	Methanol 3.5% mol	$(1.59–3.37) \times 10^{-3}$	Gurdial et al. (1993)
p-Hydroxybenzoic acid	373	20.7–41.4		$(1.33–7.49) \times 10^{-5}$	Krukonis and Kurnik (1985)
p-Hydroxybenzoic acid	318–328	10.1–20.3		$(1.4–37.2) \times 10^{-5}$	Lucien and Foster (1996)
o-Anisic acid (*o*-methoxybenzoic acid)	308–328	9.98–24.7		$(0.89–11.1) \times 10^{-2}$	Chen and Tsai (1995)
m-Anisic acid (*m*-methoxybenzoic acid)	308–328	9.98–24.7		$(1.65–17.3) \times 10^{-2}$	Chen and Tsai (1995)
p-Anisic acid (*p*-methoxybenzoic acid)	308–328	9.98–24.7		$(2.81–9.06) \times 10^{-2}$	Chen and Tsai (1995)
p-Anisic acid (*p*-methoxybenzoic acid)	323–373	5.5–15.1		$(2.87–137.3) \times 10^{-6}$	Mukhopadhyay and De (1995)
Protocatechuic acid (3,4-dihydroxybenzoic acid)	313–333	10–50		$(0.023–25.9) \times 10^{-7}$	Murga et al. (2002)
Vanillic acid	313–328	8.6–25		$(0.48–23.9) \times 10^{-6}$	Stassi et al. (2000)
Vanillic acid	313–333	8.5–50		$(0.937–60.71) \times 10^{-6}$	Murga et al. (2004)
Gallic acid	313	20	Ethanol 2%–10% v/v	$(0.30–20.4) \times 10^{-7}$	Murga et al. (2000)
Gallic acid	313–333	10–40	Ethanol 0.7%–6%	$(1.06–19.72) \times 10^{-7}$	Cháfer et al. (2007)
Methyl gallate	313–333	10–50		$(0.019–4.24) \times 10^{-6}$	Murga et al. (2002)
Propyl gallate	313–333	15–25		$(5.02–18.7) \times 10^{-6}$	Cortesi et al. (1999)

(*Continued*)

TABLE 5.1 (*Continued*)
Solubility of Selected Phenolic Compounds in Pure or Modified Supercritical CO_2

Phenolic Compound	*T* (K)	*P* (MPa)	Modifier (%)	y_2 (–)	References
Phenolic acids					
Dodecyl gallate	313–333	15–25		$(3.1–20.7) \times 10^{-6}$	Cortesi et al. (1999)
Syringic acid	313–333	8.5–50		$(0.01–127.1) \times 10^{-7}$	Murga et al. (2004)
Hydroxycinnamic acids					
o-Coumaric acid	308–323	8.5–25		$(0.21–35.4) \times 10^{-8}$	Choi et al. (1998)
m-Coumaric acid	308–323	8.5–25		$(0.16–105.5) \times 10^{-8}$	Choi et al. (1998)
p-Coumaric acid	308–323	8.5–25		$(0.52–38.5) \times 10^{-8}$	Choi et al. (1998)
p-Coumaric acid	313–333	8.5–50		$(0.3–255) \times 10^{-8}$	Murga et al. (2003)
Caffeic acid	313–333	8.5–50		$(0.08–4.7) \times 10^{-8}$	Murga et al. (2003)
Ferulic acid	301–333	12–28		$(1.55–11.8) \times 10^{-6}$	Sovová (2001)
Ferulic acid	313–333	12–28	Ethanol 0.27%–3.35% w/w	$(2.4–79.8) \times 10^{-6}$	Sovová (2001)
Ferulic acid	313–333	8.5–50		$(0.065–43.32) \times 10^{-6}$	Murga et al. (2003)
Ferulic acid	308–338	10–35		$(0.219–2.65) \times 10^{-6}$	Sun et al. (2005)
Naphthoquinone					
1,4-Naphthoquinone	318–343	10.1–36.4		$(0.18–13.5) \times 10^{-3}$	Schmitt and Reid (1986)
Stilbene					
Resveratrol	313	8–14	Ethanol 5%–15%	$(4.14–27.01) \times 10^{-5}$	Berna et al. (2001)
Xanthones					
1-Hydroxyxanthone	305–348	7.4–35.5		$(1.49–22.20) \times 10^{-4}$	Ghiasvand et al. (1999)
1-Hydroxy-3-methylxanthone	305–348	7.4–35.5		$(1.68–30.22) \times 10^{-4}$	Ghiasvand et al. (1999)
1,6-Dihydroxyxanthone	305–349	7.4–35.6		$(1.31–35.88) \times 10^{-4}$	Ghiasvand et al. (1999)
1,5,6-Trihydroxyxanthone	305–348	7.4–35.5		$(2.80–16.84) \times 10^{-4}$	Ghiasvand et al. (1999)
Anthraquinone					
1,8-Dihydroxyanthraquinone	327–357	6.9–20.1		$(0.73–14.7) \times 10^{-5}$	Galia et al. (2002)

(Continued)

TABLE 5.1 (*Continued*)
Solubility of Selected Phenolic Compounds in Pure or Modified Supercritical CO_2

Phenolic Compound	T (K)	P (MPa)	Modifier (%)	y_2 (–)	References
Anthraquinone flavone					
Flavone	308–318	9.1–25.3		$(0.35–5.61) \times 10^{-4}$	Uchiyama et al. (1997)
Flavonols					
3-Hydroxyflavone	308–318	9.1–25.3		$(0.79–5.56) \times 10^{-4}$	Uchiyama et al. (1997)
Catechin	313	20	Ethanol 2%–10% v/v	$(0.097–7.75) \times 10^{-7}$	Murga et al. (2000)
Catechin	313	8–12	Ethanol 5%–30%	$(0.51–53.6) \times 10^{-4}$	Berna et al. (2001)
Catechin	313–343	12–26		$(0.67–106.8) \times 10^{-7}$	Edi-Soetaredjo et al. (2013a)
Epicatechin	313	20	Ethanol 2%–10% v/v	$(0.079–3.56) \times 10^{-7}$	Murga et al. (2000)
Epicatechin	313	8–12	Ethanol 5%–30%	$(0.51–53.6) \times 10^{-4}$	Cháfer et al. (2002)
Epicatechin	313–343	12–26		$(6.6–175.3) \times 10^{-7}$	Edi-Soetaredjo et al. (2013b)
Quercetin	313	8–12	Ethanol 5%–30%	$(0.22–6.37) \times 10^{-3}$	Cháfer et al. (2004)
Epigallocatechin gallate	313	15–35		$(0.13–0.83) \times 10^{-4}$	Song et al. (2010)
Epigallocatechin gallate	313	15–35	Acetone 3.7% mol	$(0.30–0.78) \times 10^{-4}$	Song et al. (2010)
Epigallocatechin gallate	313–333	15–35	Ethanol 4.4%–8.4% mol	$(0.98–7.34) \times 10^{-4}$	Song et al. (2010)
Isoflavone					
Daidzein	313	25	Ethanol 0%–15% mol	$5.14 \times 10^{-10}–10^{-5}$	Nakada et al. (2009)
Genistein	313	25	Ethanol 0%–15% mol	$6.38 \times 10^{-10}–10^{-5}$	Nakada et al. (2009)
Glycitein	313	25	Ethanol 0%–15% mol	Not detected–10^{-6}	Nakada et al. (2009)

solubility and CO_2 density than that between solubility and system pressure. In this work, we propose using the methodology of Ziger and Eckert (1983). This framework relates a solubility enhancement (*E*), Equation 5.1, with a factor (*f*) defined by Equation 5.2 that depends on the ratio of the solubility parameters of SC-CO_2 and the solute (Δ) and other solute-dependent properties:

$$E = \frac{y_2 P}{P_2^v(T)}, \tag{5.1}$$

and

$$f = \varepsilon_2 \frac{\Delta}{y_1}\left(2 - \frac{\Delta}{y_1}\right) - \log\left[1 + \frac{(\delta_1)^2}{P}\right], \tag{5.2}$$

where

$$\varepsilon_2 = \frac{(\delta_2)^2 v_2^L(T)}{2.303RT}, \tag{5.3}$$

and

$$\Delta = \frac{\delta_1}{\delta_2}. \tag{5.4}$$

In Equation 5.1, *E* indicates how many times larger is the solubility in SC-CO_2 that is expected from the vapor/sublimation pressure of the solute, P_2^v (depending on whether it is in a liquid or solid state, respectively), which depends on the absolute temperature of the system (*T*). The solubility parameter of SC-CO_2 (δ_1) depends mainly on its density, whereas that of the solute (δ_2) depends on its chemical structure. Besides P_2^v and δ_2, factor *f* depends on the molar volume of the solute, v_2^L, that is, a moderate function of *T*. Figure 5.2 illustrates the use of Equations 5.1 through 5.4 to model the solubility of selected compounds in Table 5.1 for which there was information about model parameters. Ziger and Eckert (1983) proposed that all isotherms collapsed to a single line and that there was a linear relationship between log(*E*) and *f*:

$$\log(E) = \nu_2 + \eta_1 \cdot f, \tag{5.5}$$

with a solvent-dependent slope η_1 and a solute-dependent intercept ν_2, which is suggested by experimental data in Figure 5.2. There was a single value of η_1 for hydroquinone, salicylic acid, vanillic acid, flavone, propyl gallate, and dodecyl gallate (0.293), a value slightly lower for 3-hydroxyflavone (0.114), and a value slightly higher for benzoic acid (0.687). Figure 5.2 shows that SC-CO_2 causes very significant increases in the tendency of high-molecular-weight and polar solutes such as dodecyl gallate to pass to the gaseous phase (7-to-8 order of magnitude increase), with a more moderate effect for lighter or less polar solutes such as hydroquinone, benzoic acid, salicylic acid, and vanillic acid. The most impressive aspect of the

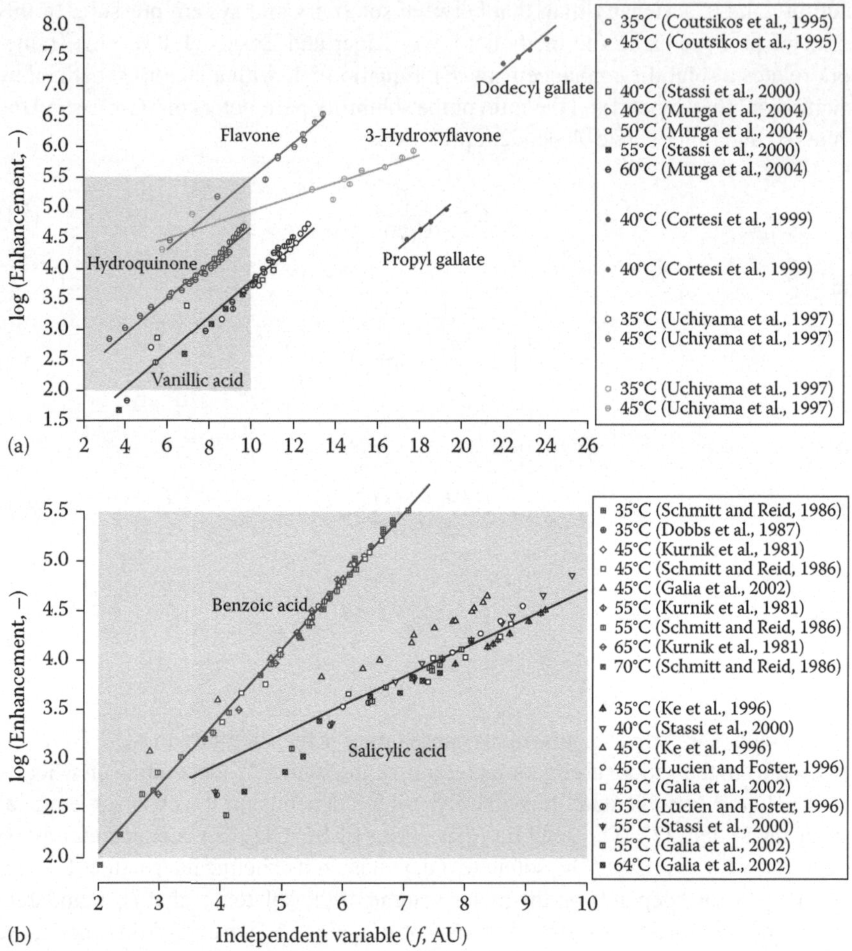

FIGURE 5.2 Solubility enhancement [log(E), Equation 5.1] as a function of independent factor f, Equation 5.2, for the solubility of phenolic compounds in supercritical CO_2. Gray areas in (a) and (b) are the same.

simple relationship described by Equation 5.5 is that it can predict the very complex relationship between y_2 and T and P shown in Figure 5.1. Indeed, the changes in solubility of benzoic and salicylic acids as a function of state condition informed in Figure 5.1 were predicted using best-fit parameters derived from Figure 5.2.

Construction of Figure 5.2 was limited to those solutes and conditions in Table 5.1 for which there was information of vapor/sublimation pressure (and molar volume) as a function temperature. However, it may be possible to expand the analysis to all solutes using group-contribution estimation methods or alternative approaches.

In the phase equilibrium of a solute and SC-CO_2, the presence of a cosolvent enhances solute solubility (Kurnik and Reid, 1982; Lucien and Foster, 1996). This positive effect is so-called cosolvency in phase equilibrium literature and is not due

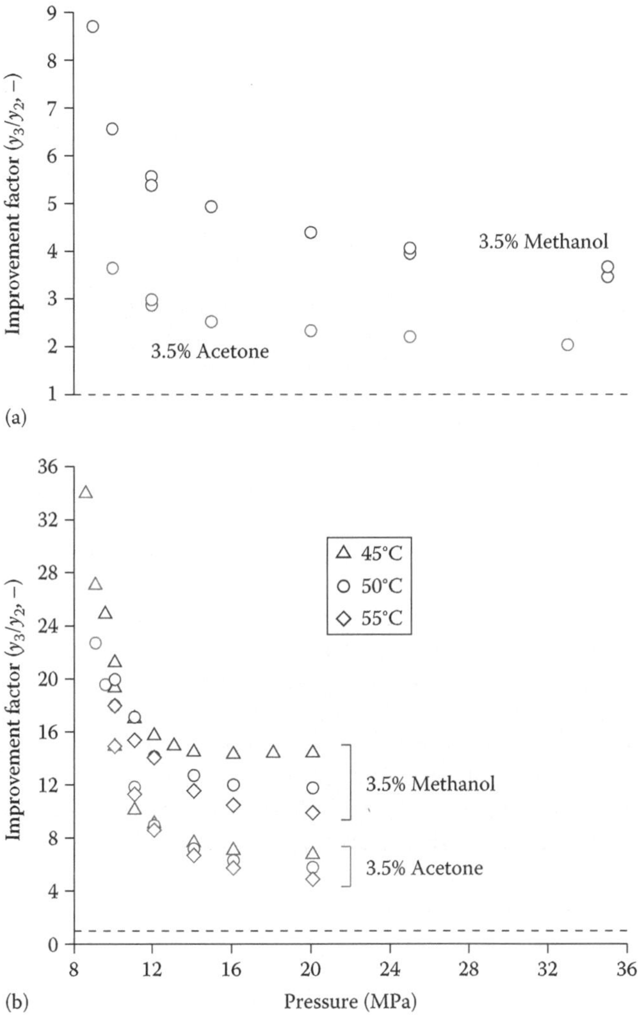

FIGURE 5.3 Improvement factor for the solubility in supercritical CO_2 of (a) benzoic acid at 35°C and (b) salicylic acid at the noted temperature as a function of system pressure and an added polar cosolvent (acetone or methanol).

solely to specific interactions having the cosolvent or modifier as a bridge between CO_2 and the target solute (Dobbs et al., 1987), but also to an increase in the density or polarizability of the solvent mixture as compared to pure CO_2 under the same system conditions (Dobbs et al., 1986). The main cosolvents for polar solutes such as polyphenols include ethanol, acetone, methanol, and water. Figure 5.3 illustrates the effects of a small amount of added acetone or methanol on the solubility of benzoic acid (a) and salicylic acid (b) in SC-CO_2 as a function of system pressure (a, b) and temperature (b). This is shown as an improvement factor or the ratio of solute solubility in the ternary system CO_2 + cosolvent + solute (y_3) and in the binary system CO_2 + solute

(y_2) under the same system conditions. For calculations, y_2 was estimated using the model of Ziger and Eckert (1983) with model parameters derived from Figure 5.2b. It can be appreciated that both acetone and methanol have a positive cosolvent effect (improvement factor above 1, signaled by the segmented line), that methanol has a more positive cosolvent effect than acetone, and that the cosolvent effect decreases as T or, to a larger extent, P increases. Thus, the cosolvent effect is more pronounced at near-critical conditions (35°C–45°C and 8.5–9.0 MPa), where the solubility of benzoic and salicylic acids in pure SC-CO_2 is small (Figure 5.1). The cosolvent effect is larger for salicylic acid (in the case of methanol, the improvement factor approaches values of about 10 for high pressures above 20 MPa) than for benzoic acid (where it approaches values of about 4 under equivalent conditions), thus confirming that high cosolvent effects are associated with low solubilities in pure SC-CO_2. The cosolvent effect increases also with the concentration of the intermediate-volatility component as observed for the solubilization of gallic acid (GA) (Murga et al., 2000), catechin (Berna et al., 2001; Murga et al., 2000), epicatechin (EC) (Cháfer et al., 2002; Murga et al., 2000), quercetin (Cháfer et al., 2004), and epigallocatechin gallate (EGCg) (Song et al., 2010) in SC-CO_2 modified using ethanol as the cosolvent.

In multicomponent systems, the solubility of a solute in SC-CO_2 depends on the thermodynamic solubility and the matrix–solute affinity. The presence of more solutes changes the phase equilibrium, thus increasing solubility.

Murga et al. (2000) measured the solubility of GA, catechin, and EC in SC-CO_2 at 20 MPa and 40°C using ethanol as the cosolvent and reported the yields of extraction of a solute mixture (grape seed concentrate) under the same conditions following a 3 h treatment. The ratio between extraction yield of polyphenols (mg/g concentrate) and specific consumption of solvent (kg CO_2/g concentrate) was used as an estimate of the solubility. From these yields, we estimated lower bound solubilities for the compounds (under the assumption that extraction yield was fully determined by solubility) that are reported in Table 5.2. Table 5.2 clearly demonstrates that the solubility of each of the solutes in a mixture (except for GA using 5% ethanol v/v) is higher than

TABLE 5.2
Solubility of Three Natural Phenols in Supercritical CO_2 Modified with Ethanol at 40°C and 20 MPa

	Solubility (mg Solute/kg SC Solvent) of Pure Compounds		Solubility (mg Component/kg SC Solvent) of Compounds in Commercial Grape Seed Concentrate	
Compound	**5% Ethanol (v/v)**	**10% Ethanol (v/v)**	**5% Ethanol (v/v)**	**10% Ethanol (v/v)**
Gallic acid	0.55	7.48	0.370	441.9
Catechin	0.47	4.80	0.926	26.1
Epicatechin	0.45	2.23	0.556	7.6

Source: Murga, R. et al., *J. Agric. Food Chem.*, 48, 3408, 2000.

when isolated (binary CO_2 + solute system). For instance, using 10% (v/v) ethanol as the modifier, the solubility of GA increased >59-fold in the complex mixture as compared to the binary system, followed by catechin (5.4-fold) and EC (3.4-fold).

5.4 APPLICATIONS

In this section, we will exemplify SC-CO_2 technologies for three important plant phenolic examples, namely, the isolation of diterpenes from rosemary leaves (Section 5.4.1), the isolation of phenolics from winery residues (grape marc, but particularly grape seeds) (Section 5.4.2), and the isolation of phenolics from tea (or alternatively, the enrichment of tea on phenolics by decaffeination) (Section 5.4.3). These processes are representative of many schemes that are summarized in Section 5.4.4 and can be applied to alternative substrates such as herbs and spices such as turmeric; fruits and vegetables; seeds, beans, and grains including coffee beans and cocoa kernels; medicinal plants such as *Tetraclinis articulata*; barks and leaves of terrestrial and marine plants such as olive and eelgrass leaves; and propolis.

5.4.1 Extraction of Phenolics from Rosemary

There are many examples on the extraction of diterpene phenolics from rosemary (*Rosmarinus officinalis* L.) and other Lamiaceae herbs. A key point of this particular application is to obtain the essential oils that are responsible for the characteristic aroma of the herb, on one hand, and the phenolic compounds responsible for the antioxidant and other biologic activities of the plant, on the other hand. A clear advantage of SC-CO_2 extraction over alternative technologies is that it allows for the fractionation of rosemary during extraction, during separation, or as a function of time to allow for obtaining extracts with increased content of antioxidants and reduced content of volatiles. This is possible by taking advantage of large differences in solubilities in SC-CO_2 between essential oils and other volatile, low-molecular-weight, and/or nonpolar, high-solubility compounds, and phenolic diterpenes and other nonvolatile, high-molecular-weight, and/or polar, low-solubility compounds. Separation processes for rosemary phenolic antioxidants are summarized in Table 5.3.

There have been several studies aimed at optimizing the extraction of rosemary antioxidants at the analytical level. Tena et al. (1997) maximized the yield of carnosic acid and purity (no cleanup required prior to high-performance liquid chromatography [HPLC] analysis) of SC-CO_2 rosemary extracts by using the highest tested temperature (120°C) and pressure (38.3 MPa) and an intermediate extraction time (20 min; corresponding to 2.71 kg g^{-1} CO_2/substrate). Bicchi et al. (2000) consecutively extracted volatiles, phenolic diterpenes, and other phenolics using 60 g g^{-1} pure CO_2/substrate, 90 g g^{-1} CO_2/substrate modified with 5% methanol, and 90 g g^{-1} CO_2/substrate modified with 10% methanol all at 60°C and 25.3 MPa. The rosemary extracts produced by Tena et al. (1997) and Bicchi et al. (2000) contained higher concentrations of carnosic acid and other phenolic diterpenes than extracts obtained using conventional organic solvents. Something similar was previously reported by Schwarz et al. (1992), but the authors did not specify the extraction conditions for the extract. Ibáñez et al. (1999) extracted the volatiles from air-dried, freeze-dried,

TABLE 5.3
Supercritical Extraction Conditions from Rosemary Using CO_2

Pretreatment	Particle Size (mm)	*T* (°C)	*P* (MPa)	Time Extraction/ Static Time (min)	Specific Solvent Rate (kg CO_2/L·h)	Modifier	References
Ground leaves	≤0.5	40–120	7.7–38.3	60/10	8.17–32.67		Tena et al. (1997)
Ground leaves		60	25.3	45		5%–10% MeOH	Bicchi et al. (2000)
Ground leaves		50–60	30–35	30		0%–2% EtOH	Señoráns et al. (2000)
Comminuted leaves	0.30–0.71	30–50	20–30	450	50.4	6%–13.8% EtOH, 2-propanol	Leal et al. (2003)
Ground leaves	0.297–0.707	30–40	10–30	180	1.36		Carvalho et al. (2005)
Ground leaves	0.5–0.999	40–60	15–35	30		0%–7% EtOH	Cavero et al. (2005)
Ground leaves		40	10	60/5	22.53		Ramírez et al. (2005)
Leaves		100	35	35/5	4.29	2%–5% MeOH, EtOH, 2-propanol	Yesil-Celiktas et al. (2007)
Ground leaves	0.841	40–80	27–34.5	60/60			Chang et al. (2008)
Ground	0.400	40	11.5	360	2		Babovic et al. (2010)
Ground	0.400	100	35	360	2		Babovic et al. (2010)
Ground leaves	0.841	40–80	34.5	90/30			Kuo et al. (2011)

(*Continued*)

TABLE 5.3 (*Continued*)
Supercritical Extraction Conditions from Rosemary Using CO_2

Pretreatment	Particle Size (mm)	T (°C)	P (MPa)	Time Extraction/ Static Time (min)	Specific Solvent Rate (kg CO_2/L·h)	Modifier	References
Ground leaves	0.2–0.6	40	15–30	60–180	1.80	5%–10% EtOH	Vicente et al. (2013)
Milled		40–60	10–17		17.56–97.1		Omar et al. (2013)
Ground leaves		40	15–30	60–120	1.80	7% EtOH	Sánchez-Camargo et al. (2014)
Extracts by steam distillation and Soxhlet extraction		40–80	10–35	60	0.06–0.09		López-Sebastián et al. (1998)
Ground leaves		40, 60	10, 40	30/5	0.06–0.09		Ibáñez et al. (1999)
Ground leaves	0.2–0.6	40	30	300	1.8		Vicente et al. (2012)
Ground leaves	0.2–0.6	40	15	180	1.8	5%–10% EtOH	Vicente et al. (2012)
Ground leaves	0.2–0.6	40	30	360	1.8		Vicente et al. (2012)
Ground leaves	0.4	40, 40, and 100	11.5, 30	250	2		Ivanović et al. (2009)
Ground leaves	0.4			250	2		Ivanović et al. (2009)

and commercial samples of dried rosemary using 66.6–88.7 g g^{-1} pure CO_2/substrate at 40°C and 10 MPa preceded by 5 min static extraction under the same conditions and then extracted the antioxidant fraction using 94.3–126 g g^{-1} pure CO_2/substrate at 60°C and 40 MPa preceded by 5 min static extraction under the same conditions and found that the second fraction of the commercial samples had the highest antioxidant activity. Yesil-Celiktas et al. (2007) extracted phenolic diterpenes using ca. 128 g g^{-1} CO_2/substrate modified with 5% methanol at 100°C and 35 MPa following 5 min static extraction under the same conditions. The use of 5% methanol as the modifier increased the yield of carnosic acid 16.3% as compared to pure SC-CO_2 (Yesil-Celiktas et al., 2007). Chang et al. (2008) maximized the yield of carnosic acid and carnosol of rosemary extracts by using 9.42 g g^{-1} CO_2/substrate (1 h extraction) at the highest tested temperature (80°C) and pressure (34.5 MPa) following 1 h of static extraction under the same conditions.

In fractional extraction, volatiles are fully extracted at a moderate pressure of 10–15 MPa and then antioxidants are extracted by increasing the solvation power of SC-CO_2. This increase in solvent power is effected by increasing the extraction pressure to 30–40 MPa (Ibáñez et al., 1999; Ivanovic et al., 2009), increasing the extraction temperature to 100°C (Ivanovic et al., 2009), adding 5%–10% of a polar modifier such as ethanol (Sánchez-Camargo et al., 2014; Vicente et al., 2013), or a combination of these factors. Extracting rosemary leaves with SC-CO_2 at 35°C–60°C and 8–15 MPa, Reverchon and Senatore (1992) found that the best extraction conditions were 10 MPa and 40°C with selective precipitation of unwanted cuticular waxes in a first separator operated at 8 MPa and 10°C and of volatiles in a second separator at 0°C and 2.5 MPa. The volatiles recovered in the second separator had an increased content of oxygenated monoterpenes, decreased content of monoterpene hydrocarbons, and a more intense aroma that resembled more closely the aroma of the original plant material than hydrodistilled oil (Reverchon and Senatore, 1992). Extracting other Lamiaceae herbs with SC-CO_2 at 40°C and 10–18 MPa, the group of Bernardo-Gil found that the extraction yield of summer savory (*Satureja hortensis* L.) did not increase at ≥12 MPa, that the extract obtained at 12 MPa had virtually no antioxidant activity, and that the residue from the SC-CO_2 extraction at 12 MPa had an increased antioxidant activity as compared to the substrate, possibly due to the removal of prooxidant essential oils and waxes (Esquivel et al., 1999). Ribeiro et al. (2001) reached similar conclusions when extracting lemon balm (*Melissa officinalis* L.); the residue of the SC-CO_2 extraction carried out at 10 MPa and 35°C had an increased phenol content and improved antioxidant activity as compared to the original substrate. Consequently, Ibáñez et al. (1999) dearomatized rosemary at 40°C and 10 MPa and Ivanovic et al. (2009) at 40°C and 11.5 MPa prior to extracting the antioxidant fraction under more intense conditions of 60°C and 40 MPa and 40°C or 100°C and 30 MPa, respectively.

In fractional separation, SC-CO_2 with a high solvent power (40°C–60°C, 30–35 MPa, and/or modified with 2%–7% ethanol) is used to extract rosemary and the antioxidant and volatile fraction are precipitated out in a cascade of two or more separators where pressure is reduced progressively. Señoráns et al. (2000) attempted extraction under different conditions with pure SC-CO_2 (50°C and 35 MPa) or modified with 2% ethanol (40°C and 30 MPa or 60°C and 35 MPa) and different fractional separation schemes using three stages. Authors found that the addition of ethanol coupled with the recovery of a fraction at high pressures (15–20 MPa) allowed for the separate collection of phenolic compounds with high antioxidant activity. Cavero et al. (2005) extended the work of Señoráns et al. (2000) by performing extractions at 40°C or 60°C and 15, 25, or 35 MPa using pure CO_2 or modified with 4% or 7% ethanol. In these experiments, the temperature and pressure in the first separator were set to the extraction temperature and 40% of the extraction pressure, and the temperature and pressure in the second separator were set at 20°C and 20 MPa. The antioxidant activity of extracts recovered in the first separator was higher when extractions where carried out using ethanol as the modifier and was related to the level of carnosic acid in the samples. In the experiments of Cavero et al. (2005), essential oils were extracted first and precipitated in the second separator but, when continuing extractions, some lower-solubility compounds having antioxidant activity were precipitated in the second separator.

Advantage can be taken also of changes in extract composition with time for the purpose of fractionation rosemary extracts. Indeed, when performing SC-CO_2 extraction with fractional separation, volatiles are extracted first and precipitated in the second separator. However, continuation of extraction causes the coprecipitation of some lower-solubility, antioxidant compounds also in the second separator. To test this, Vicente et al. (2012) carried out extractions at 40°C and 30 MPa for 5 h and effected separation in a first separator at 40°C and 10 MPa for a variable period between 0 h (no fractionation) and 5 h (continuous fractionation) and in a second separator at 40°C and 5 MPa (the recirculation pressure) obtaining the same total recovery but different compositions and antioxidant activities of the two fractions depending on the fractionation time. Based on these results, the group of CIAL (CSIC, UAM) in Madrid, Spain, attempted optimizing process conditions with extractions being carried out in a single stage or two stages using pure SC-CO_2 or modified with small amounts (≤7%) of ethanol at low (15 MPa) or high pressure (30 MPa) (Sánchez-Camargo et al., 2014; Vicente et al., 2012, 2013).

5.4.2 Extraction of Phenolics from Grape Peels, Seeds, Stalks, and Winery By-Products

There are many examples of the extraction of phenolic compounds from grapes (*Vitis vinifera* L.). In the particular case of the seeds, a key point in process development is obtaining separately the prooxidant oil (triglyceride) fraction from the meristematic cells and the antioxidant fraction from the seed testa. This work will not address the extraction of the oil, unless this is done as pretreatment of the seeds prior to extracting the antioxidant fraction. Only two works have attempted using pure SC-CO_2 to extract the antioxidant fraction. Agostini et al. (2012) extracted very low levels of procyanidins and phenolics using 41 g g^{-1} CO_2/milled seeds (1 h extraction) at 80°C and 25 MPa together with α-tocopherol and other antioxidant compounds. On the other hand, Rombaut et al. (2014) assisted the mechanical extraction of oil using CO_2. Gas-assisted mechanical expression relies on the displacement of oil by CO_2 during pressing, resulting in an increase in yield and improved coextraction of seed testa polyphenols (Rombaut et al., 2014). Separation processes for grape phenolic antioxidants are summarized in Table 5.4.

TABLE 5.4
Supercritical Extraction Conditions from *Vitis vinifera* L. (Grape) Using CO_2

Pretreatment	Particle Size (mm)	*T* (°C)	*P* (MPa)	Time Extraction/ Static Time (min)	Specific Solvent Rate (kg CO_2/L·h)	Modifier	References
Skin		40	15	15/0.5		7.5% EtOH	Pascual-Martí et al. (2001)
Skin		60	25	15/3		20% EtOH	Chafer et al. (2005)
Ground peel (*Vitis labrusca* B.)	0.5	37–46	14–17	30		5%–8% EtOH	Ghafoor et al. (2010)
Ground seed		35	45.6	–/15		20% MeOH	Palma et al. (1999)
Crushed seed				180/20		0.5 mL/min MeOH	Palma and Taylor (1999)
Crushed seed		80	65.5	60	11.22	30%–40% MeOH	Ashraf-Khorassani and Taylor (2004)
Milled defatted seed (from seed oil industry)	0.310–0.681	40	20–30			2%–15% EtOH	Murga et al. (2000)
Crushed seed	0.23–0.425	30 and 50	25–30	60	1	5%–20% EtOH	Yilmaz et al. (2011)
Ground seed	0.32–0.84	80	25	60			Agostini et al. (2012)
Ground seed (*Vitis labrusca* B.)	<0.5 mm	37–46	13.7–16.7	30		5%–8% EtOH	Ghafoor et al. (2012)
Ground seed	0.375	35–60	8–15		4.76		Marqués et al. (2013)
Ground seed	0.67	50–60	15–30	240		12.5%–17.5% EtOH	Oliveira et al. (2013)
Ground seed	0.6–1.18	75–120	23–53.8	120	2.5–8.5		Rombaut et al. (2014)
Ground marc		50	35	15		5% MeOH	Arce et al. (2001)
Ground marc		50	35	15		10% H_2O	Palenzuela et al. (2002)
Ground marc		50	35	15		3% MeOH	Palenzuela et al. (2004)
Ground seed and other waste materials by hand		40–60	10–50	60–120	3.17–5.81	5%–20% mol H_2O; MeOH	Mantell et al. (2003)

(Continued)

TABLE 5.4 (*Continued*)
Supercritical Extraction Conditions from *Vitis vinifera* L. (Grape) Using CO_2

Pretreatment	Particle Size (mm)	T (°C)	P (MPa)	Time Extraction/ Static Time (min)	Specific Solvent Rate (kg CO_2/L·h)	Modifier	References
Ground pomace (wine industry by-products)		45	10–25		2.75	5% MeOH	Louli et al. (2004)
Skins, seeds, and small amount of stems		35–50	8–35	60		0%–8% EtOH	Pinelo et al. (2007)
Milled marc (composed of skins, seeds, and stalks)	0.35–0.55	45	20–50	200/30	9.6	0.90 mL/min EtOH	Fiori et al. (2009)
Pulverized whole grape pomace (seeds, stems, and skin)	0.165–0.319	35–55	10–40	180	4.80	5% EtOH	Casas et al. (2010)
Supercritical extract		25–30	10–13				Bleve et al. (2008)
Milled marc							Vatai et al. (2008)
Residues lyophilized		40	20	90	2.68	0.7 mL/min MeOH	Floris et al. (2010)
Ground bagasse	0.83	40	20–35	300/10	0.85	10% EtOH	Farías-Campomanes et al. (2013)
Ground marc	0.83	40–60	10–28	180	6–10	15% EtOH, H_2O	Da Porto et al. (2014a)
Ground marc	0.83	40	8–30	180	4–6	7.5%–15% EtOH, H_2O	Da Porto et al. (2014b)
Ground marc	0.83	45	28	180	10		Da Porto et al. (2015)
Ground wine waste		35	25 and 10.3	5		40% MeOH	Aizpurua-Olaizola et al. (2015)

Seminal work on the fractional extraction of grape seeds is due to Prof. Taylor's group in Virginia Polytechnic Institute and State University focusing on analytic aspects. Palma et al. (1999) studied the effect of system temperature (35°C or 55°C), CO_2 density (850 or 950 kg/m^3; pressure ranged from 18.2 MPa at 35°C for the lowest density to 50.6 MPa at 55°C for the highest density), and the concentration (10% or 40%) and type (methanol or ethanol) of polar modifier on the recovery and reproducibility of the extraction of catechin for ground grape seeds. The best extraction conditions were 55°C and 50.6 MPa (CO_2 density = 950 kg/m^3) using CO_2 modified with 10% methanol, which resulted in higher recoveries of catechin than those of solid–liquid extractions or sonication-assisted solid–liquid extraction using aqueous methanol (20%) solutions. The authors used 667 g g^{-1} CO_2/substrate during dynamic extraction (21–23 min depending on CO_2 density) following 20 min of static extraction. Palma and Taylor (1999) extracted fatty acids, aliphatic aldehydes, and sterols using 16.7 g g^{-1} CO_2/ground seeds and then extracted GA, catechin, EC, and other phenolic compounds using 16.7 g g^{-1} CO_2/ground seeds modified with 16.7% methanol. Extraction conditions at both stages were 35°C and 45.6 MPa and followed 15 min static extraction under the same conditions employed for dynamic extractions. Ashraf-Khorassani and Taylor (2004) extracted oil at 80°C and 65.5 MPa using 22.4 g g^{-1} CO_2/substrate of pure SC-CO_2 (60 min), then extracted monomeric polyphenols at 80°C and 65.5 MPa using 22.4 g g^{-1} CO_2/substrate of CO_2 modified with 40% methanol (60 min), and finally extracted polyphenolic dimers/trimers using pressurized methanol at 80°C and 13.8 MPa (30 min static extraction followed by 10 cm^3 solvent flush, and 5 min of CO_2 purge at 13.8 MPa).

Sometimes, SC-CO_2 extraction conditions selected for analytical purposes, such as those proposed by Prof. Taylor and his coauthors (Ashraf-Khorassani and Taylor, 2004; Palma et al., 1999; Palma and Taylor, 1999), do not find counterparts for industrial applications because they rely on extreme solvent-to-substrate ratios, pressures, and/or modifier addition and/or questionable modifier selection (e.g., toxic methanol instead of environment-friendly ethanol). Another example of this is the two-step procedure recommended by Aizpurua-Olaizola et al. (2015) for the separate recovery of oils (5 min extraction at 35°C and 25.3 MPa using 2 cm^3/min of pure SC-CO_2) and polyphenols (5 min extraction at 35°C and 10.3 MPa using 2 cm^3/min of CO_2 modified with 40% methanol) from 0.25 g samples of seeds or mixtures of seeds and skins from winery waste. Another problem when applying the results in the literature relates to the interpretation of the results from statistical analysis when using a small or impractical experimental region, which sometimes occurs when applying response surface methodology. As an example, Ghafoor et al. (2012) studied the effects of extraction temperature (37°C–46°C), extraction pressure (13.7–15.7 MPa), and amount of modifier added (5%–8% ethanol) on the yield, phenol content, and antioxidant activity of ground grape (*Vitis labrusca* B.) seed extracts. In these experiments, 3 g samples were extracted 30 min using 2 cm^3/min of CO_2. All responses improved with extraction temperature and pressure, as expected, but did not improve significantly as the amount of ethanol used increased, which was unexpected.

Murga et al. (2000) studied the effects of extraction pressure (20 or 30 MPa) and modifier type (methanol or ethanol) and concentration (2%–15%) on the yield and composition of SC-CO_2 extracts from a commercial concentrate of complex phenols and tannins and defatted grape seeds. Samples of 3 g were extracted with 0.5 cm^3/min of modified CO_2 for 3 h. As expected, they observed that the solvent power of SC-CO_2 increased with extraction pressure and the amount of added modifier so that as the polarity and molecular weight increased, a higher pressure and/or percentage of modifier were required for the particular phenolic compound to be extracted.

Yilmaz et al. (2011) studied the effect of extraction temperature (30°C or 50°C), extraction pressure (25–30 MPa), and added ethanol cosolvent (5%–20% v/v) on the SC-CO_2 extraction of different phenolics from crushed grape seeds. All extractions were extended for 1 h. Yilmaz et al. (2011) found that it was possible to fractionate the extract by varying the solvent power of SC-CO_2 mainly by increasing system pressure and concentration of ethanol as the cosolvent; the compound first extracted is EC gallate (ECg), followed by catechin and EC and finally GA, EGC, and EGC gallate (EGCg).

Mantell et al. (2003) compared water and methanol as modifiers for the SC-CO_2 extraction of anthocyanins from red grape skin waste from a vinification process. The authors studied the effects of extraction temperature (40°C or 60°C), extraction pressure (10 or 50 MPa), solvent flow rate (12 or 22 mmol/min), modifier type and application level (5% or 20%), and extraction time (1 or 2 h) for extraction yield of anthocyanins from 3 g ground samples. All variables with the exception of pressure had a significant ($p < 0.05$) positive effect on the extraction yield of anthocyanins, whereas pressure had a significant negative effect, and methanol performed significantly better than water as a modifier. Consequently, recommended extraction conditions were 2 h extraction at 60°C and 10 MPa using 22 mmol/min of CO_2 modified with 20% methanol (Mantell et al., 2003). In a related study, Casas et al. (2010) extracted resveratrol from dried and ground skins, pomace, seeds, and stems from white grape pomace using pure and ethanol-modified SC-CO_2. Samples of 4.6 g were extracted using 0.8 g/min of CO_2 for 3 h. The resveratrol content was largest in the skins, intermediate in the pomace, lower in the seeds, and lowest in the stems. The yield increased as pressure increased from 10 to 40 MPa, and as a result of the addition of 5% ethanol as a modifier, with no resveratrol being extracted using pure SC-CO_2 at 10 MPa. The resveratrol yield decreased slightly when temperature increased from 35°C to 55°C. Consequently, recommended extraction conditions were 3 h extraction at 35°C and 40 MPa using 0.8 g/min of CO_2 modified with 5% ethanol (Casas et al., 2010).

Pinelo et al. (2007) studied the effect of extraction temperature (35°C or 50°C), extraction pressure (8 or 35 MPa), and modifier concentration (0% or 8% azeotropic ethanol) on the SC-CO_2 extraction of distilled white grape pomace (55%–60% moisture) in a pilot plant with two separators set at 35°C and 60% of the extraction pressure (4.5 or 21 MPa, respectively) and 25°C and 1 MPa, respectively. Extraction yield was independent on extraction temperature and increased markedly when increasing extraction pressure and using azeotropic ethanol as a modifier. Extracts obtained at 35 MPa using CO_2 modified with 8% azeotropic ethanol contained GA, catechin, and EC, unlike solid–liquid extracts obtained using azeotropic ethanol that contained more proanthocyanidins.

Fiori et al. (2009) attempted extracting polyphenols from grape skins and seeds using ethanol-modified SC-CO_2. Dried and ground skins (53–55 g) were extracted at

45°C and 20–50 MPa using 85–94 g CO_2/substrate (5 h) containing 54–65 cm^3 kg^{-1} ethanol/CO_2. As expected, yield increased with extraction pressure. Grape seeds were extracted by combining SC-CO_2 extraction at 40°C and 50 MPa using pure CO_2 or ethanol-modified CO_2 with liquid ethanol extraction. SC-CO_2 extraction was ineffective for extracting polyphenolic compounds; it can be anyway utilized to selectively extract grape seed oil by avoiding any solvent contamination of the matrix that can be further extracted for the recovery of polyphenolic compounds by means of a liquid organic solvent.

There are several works on the SC-CO_2 extraction of grape marc using aqueous ethanol as a modifier led by Prof. Da Porto from the University of Udine. Grape marc from white grapes was dried and ground (d_p = 0.83 mm, mean Sauter). Da Porto et al. (2014a) removed the oil from 0.48 kg of ground marc using 75 g g^{-1} CO_2/substrate (3 h) at 45°C and 28 MPa. The phenolic fraction from 0.10 kg of ground, deoiled marc was subsequently extracted at 40°C–60°C, 10 or 20 MPa using 290–297 kg kg^{-1} CO_2/substrate (5 h) modified with 15% water or azeotropic ethanol as the modifier. Extraction rate and yield were not affected largely by extraction temperature but decreased with extraction pressure, presumably due to pressure-gradient-induced agglomeration of the substrate. The best results were obtained by consecutively extracting the deoiled marc for 5 h using SC-CO_2 modified with 15% water at 40°C and 10 MPa and then for an additional 5 h using SC-CO_2 modified with 15% azeotropic ethanol at 60°C and 10 MPa, and then combining the two extracts (Da Porto et al., 2014a). This work was extended by Da Porto et al. (2014b) who attempted the SC-CO_2 extraction of ground, deoiled marc using aqueous 57% ethanol solution as the modifier. Because the extraction rate and yield of polyphenols in experiments carried out at 40°C increased as extraction pressure decreased from 30 to 8 MPa, subsequent experiments were performed at 40°C and 8 MPa to minimize agglomeration of the substrate in the packed bed. The authors studied the effect of mass flow rate of CO_2 (66.7 or 100 g/min) and percentage of the modifier in CO_2 (7.5%–15%). Da Porto et al. (2014b) found that polyphenol yield increased under conditions where the molar fraction of CO_2 was lowest (66.7 g/min of SC-CO_2 modified 7.5% aqueous ethanol or 100 g/min of SC-CO_2 modified with 10% aqueous ethanol). Da Porto et al. (2015) combined ultrasound-assisted extraction of ground, deoiled marc with aqueous 57% ethanol solution (4 or 10 min at 20°C or 80°C) with extraction using 100 g/min of CO_2 modified with 10% aqueous ethanol at 40°C and 8 MPa. The yield and antioxidant activity of the extracts was maximal when combining 4 min of ultrasound-assisted extraction at 80°C with SC-CO_2 extraction using CO_2 modified with aqueous ethanol.

Louli et al. (2004) studied an alternative process for red grape pomace consisting of conventional solvent extraction followed by the concentration of phenolics in the extract by removal of nonphenolics using pure SC-CO_2. Conventional extraction was done using aqueous 3% KOH solution (static extraction at 18°C–20°C under N_2 atmosphere), methanol (Soxhlet extraction at ca. 60°C), and ethyl acetate (Soxhlet extraction at ca. 75°C). The process was scaled-up using a stem-free, crushed substrate. Extraction was done at <30°C using 0.36 m^3 of ethyl acetate (the best solvent) that percolated a packed bed with 200 kg of the substrate and was continuously recirculated (residence time = 3 h) for a period of 16 h. The extract was dried by vacuum (40–50 mmHg) evaporation. Thirty grams of dried ethyl acetate extract was treated using 200 g g^{-1} CO_2/substrate at 45°C and 10–25 MPa (5.5 h) in a pilot device equipped with two separators operating at 50°C and 8 MPa and 50°C and 2–4 MPa, respectively. This treatment increased the content of phenolics and antioxidant activity of the extracts by removing fatty material, with an associated improvement in the organoleptic properties of the product. The amount removed was higher when processing the ethyl acetate extract of a substrate containing stems. There was no need to add 5% methanol as a modifier or to increase pressure above 15 MPa for these purposes.

Other process schemes include the SC antisolvent-assisted removal of conventional solvents to obtain polyphenol-enriched powders carried out in Prof Reverchon's group in Italy (Floris et al., 2010; Marqués et al., 2013). Vatai et al. (2008) formulated a conventional grape seed extract by applying it on a solid carrier using the so-called high-pressure concentrated form technique. These processes are described more in detail in Section 5.5.4.

5.4.3 Extraction of Phenolics from Tea

Another interesting case study is the enrichment of tea on flavonoids by selective decaffeination of tea leaves, which depends critically on differences in solubility in SC-CO_2 between caffeine and tea flavonoids depending on system temperature and pressure and type and concentration of the modifier. Separation processes for tea phenolic antioxidants are summarized in Table 5.5.

Seminal work on this subject was performed in Prof. Chang's group in National Chung-Hsing University (Taichung, Taiwan). Chang et al. (2000) studied the extraction of caffeine, main tea catechins (EC, EGC, ECg, and EGCg), and GA from green tea using 2.6–10 g/min of SC-CO_2 at 50°C and 31 MPa. The authors extracted 90 g tea powder, mixed in some experiments with 90 cm^3 of different static cosolvents: water and aqueous ethanol (18%–99.8%). The extract was absorbed in aqueous ethanol (50%) at 28°C and 5 MPa. Pure SC-CO_2 extracted virtually no GA or tea catechins whose content increased when using a static modifier and when increasing the ethanol content in it up to 95%. On the other hand, caffeine could be extracted using pure SC-CO_2 and extraction yield increased with the ethanol content in the static modifier up to 70%. Consequently, the ratio of total polyphenols to caffeine in the extract increased from 0.73 when using pure SC-CO_2 to a maximum of 1.50 when using 95% ethanol as a static modifier. This work was extended by Chang et al. (2001) for oolong tea for which the polyphenol-to-caffeine ratios were smaller. Better yields and selectivities were obtained by Soxhlet extraction when using water and specially 95% ethanol (Chang et al., 2000, 2001). Huang et al. (2007) updated these two works including the use of a continuous modifier and the optimization of extraction conditions to simultaneously increase caffeine yield and reduce catechin coextraction. The authors did preliminary work to optimize conventional extraction conditions (stirred extraction with boiling water) as well as to select the SC-CO_2 decaffeination conditions. Specifically, they determined caffeine extraction rate as a function of system temperature and pressure and caffeine removal and catechin retention as a function of substrate particle size, amount and mode of addition of the

TABLE 5.5
Supercritical Extraction Conditions from Ground *Camellia sinensis* L. Leaves (Green Tea) Using CO_2

Particle Size (mm)	T (°C)	P (MPa)	Time Extraction/ Static Time (min)	Specific Solvent Rate (kg CO_2/L·h)	Modifier	References
	60	31			10% EtOH, H_2O	Chang et al. (2000)
	24–60	31–50			18%–98% EtOH	Chang et al. (2001)
0.074–0.177	40–60	20–30			0.56% EtOH, H_2O	Huang et al. (2007)
						Perva-Uzunalic et al. (2006)
0.425–0.710	70	30	51	6.25	2.3% EtOH	Lee et al. (2007)
0.5437	50–80	15–30	120	5.10	4.6 g/100 g CO_2 EtOH	Park et al. (2007a)
0.425–0.710	70	30	120	5.10		Park et al. (2007b)
0.52	40–80	20–40	300		4%–7% H_2O	Kim et al. (2008)
	50	15.1	15			Ligor et al. (2008)
0.745						Meterc et al. (2008)
0.4–1.6	40–80	30	60–180		5–30 mL EtOH, H_2O	Sun et al. (2010)
0.507–0.841	40–60	10–30	150	2.41–7.52	1 mL/2 g Substrate EtOH	Ghoreishi and Heidari (2012)
	40–60	10–30	40–120	1.10–11.62		Ghoreishi and Heidari (2013)
0.50–0.84	40–60	10–30	40–120	1.10–11.62	1 mL/2 g Substrate EtOH	Heidari and Ghoreishi (2013)
0.42	40–60	10–20	60	9.60	1–3 g/min EtOH	Maran et al. (2013)

water modifier, system temperature and pressure, and CO_2-to-substrate ratio. Based on these results, they defined the optimization protocol using response surface analysis. The authors conditioned 8 g of green tea powder (particle diameter = 74 μm) with 2 g of water for 30 min and made extraction with 7.1 g/min of CO_2 containing 0.56% water. Extractions were done using 60–100 g g^{-1} CO_2/substrate (67.5–112.5 min) at 40°C–60°C and 20–30 MPa. Optimal extraction conditions were 80 g g^{-1} CO_2/substrate (90 min) at 50°C and 27.5 MPa that resulted in the extraction of 71.9% of caffeine while retaining 67.8% of catechins in the treated tea.

There have been several works on tea decaffeination led by Prof. K.H. Kim from Korea University and Prof. Y.S. Kim from Ewha Womans University, both in Seoul. Park et al. (2007a) studied the effects of sample particle diameter, CO_2 mass flow rate, and extraction temperature on the relative extraction of caffeine and catechins from ground or intact green tea leaves. The authors mixed the tea (10 g) with a static modifier (95% ethanol) and interrupted extraction every 40 min to replace the modifier. The amount of added modifier was kept constant on a per-CO_2 basis: 46 g kg^{-1} aqueous ethanol/CO_2. As an example, in experiments using 8.5 g/min of CO_2, the authors complimented the substrate with 15.6 g portions of 95% ethanol at the beginning and following every interruption of the extraction process. Total extraction time was 2 h. In extractions at 30 MPa, caffeine yield increased as temperature increased from 50°C to 80°C, and in extractions at 70°C caffeine yield increased as pressure increased from 15 to 30 MPa. Caffeine yield also increased slightly but caffeine concentration in CO_2 decreased markedly as the mass flow rate of CO_2 increased from 5.5 to 14.5 g/min. Caffeine yield decreased slightly as the particle diameter of ground tea increased from 237 or 544 μm, but the reduction was substantial when using unground tea leaves (81%). Caffeine yield was above 90% when extracting ground tea using ≥8.5 g/min of SC-CO_2 at ≥70°C and ≥20 MPa. Caffeine yield during extraction of 237 μm particles using 11.5 g/min of SC-CO_2 at 70°C and 30 MPa increased from 76% after 40 min, to 94% after 80 min, and to 98% after 120 min. Unfortunately, catechins concurrently extracted with caffeine depending on their comparative solubilities: EC = 48%, EGCg = 71%, ECg = 77%, and EGC = 81%. When using unground tea leaves, caffeine extraction rate decreased but that of catechins decreased even further with a resulting improvement in the selectivity of the decaffeination process. Park et al. (2007a) however claimed that further improvements were required to improve the selectivity of tea decaffeination. Park et al. (2007b) extended the previous work to consider also the effect of the type (water or 95% ethanol) and concentration of the cosolvent (29–88 g kg^{-1} modifier/CO_2) of modifier on the selectivity of the decaffeination process. The authors included also pure SC-CO_2 as a reference solvent. The authors found that caffeine yield increased with the amount of added modifier. At the same time, catechin retention decreased with the amount of added modifier. Caffeine yield was higher when using 95% ethanol than water, whereas differences in retention of catechins between the two modifiers depended on their polarity. Retention increased for the less polar EC, EGCg, and ECg and decreased for the more polar ECg when replacing water by 95% ethanol. Park et al. (2007b) concluded that 95% ethanol was more selective than water for tea decaffeination. Lee et al. (2007) studied the effect of the decaffeination processes suggested by Park et al. (2007a,b) on the aroma of tea. For that, they mixed

1.88 kg of ground green tea with different amounts of 95% ethanol (static modifier) so as to reach 23 (sample I) or 70 g kg^{-1} 95% ethanol/CO_2 (sample II) and extracted it with 1.25 kg/min of SC-CO_2 at 70°C and 30 MPa. Every 51 min extraction was interrupted to add an amount of static modifier equivalent to that added initially. The total extraction time was 204 min (four stages; 135 kg kg^{-1} CO_2/substrate). In addition, a sample treated with 70 g kg^{-1} 95% ethanol/CO_2 was decaffeinated for 714 min (14 stages; 135 kg kg^{-1} CO_2/substrate) (sample III). The levels of decaffeination were 38.6%, 64.8%, and 87.4% for samples I, II, and III, respectively. Lee et al. (2007) found that most volatile components decreased to an extent that depended on the level of decaffeination up to the point of being completely removed in sample III, particularly nonpolar terpenes such as linalool and phenylacetaldehyde and floral compounds such as hexanal and (*E*)-2-hexenal having a large influence on the characteristic aroma of green tea.

Kim et al. (2008) attempted also to optimize the selectivity of tea decaffeination process by increasing the caffeine-to-EGCg ratio (β) in the treated sample. To do that, they extracted a 50 g sample of ground green tea with its initial moisture content adjusted to 4%–7% (water as the static modifier) using 24 g/min of SC-CO_2 at 40°C–80°C and 20–40 MPa for 5 h. Extraction yields of caffeine and EGCg of a sample containing 7% water increased as temperature increased from 40°C to 80°C at 40 MPa, and as pressure increased from 20 to 40 MPa at 50°C. They increased also as moisture increased for 4%–7% at 50°C and 50 MPa. The selectivity reached a maximum of 2.57 when extracting the sample adjusted to 7% water at 40°C and 50 MPa, which was higher than performing extraction in an agitated flask using water ($\beta = 0.88$) or ethanol ($\beta = 0.24$). Kim et al. (2008) attempted to explain the optimal decaffeination conditions by computing solubility parameters for caffeine, EGCg, and water-modified CO_2 as a function of extraction conditions but their attempt was flawed in that they did not consider the actual amounts of water dissolved in SC-CO_2 during extraction.

There are a couple of works on tea decaffeination that emphasize statistical aspects. Sun et al. (2010) compared the performance of water, ethanol, and aqueous 50% ethanol solution as the continuous modifier (93 cm^3 kg^{-1} modifier/CO_2) for the selectivity of the decaffeination of a 10 g sample of ground green tea with 2.76 g/min of SC-CO_2 at 60°C and 30 MPa. Residual levels of caffeine were below 1% in all cases, but residual catechins were three to four times higher in the sample treated with ethanol (15.6%) than the other modifiers. Thus, the authors selected ethanol as the continuous modifier despite its reduced capacity to extract caffeine from tea. Sun et al. (2010) then studied the effect of particle diameter (0.5–1.6 mm), extraction temperature (40°C–60°C), extraction time (1–3 h), and the amount of added modifier (5–30 cm^3) on the extraction of caffeine and retention of EGCg and catechins when extracting 10 g samples with 2.76 g/min of CO_2 at 30 MPa. All independent variables affected ($p < 0.05$) caffeine extraction significantly but only particle diameter affected ($p < 0.05$) catechin retention significantly. Optimal decaffeination conditions were small particles (0.4 mm diameter), high temperature (80°C), intermediate extraction time (2 h), and intermediate amount of modifier (15 cm^3, corresponding to 47 cm^3 kg^{-1} modifier/CO_2), for which decaffeination reached 70% and catechin retention reached 94%. Maran et al. (2013) used response surface methodology to analyze the effect of extraction temperature (40°C–60°C), extraction pressure (10–20 MPa), and modifier mass flow rate (1–3 g/min of ethanol) on the phenolic content, flavonoid content, tannin content, and antioxidant activity of ground tea leave (>mesh) extracts. The authors did not measure the caffeine content of treated leaves. In the experiments, 5 g samples were extracted with 8 g/min of CO_2 for 1 h (96 g/g CO_2/substrate, 11.1%–27.3% ethanol). This is an experiment with low extraction pressures and a high level of modifier. The authors determined an optimal extraction condition that validated experimentally.

The focus of Perva-Uzulanic et al.'s (2006) work was the extraction of catechins using conventional extraction as a function of the solvent (aqueous solutions with a different concentration of acetone, methanol, ethanol, or acetonitrile, or pure) and extraction temperature (boiling or controlled temperature in the case of pure water), but they included SC-CO_2 extraction using water as a modifier (sample moistened to 15%–50%, extracted with wet CO_2 at 50°C–80°C and 22.5–35.0 MPa, and air-dried in an oven at 40°C to 4% water) as a pretreatment for the selective extraction of caffeine.

5.4.4 Classification of Applications

The review of the three examples above highlights some general features about the extraction of plant polyphenols using SC-CO_2. Although pure SC-CO_2 has shown little promise for the direct extraction of these solutes from solid substrates, because of their limited solubility, particularly of the most interesting high-molecular-weight and polar dimer, trimer, and glucoside types, SC-CO_2 can be used as part of multiple unit approaches. This includes the removal of volatile compounds responsible for the characteristic aroma of rosemary and other Lamiaceae herbs, tea, and vegetable substrates, vegetable oils in grape seeds and other oilseeds, and alkaloids and other toxic or questionable metabolites such as caffeine from tea, thus enriching the residue in desirable phenolic compounds. Alternatively, the solvent power of SC-CO_2 can be increased by supplementing it with a polar modifier that for all practical purposes can be limited to ethanol, water, and mixtures of the two. Based on similar observations, Díaz-Reinoso et al. (2006) classified SC-CO_2 extraction processes for plant polyphenols in four types according to two criteria, namely, the solvent power of SC-CO_2 (low or high) and the order in which SC-CO_2 was applied (first or second).

Low- to intermediate-power SC-CO_2 can be used to remove prooxidant essential oils and triglycerides or other undesirable compounds such as caffeine. It can be applied first to produce a solid raffinate that can be treated then with other solvents to produce an extract that is enriched with plant phenolics. The other (second) solvent can also be SC-CO_2, but in this case, it should have a very high power, which demands high concentrations of polar modifiers. This is difficult in the case of water, because of its limited solubility in SC-CO_2. Prospects are better for ethanol because it is completely miscible with CO_2 at pressures above ca. 12 MPa at 60°C (or ever a lower pressure at a lower temperature) (Gutiérrez et al., 2010), but ethanol is less polar than water (King, 2014).

Low- and intermediate-power SC-CO_2 can also be applied in second to remove the prooxidants or other undesirable compounds from a conventional extract, thus enriching it in the required plant polyphenols. The conventional extract can be initially in a liquid or solid state, depending on the application of a complimentary desolventizing process. If it is in the solid state, the process will resemble a SC-CO_2 extraction (Section 5.5.1). This process will be simpler in that no additional steps will be required to separate higher-solubility solutes and conventional solvents that are typically very soluble in CO_2 (an exception would be aqueous extracts). The case of liquid extracts allows alternative high-selectivity separation processes such as CounterCurrent (CC) extraction (Section 5.5.2) or preparative chromatography (Section 5.5.3). These two processes result in a liquid product. However, as a final alternative, the process could result in a solid product if the main purpose were the removal of the solvent and volatile compounds rather than the enrichment of the extract (or raffinate) in specific compounds. This is reviewed in Section 5.5.4 that is devoted to the production of powders.

Readers interested in the multiple unit processing of plant material using platforms incorporating SC-CO_2 as well as an alternative solvent can refer to the reviews of King and Srivinas (2009, 2015) and King (2014).

5.5 SEPARATION PROCESSES USING SUPERCRITICAL CO_2

5.5.1 Batch Extraction from Solid Substrates

As presented in Section 5.4, most examples of application of SC-CO_2 technologies to the isolation of polyphenols from biologic matrices relate to extraction from solid substrates. Because of the technical difficulties of continuously feeding or retiring solid particles from a high-pressure vessel, SC-CO_2 extraction processes of solid substrates are typically carried out batchwise, using porous baskets with a packed bed of substrate particles and quick-opening extraction vessels. Industrial plants typically use several extraction vessels that can be operated using a simulated CC flow pattern of SC-CO_2 (del Valle, 2015). Exceptions include the use of intermediate-pressure blow cases (Lack and Seidlitz, 1993) and extruders (Eggers, 1996) both to feed fresh substrate and to retrieve exhausted substrate continuously from an extraction vessel, but to the best of the authors' knowledge, patents for blow cases and extruders have not been applied commercially. The batch extraction of solid substrates is the most common application of SC-CO_2 as a separating agent; it is dealt with in book chapters of authoritative books on SCF separation processes (Brunner, 1994; Clifford, 1999); book chapters on applications in food engineering (Dunford et al., 2003), processes, components, and costs of industrial plants (Martínez and Vance, 2008), principles and applications to the food, beverage, and nutraceutical industries (Knez et al., 2010); and book sections on extraction from solids (Lack et al., 2001), principles of separation processes using SCFs (Nagahama, 2002; Rosa and Meireles, 2009), and extraction of natural products (Goto, 2002). Very comprehensive reviews on the extraction of solid substrates, published recently by Sovová and Stateva (2011) and de Melo et al. (2014), can be consulted by the interested reader. This section will review the effect of substrate pretreatment and extraction conditions on the SC-CO_2 extraction of plant polyphenols, the modeling of the process, and the application of models for process simulation, plant design, and economics.

The SC-CO_2 extraction rate and yield of a solid substrate is affected by the substrate and its pretreatment, on one hand, and the conditions of the SC-CO_2 including system temperature and pressure, type and concentration of added polar cosolvent, and velocity. The experimental results are typically plotted as a cumulative extraction curve presenting the yield of a solute or solute mixture (extract) *versus* extraction time or specific solvent consumption (mass of CO_2 per unit mass of a solid substrate). Cumulative extraction curves are typically simple with an initial straight portion (constant extraction rate) and then declining progressively up to reaching a horizontal asymptote representing the content of solute or solute mixture (extract) in the substrate that can be extracted by SC-CO_2 under the selected experimental conditions.

An important operating condition that is generally improperly analyzed is the velocity of CO_2 flowing through the extraction vessel, typically characterized by the superficial velocity (U), or the velocity of the CO_2 stream when flowing through an empty vessel with the same inner diameter as the packed bed. The effect of U on extraction rate is improperly analyzed when the flow rate (Q) of CO_2 is used as an independent variable instead, because U depends on Q and the volume (V_E) of the extraction vessel (the larger the vessel, the smaller the U for any given Q). Because Q can correspond to mass flow rate instead of volumetric flow rate (volume of CO_2 per unit time), U may also depend on the extraction conditions (temperature and pressure) that affect CO_2 density (ρ), as informed in the following equation:

$$U = \sqrt[3]{\frac{4}{\pi}\left(\frac{H/D}{V_E}\right)^2 \cdot \frac{Q}{\rho}}, \qquad (5.6)$$

where H/D represents the height-to-diameter ratio of the extraction vessel. Another source of confusion is that some reports inform the volumetric flow rate of expanded CO_2 (Q_{VE}) at the exit of a one-pass laboratory (or benchtop) unit, which has no relationship to the flow condition inside the extraction vessel. Considering that expanded CO_2 may typically reach normal conditions (25°C and 1 atm), Q relates to Q_{VE} according to the following equation:

$$Q = 1.808 \cdot Q_{VE}, \qquad (5.7)$$

where

$[Q]$ = g/min

$[Q_{VE}]$ = NL/min

ρ = 1.808 g/L is the density of CO_2 at 25°C and 1 atm

To make the results independent of the size of the extraction system, we decided to inform the specific solvent rate (in kilograms of CO_2 per liter of empty extraction vessel and per hour) in Tables 5.3 through 5.5. The specific solvent rate gives an idea about the superficial CO_2 velocity without requiring corrections for the inner diameter of the extraction vessel (which, the same as the H/D ratio, is not always informed) or the density of CO_2 under extraction conditions (which may exhibit experiment-to-experiment variations).

An increase in U has a positive effect on the SC-CO_2 of both essential oil from sandarac tree (*Tetraclinis articulata*) at 40°C and 9 MPa (Figure 5.4a) (Herzi et al., 2013) and oil from turmeric (*Curcuma longa*) at 40°C and 30 MPa (Figure 5.4c) (Gopalan at al., 2000), which may be due to either an increase in the amount of CO_2 passed through the extraction vessel or an improvement in the external (film) mass-transfer coefficient (k_f) across the static film (boundary layer) of CO_2 next to the substrate particles. To establish the nexus between improvements in extraction rate and improvements in external mass transfer, it is better to plot cumulative extraction yield against specific CO_2 consumption (Figure 5.4b and d) than time (Figure 5.4a and c, respectively). When extraction rate is defined by equilibrium (solute or extract solubility) rather than kinetic (external mass transfer) factors, the initial slope of cumulative extraction plots of extract yield *versus* specific CO_2 consumption, which represents the "apparent" solubility of the extract in SC-CO_2 under process conditions (g kg^{-1} extract/CO_2) (del Valle and Urrego, 2012), does not change (cf. Figure 5.4a). Increasing U increases k_f, thus increasing extraction rate, but only when external mass transfer influences it to some extent, such as in the initial stages of the extraction of small particles of a substrate having a large specific surface (surface-to-volume ratio) and containing a large amount of free, high-solubility solute (e.g., essential oil in sandarac tree). Using large values of U is not justified when extraction is internally

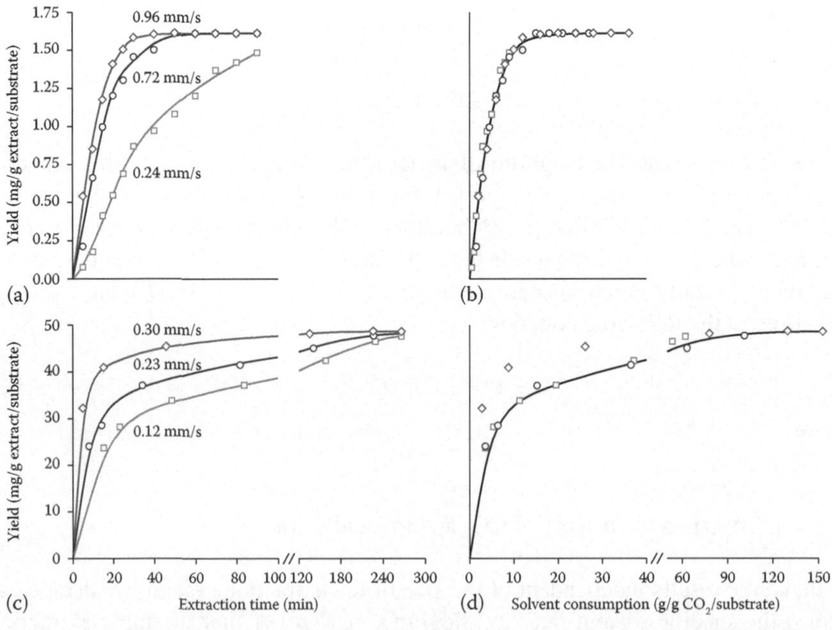

FIGURE 5.4 Effect of supercritical CO_2 superficial velocity (U, noted values in the graphs) on the extraction yield as a function of (a, c) extraction time or (b, d) solvent consumption for (a, b) ground *Tetraclinis articulata* leaves treated with CO_2 at 40°C and 9 MPa. (From Herzi, N. et al., *J. Supercrit. Fluids*, 82, 72, 2013.) (c, d) Ground turmeric (*Curcuma longa*) rhizomes treated at 40°C and 30 MPa. (From Gopalan, B. et al., *J. Agric. Food Chem.*, 48, 2189, 2000.)

controlled (e.g., when extracting large particles of a densely compacted substrate having small specific surface). There are practical limits to the reduction in U because values of k_f during SC-CO_2 extraction of high-solubility solutes, under solvent upflow conditions, and using low values of U can diminish to very small levels due to the negative influence of natural convection phenomena as illustrated by Germain et al. (2005) for the extraction of essential oils and triglycerides from vegetable substrates.

Equilibrium and/or kinetic factors determine the effect of extraction conditions on extraction rate and yield depending on the binding of a specific solute and the components of a complex extract to the solid matrix, the solubility of this solute or solute mixture (extract), and the mass transport properties. Little is known about the effect of CO_2 conditions on sorption phenomena affecting the binding of a solute or solute mixture (extract) to the solid matrix (Urrego et al., 2015). In experiments where both the substrate and CO_2 velocity remain unchanged, the internal (solid phase) and external (fluid phase) mass-transfer rates depend solely on the physical properties of the loaded SC-CO_2 phase. Extraction T and P have a pronounced effect on the high-pressure phase equilibrium (Section 5.3) and the transport properties (viscosity, diffusivity) of the SC-CO_2 phase. Extraction rates may increase as the solubility of the solute or its diffusivity in SC-CO_2 (D_{12}) increases depending on the controlling mechanism (phase equilibrium or kinetic, respectively) of the extraction process. The positive effect of an increase in P on the extraction rate of turmeric at 40°C (Figure 5.5) is due possibly to an increase in oil solubility with the pressure and solvent power of SC-CO_2 (Gopalan et al., 2000).

As explained in Section 5.3, near the critical point of CO_2 (or at least below the crossover pressure), density effects dominate vapor pressure effects, and solubility

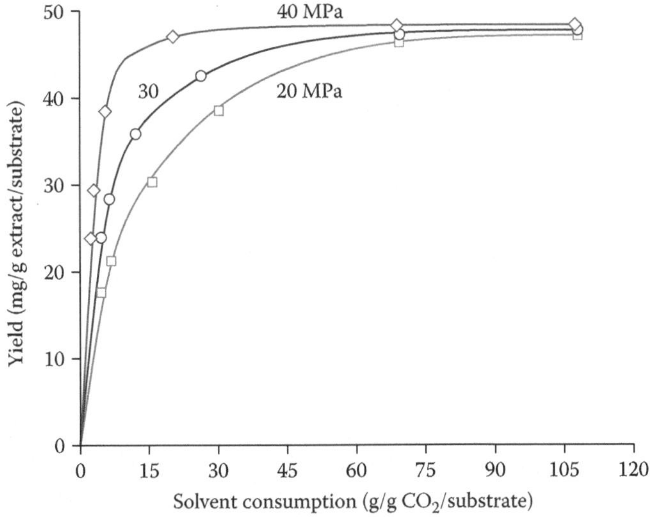

FIGURE 5.5 Effect of extraction pressure on cumulative extraction curves for ground turmeric (*Curcuma longa*) rhizhomes treated with supercritical CO_2 at 40°C. (From Gopalan, B. et al., *J. Agric. Food Chem.*, 48, 2189, 2000.)

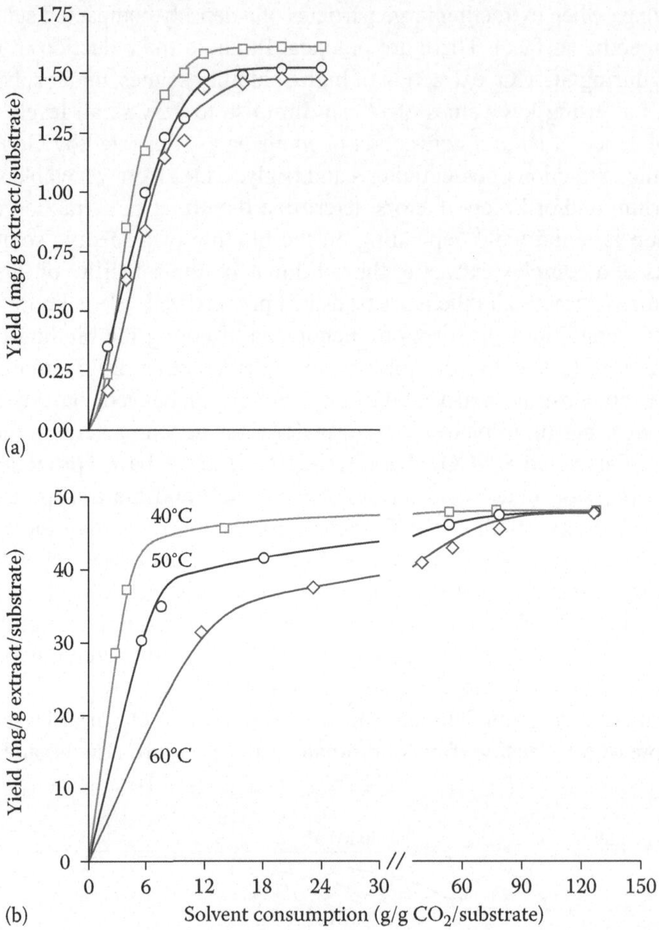

FIGURE 5.6 Effect of extraction temperature on cumulative extraction curves for (a) ground *Tetraclinis articulata* leaves (d_p = 1.5 mm) treated with 20 g/min of CO_2 at 9 MPa. (From Herzi, N. et al., *J. Supercrit. Fluids*, 82, 72, 2013.) (b) Ground turmeric (*Curcuma longa*) rhizomes treated with CO_2 at 30 MPa. (From Gopalan, B. et al., *J. Agric. Food Chem.*, 48, 2189, 2000.)

decreases as *T* increases isobarically. This is the case in the extraction of sandarac tree essential oil at 9 MPa (Figure 5.6a) (Herzi et al., 2013) as well as turmeric oil at 30 MPa (Figure 5.6b) (Gopalan et al., 2000), where operational solubilities decrease when temperature increases from 40°C to 60°C. Using the equation of del Valle et al. (2012) for the solubility of vegetable oils in SC-CO_2, it is possible estimating that the crossover pressure increases slightly with temperature from ca. 28 MPa at near-critical 35°C to ca. 42 MPa at 80°C (Table 5.6). Thus, it is possible that the crossover pressure for turmeric extract was above 30 MPa in the 40°C–60°C interval. On the other hand, vapor pressure effects dominate away from the critical point,

TABLE 5.6
Crossover Pressure (or Pressure at Which the Solubility Does Not Decrease or Increase as the Temperature Increases) as a Function of Temperature for the Solubility of Vegetable Oil in Supercritical CO_2, as It Can Be Estimated Using the Equation of del Valle et al. (2012a)

Temperature Interval (°C)	Approximate Crossover Pressure (MPa)
35–40	27.8
45–50	31.9
55–60	35.6
65–70	39.0
75–80	42.0

where CO_2 is less compressible and density changes moderately as a result from large changes in T. This is the case in the extraction of phenolics from olive leaves at 33.4 bar (Figure 5.7), where the operational solubility and yield increases markedly as temperature increases from 80°C to 120°C (Le Floch et al., 1998). Another explanation for the significant effect of T on the extraction rate and yield of olive leaves may be due to improved desorption of phenolics from the solid matrix as previously observed for the extraction of organics from soil samples (Hawthorne and Miller, 1994; Langenfeld et al., 1993).

Generally, a change in extraction conditions has opposite effects on the phase equilibrium and kinetic factors, in that the solubility increases and D_{12} decreases as T decreases, or as P increases. The only exception occurs when T increases and P is above the crossover value, where both D_{12} and solubility increase. This may be the case for the SC-CO_2 extraction of phenolics from olive leaves (Figure 5.7).

The binding of a solute or solute mixture (extract) to the solid matrix, its solubility, and the mass transport properties are affected not only by T and P but also by the addition of a polar modifier such as ethanol or water and its concentration in CO_2. Several factors contribute to the positive effect of a modifier on extraction rate and yield including specific interactions of the added substance with CO_2, the target solute(s), and/or functional groups on the surface of the solid matrix and/or physical modification of the solid matrix. We add a substance to CO_2 to increase the solubility of a nonvolatile target solute in "modified" CO_2 as compared to that in pure CO_2 under the same system (T and P) conditions, or the so-called cosolvency effect described in Section 5.3. The added substance may also affect the solid matrix where the target solute is initially located so that it is more easily extracted using "modified" than pure CO_2 under the same system conditions. The so-called entrainer effects of the modifier responsible for improving extraction include swelling of the solid matrix (Fahmy et al., 1993) and competing with the target solute for binding sites on the surface of the solid (King and France, 1992). The names entrainer and modifier were coined for adsorptive separation processes such as chromatography and relate to the effect of the added modifier on the structure and surface chemistry of the adsorbents. The action of modifiers is extremely specific so that they increase or decrease the

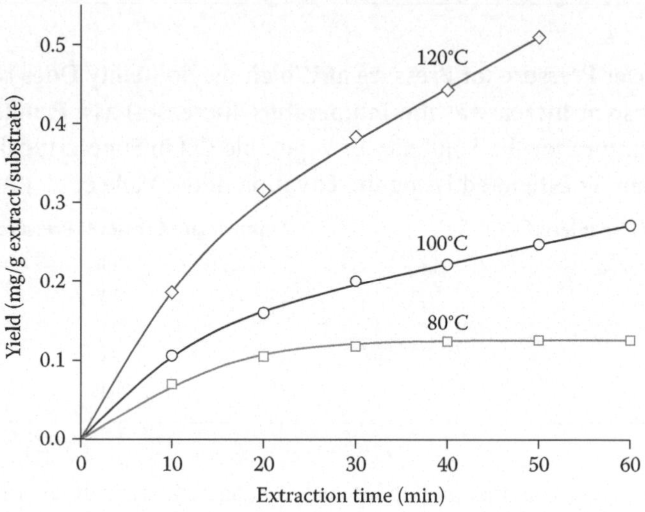

FIGURE 5.7 Effect of temperature on cumulative extraction curves of phenolic compounds (caffeic acid equivalent) from ground olive (*Olea europaea* L.) leaves treated with 2 cm^3/min of CO_2 at 33.4 MPa. (From Le Floch, F. et al., *Talanta*, 46, 1123, 1998.)

separation factor for different solutes. Ethanol unsurprisingly dominates the application of modifiers in SC-CO_2 extraction (de Melo et al., 2014), particularly of polyphenols (Tables 5.3 through 5.5). With ethanol being polar, adding small amounts of ethanol to CO_2 may increase polarity markedly. Despite being more polar than ethanol, water is not as effective because it is much less soluble in CO_2 than ethanol.

As exemplified in Section 5.4, supercritical extraction with modified CO_2 can be implemented in two distinct ways: continuous and pulsed (de Melo et al., 2014). In the more conventional continuous procedure, SC-CO_2 is mixed with the modifier in a fixed proportion along time, which requires a pump and independent feed line for the two. In the alternative static procedure, the solid substrate is impregnated with the modifier at the onset of the experiment, prior to its extraction with pure SC-CO_2. Reproducibility of the results improves when using the continuous procedure, because when applying the pulsed procedure, the concentration decreases continuously during extraction. In these cases, the experiment can be stopped to reimpregnate the solid substrate with the modifier as done in the works led by K. H. Kim and Y. S. Kim on tea decaffeination described in Section 5.4.3 (Lee et al., 2007; Park et al., 2007a,b). An alternative to using independent pumps and feed lines for CO_2 and the modifier in the case of a limited-solubility modifier, such as water, is to use a saturator (high-pressure vessel filled with a modifier) in the feed line of CO_2 before the extraction vessel, as done by Saldaña et al. (2002) for the decaffeination of guarana seeds using water-saturated CO_2. The typical approach to assess the impact of a modifier involves comparing cumulative extraction curves between the modified and pure SC-CO_2 results. For example, Luengthanaphol et al. (2004) observed a notable increase in the yield of EC from tamarind seed coat using SC-CO_2 at 80°C

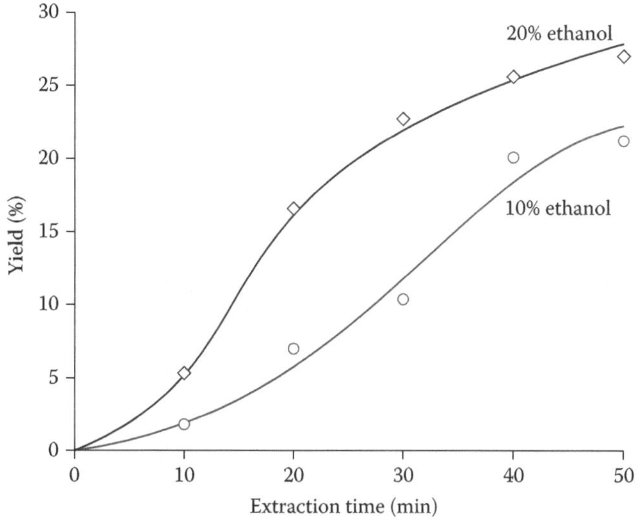

FIGURE 5.8 Effect of ethanol as the modifier on cumulative extraction curves of phenolic compounds from ground eelgrass (*Zostera marina*) leaves using supercritical CO_2 at 80°C and 25 MPa. (From Pilavtepe, M. et al., *J. Supercrit. Fluids*, 68, 87, 2012.)

and 30 MPa as a result of adding 10% ethanol as the modifier, from virtually nothing (<0.5 ppm) when using pure CO_2 to about 28 ppm when using ethanol-modified CO_2. Figure 5.8 exemplifies the positive effect of ethanol as a modifier for the SC-CO_2 extraction of ground eelgrass leaves at 80°C and 25 MPa (Pilavtepe et al., 2012). Similar observations were made for total and individual phenolic acids (i.e., benzoic, caffeic, *p*-coumaric, chicoric, ferulic, rosmarinic acids) (Pilavtepe et al., 2012).

Commonly, vegetable substrates are mechanically pretreated to improve the subsequent SC-CO_2 extraction. Pretreatments have multiple purposes, including release of solutes from cells (Eggers, 1996; del Valle et al., 2003b; Williams, 1997), ease solvent flow through the packed bed (Williams, 1997), and increase the load of the extraction vessel (del Valle and Aguilera, 1989; del Valle et al., 2003b; Hallberg et al., 1989). In the case of low-oil seeds such as those of grapes, alternative pretreatments include flaking and size reduction (Williams, 1997). Size reduction processes such as crushing, chopping, and grinding rely on impact and attrition mechanisms to break superficial cells, so that the amount of released solute is inversely proportional to the final particle diameter (d_p) (Reverchon and Marrone, 2001). In addition to increasing the fraction of easily available superficial oil, a reduction in d_p increases the specific surface (surface-to-volume ratio) of the substrate (Reverchon and Marrone, 2001). Thus, size reduction improves extraction rate and/or yield as illustrated in Figure 5.9 for the extraction of turmeric oil at 40°C and 30 MPa. The use of packed beds with small particles is not encouraged, however, because of the low packing density and agglomerating tendency of the substrate (del Valle et al., 2012b), and the poor fluid flow of the solvent that may cause channeling and/or blockage of supporting screens (del Valle et al., 2006).

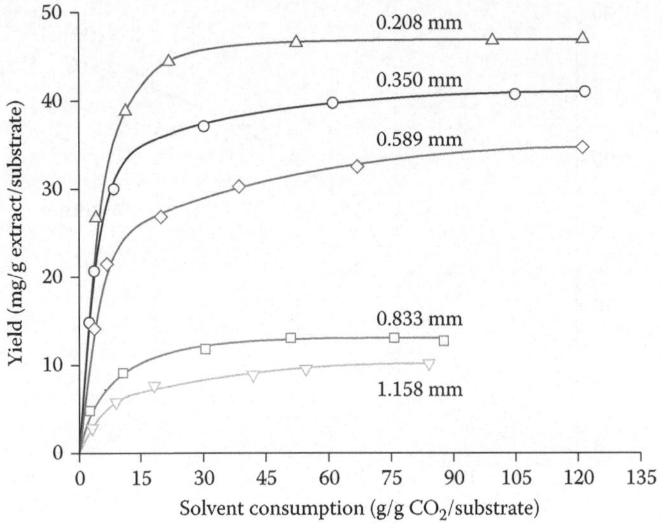

FIGURE 5.9 Effect of particle size (d_p, noted values in the graph) on cumulative extraction curves of ground turmeric (*Curcuma longa*) rhizhomes treated with 3 cm^3/min of CO_2 at 40°C and 30 MPa. (From Gopalan, B. et al., *J. Agric. Food Chem.*, 48, 2189, 2000.)

The solute can also be released from the cells using high-shear-stress processes such as prepressing, flaking, and pelletization, which destroy cell walls and other barriers to mass transfer within the solid without resorting to a reduction in particle size (Aguilera and Stanley, 1999). Sheared vegetable substrates are optimally conditioned for a fast extraction because of the destruction of natural barriers to mass transfer (cell walls) and the opening up of diffusion paths inside the tissue (Eggers et al., 1985). High-shear-stress pretreatments may have the added advantage of increasing vessel load density (del Valle and Aguilera, 1989; del Valle et al., 2003b; Hallberg et al., 1989; Sielfeld et al., in press). Indeed, during the densification process of a powdery material, particles are rearranged, fractured, and/or plastically deformed to fill the void fraction (Stanley-Wood, 1990). These various events during uniaxial, confined pressing increased density of a milled substrate at the expense of partial solute trapping within the solid matrix (del Valle and Aguilera, 1989). There are compaction processes that are better suited for SC-CO_2 extraction of solid substrates at the industrial level than uniaxial, confined pressing, such as the multistage pretreatment consisting of fine milling, high-pressure pelletization, and particle size reduction applied by del Valle et al. (2003a) to jalapeño pepper flakes that fractured cell walls extensively and liberated intracellular solutes effectively. An advantage of these high-shear-stress processes as compared to size reduction processes is that the fraction of easily available solute in the sheared substrate is related less to superficial fractures caused by impact efforts (case of milled substrates) than to inner fractures caused by shear efforts (del Valle et al., 2003a), thus circumventing typical operational problems associated with small particle processing. Sielfeld et al. (in press) compared pelletization and milling as pretreatments for the extraction of carnosic acid and other phenolic antioxidants from rosemary and found that pelletization increased the volumetric yield (or the mass of recovered solute per unit time and per unit volume of extraction vessel) of the SC-CO_2 extraction process despite a reduction in extraction yield. They hypothesized that this was a consequence of the increase in bulk density and improved destruction of inner barriers to mass transfer in the pretreated substrate. Sielfeld et al. (in press) claimed that pelletizing could increase SC-CO_2 extraction rate and yield of rosemary antioxidants if it were undertaken under optimized conditions, producing smaller pellets (to increase the specific surface of the substrate) at lower temperature (to prevent deleterious oxidative reactions).

Mass-transfer models on SC-CO_2 extraction from solid substrate may facilitate the use of laboratory data for commercial process design purposes. This subject was reviewed by Al-Jabari (2002), del Valle and de la Fuente (2006), Oliveira et al. (2011), and Huang et al. (2012). From the standpoint of using laboratory data for the design of industrial processes, the most interesting mass-transfer models are those based on mass balance equations for thin sections of a packed bed, with different models in literature using different assumptions about the mass-transfer mechanism within the particles. The diffusional model assumes internal mass transfer by diffusion within solid particles, external mass transfer by convection through a static film of SC-CO_2 next to the particles, and axial dispersion of dissolved solute in the SC-CO_2 phase along the bed. When the inner structure of the solid particle where mass transfer takes place can be assumed to be a network of interconnected pores, two important models that can be applied are the so-called shrinking-core and desorption–dissolution–diffusion models. The shrinking-core hypothesis assumes that the solute is retained in the pores of the solid matrix by mechanical or capillary forces, so that the pore space is divided into an inner core filled with a condensed solute and an outer region containing a solution of the solute in SC-CO_2 that are separated by a moving boundary (Goto et al., 1996; King and Catchpole, 1993). Meanwhile, the desorption–dissolution–diffusion hypothesis assumes that the solute is partially adsorbed on the solid matrix within the pores, so that a fraction of the solute is adsorbed on the solid matrix and the rest is dissolved in the SC-CO_2 phase within the inner pores of the solid, which are related by an equilibrium sorption isotherm and isobar (Goto et al., 1993). del Valle and de la Fuente (2006) analyzed simplifications of this diffusional model including external control of mass-transfer rates, internal control of mass-transfer rates, and linear driving force, among others.

Solving the equations of the diffusional mass-transfer model demands information on parameters describing mass-transfer and equilibrium aspects of the problem. Important model parameters include (1) a solubility isotherm and isobar describing the partition of solute or solute mixture (extract) between the solid substrate and SC-CO_2 under process conditions, (2) the effective diffusivity of the solute in the solid (D_e), and (3) the external mass-transfer coefficient (k_f) and axial dispersion coefficient (D_L) describing the transport of the extract in the SC-CO_2 phase (del Valle, 2015). The solubility isotherm and isobar can be measured experimentally (Urrego et al., 2015) or best fitted to experimental cumulative extraction curves. The effective diffusivity can be estimated from a quantitative description of the microstructure of the pretreated solid matrix (del Valle et al., 2006; Uquiche et al., 2004, 2005) or best fitted to experimental cumulative extraction curves. Finally, k_f and D_L can be estimated from dimensionless correlations for mass transfer in packed beds operating with SCFs (del Valle, 2015; del Valle and de la Fuente, 2006; del Valle et al., 2011). D_e, k_f, and D_L influence the concentration gradient-driven mass-transfer rates within the solid, in the stationary SC-CO_2 film surrounding each particle, and along the bed, respectively.

The state conditions (P, T) of SC-CO_2 together with the specification of a key component of the extract allows for the estimation of the physical properties (ρ, μ, D_{12}) of the solvent under process conditions. Indeed, for low-solubility solutes, ρ and μ can be estimated as a function of P and T using NIST Database for pure CO_2 (Lemmon et al., 2010) under the assumption that the dissolved solute does not affect them. The value of D_{12}, on the other hand, can be estimated as a function of the reduced temperature and reduced density of SC-CO_2 and the molecular weight (MW) and critical volume (V_{c2}) of a representative solute using the correlation of Catchpole and King (1994).

Once the process parameters are estimated, dimensionless Reynolds (Re) and Schmidt (Sc) numbers and H/d_p can be used in turn to estimate D_L and k_f. Indeed, D_L can be estimated using the dimensionless correlation of del Valle et al. (2011) for D_L/D_{12} as a function of the product $Re{\cdot}Sc$ and the ratio H/d_p. According to del Valle (2015), predictions of $Sh{\cdot}Sc^{-1/3}$ as a function of Re using the correlation of King and Catchpole (1993) are higher than those of the one of Tan et al. (1988), which are in turn higher than values predicted by the correlation of Puiggené et al. (1997). These correlations, based on experimental data on dissolution of pure solids, or evaporation of liquids on nonadsorbing solid particles in packed beds in the presence of a SCF, predict smaller values of $Sh{\cdot}Sc^{-1/3}$ as a function of Re than a more general equation proposed by Wakao and Kaguei (1982) that applies for mass transfer in packed beds with low-pressure liquids and gases. Once an appropriate correlation is selected to estimate Sh as a function of Re and Sc, k_f can be estimated from the definition of Sh.

The effective diffusivity of the solute is in the treated substrate (D_e) that can be estimated using a microstructural factor (F_M), which embodies all effects of the substrate and its pretreatment on inner mass transfer (Aguilera and Stanley, 1999), using the following equation:

$$D_e = \frac{D_{12}}{F_M}. \tag{5.8}$$

The term F_M gives the retarding effect of the solid matrix on the mass-transfer rate by indicating how many times smaller the effective diffusivity of the solute is in the treated plant material than its binary diffusion coefficient in SC-CO_2. It is convenient to use F_M instead of D_e in modeling a SC-CO_2 extraction process because it is independent of T, P, U, and d_p (del Valle, 2015; del Valle and de la Fuente, 2006; del Valle et al., 2006).

Table 5.7 summarizes works on the SC-CO_2 of polyphenols from solid substrates that were mathematically modeled using differential mass-transfer models such as the one described in this section. Rahimi et al. (2011) assumed that mass transfer between solid particles in a packed bed and SC-CO_2 was defined by a

TABLE 5.7
Substrate and Extractor Vessel Characteristics

Raw Material	d_p (mm)	V (cm³)	L/d_E (–)	ε (–)	Q/M (g CO_2/g·h)	References
Ground *C. recutita* Rauschert flowers	0.3	289	2.31	0.83	1.55–5.39	Rahimi et al. (2011)
Ground *Z. marina* leaves L.	0.3–0.5	100	45	0.32	90	Pilavtepe et al. (2012)
C. sinensis leaves	0.674	8	14	0.4532	13.1–44.7	Ghoreishi and Heidari (2012)
Ground *A. satureioides* flowers	0.819	197	0.1–1.55	0.8455	9.12–432	Hatami et al. (2012)

desorption–dissolution–diffusion mechanism. Particles of the substrate were considered as porous solids. The physical properties of chamomile extract were estimated by applying Kay's mixing rule to eight components quantified using a Gas Chromatography with Flame Ionization Detection (GC-FID) analysis. The equilibrium relationship between the solute concentration in the solid phase (c_s) and the solute concentration in the SC-CO_2 phase near the surface of the particle (c_f^*) was described using a distribution coefficient K that was established as a function of T, P, and U. K was used as an adjustable parameter to minimize the discrepancies between experimental data and model predictions.

The model and procedure used by Hatami et al. (2012) were similar to those of Rahimi et al. (2011). The physical properties and mass-transfer parameters of the extract were estimated by applying Kay's mixing rule to four compounds in the extract that were quantified using a GC-FID analysis. The distribution coefficient K was considered a linear function of time and best fitted to experimental data.

Ghoreishi and Heidari (2012) described mass transfer as an irreversible desorption process from porous substrate particles. Physical properties and mass-transfer parameters were estimated using EGCg as a representative component of the extract. A local linear equilibrium was assumed within the pores of the substrate. The distribution coefficient K was determined by thermodynamic modeling of solubilities.

Pilavtepe et al. (2012) defined mass transfer using a linear driving force model based on the difference between the solute concentration in the solid phase and the solute concentration in the SC-CO_2 phase next to the surface of the solid particle. Particles of the substrate were considered as nonporous solids. Physical properties and mass-transfer parameters were estimated using the mole fractions and molecular mass of phenolic compounds in the extract quantified by Ultra-Performance Liquid Chromatography (UPLC). The equilibrium relationship between the solute concentration in the solid phase and the fluid phase on the solid was described with Henry constants, which were used as adjustable parameters to minimize the error between experimental data and model predictions.

Table 5.8 summarizes extraction conditions, and Table 5.9 presents the values of k_f, D_L, and D_e for the works listed in Table 5.7.

TABLE 5.8
Extraction Conditions

Raw Material	T (°C)	P (MPa)	ρ (kg/m^3)	$D_{12} \times 10^8$ (m^2/s)	U (mm/s)	Re (–)	Sc (–)	References
C. recutita [L.] Rauschert	40	25	565–893	0.63–1.36	0.016–0.057	0.13–0.51	5.64–14.88	Rahimi et al. (2011)
Z. marina L.	30–80	15–35	691–967	0.428–1.042	1.65–2.30	4.32–8.71	10.23–35.66	Pilavtepe et al. (2012)
C. sinensis	40–60	10–30	543–912	0.807–1.994	0.131–0.445	0.88–2.98	16.3–30.7	Ghoreishi and Heidari (2012)
A. satureioides	30–45	20–30	827–977	1.08–2.01	0.042–0.113	0.57–4.54	4.86–13.91	Hatami et al. (2012)

TABLE 5.9
Mass-Transfer Parameters and Distribution Coefficient

Raw Material	$D_L \times 10^8$ (m²/s)	*Sh* (–)	$k_f \times 10^5$ (m/s)	*K* (–)	References
C. recutita [L.] Rauschert	1.90–6.97	0.12–0.17	2.51–7.78	8.6–43.1	Rahimi et al. (2011)
Z. marina L.	397–450	1.84–2.16	2.30–4.78	2–10	Pilavtepe et al. (2012)
C. sinensis	16.56–62.77	0.0096–0.017	0.0097–0.0283	0.484–3.126	Ghoreishi and Heidari (2012)
A. satureioides	13.79–184.92	0.04–2.25	0.0757–5.52	15.7–38.0	Hatami et al. (2012)

Mathematical modeling of extraction curves is at the core of process design and economics for industrial SCFE plants for the treatment of solid substrates. del Valle (2015) summarized this for SC-CO_2 extraction of vegetable oil from prepressed seeds. Indeed, the mathematical model can be used to simulate the evolution of extraction yield as a function of process time in an industrial plant having ≥3 extraction vessels with a simulated CC contact between SC-CO_2 and the solid substrate that single-extraction-vessel laboratory or pilot plant units cannot produce experimentally. This is important because it is the relationship between the extract yield and extraction time that anchors the precise estimations of extraction cost provided by del Valle (2015) in the case of the extraction of vegetable oil from prepressed seeds. We do not believe, however, that we should dispose of a robust, predictive mathematical model for the extraction of plant phenolics that can be applied to simulate the extraction of a substrate having, for example, different size and packing characteristics than those assayed in laboratory experiments. Reducing production cost would rely on the economies of the scale associated with multivessel plants having large extraction vessels operating with large superficial CO_2 velocities and at high pressure. Mathematical simulation can guide the economic optimization task, but pilot plant experimentation will be required at the end to validate simulated results (del Valle, 2015).

5.5.2 Countercurrent Fractionation from Liquid Substrates

The subject of CC SCF Fractionation (SCFF) of liquid mixtures in packed columns has been periodically reviewed by Brunner either specifically (Brunner, 1998, 2009) or as part of several separation processes (Brunner, 2005, 2010). We recently published a review on the use of SC-CO_2 for the fractionation of liquid mixtures using not only CC contacting in packed columns but also alternative contacting devices such as membrane contactors, mixer–settler arrangements, and spray processes (Bejarano et al., 2016). This subject is also included as chapters in the books of Brunner (1994) and Clifford (1999) and book sections on pressure distillation (Laurent, 2001a) and packed towers (Laurent, 2001b).

CC-SCFF of liquid mixtures can be considered as an end-of-line process, typically used to fractionate materials extracted by conventional methods (Tabera et al., 2004). Basically, CC-SCFF is a mass-transfer operation that can be considered as a stripping process and is usually carried out in a packed column. However, other modes of operation exist (reflux mode, semibatch mode) (Bejarano et al., 2016).

In an ordinary stripping process in a packed column, liquid–gas contacting is continuous and nonstaged, unlike in plate or other types of columns. The solute of interest is transferred from a descending liquid stream to an ascending SC-CO_2 stream. In CC-SCFF, due the liquid-like density of the SC-CO_2 gas, its velocity through the packing conduits can be about 100 times lower (around cm/s instead of m/s) than in distillation (de Haan and de Graauw, 1991). In order to achieve the separation, it is essential that operational conditions warrant the presence of a two-phase region inside the column. Therefore, phase equilibrium, interfacial tension, and differences in density between phases are of great importance to ensure the partial immiscibility and separation of the contacting streams.

In general, for the design of packed columns, the separation is treated as a staged process and uses the packed bed Height Equivalent to a Theoretical Plate (HETP) to convert the required number of ideal stages to a packing height. The basic model for the mass-transfer performance of a packed column in distillation or absorption/stripping processes is often expressed by the HETP. According to the double-film theory, the relationship between the HETP and the height of the mass-transfer unit for the gas and liquid phases is given by the following equations (Sinnott, 1999; Wang et al., 2005):

$$\mathrm{HETP} = \frac{Z}{N}, \tag{5.9}$$

and

$$\mathrm{HETP} = \frac{\ln(\lambda)}{\lambda - 1} \cdot \left(\frac{G_\mathrm{m}}{k_\mathrm{G} \cdot a_\mathrm{e}} + \lambda \cdot \frac{L_\mathrm{m}}{k_\mathrm{L} \cdot a_\mathrm{e}} \right), \tag{5.10}$$

where

Z is the height of packing

N is the number of theoretical stages

λ is the stripping factor or ratio between the slope of the equilibrium line and that of the operating line (mass balance equation)

G_m and L_m are the molar flow rates per unit cross-sectional area of the gas and liquid phases, respectively

k_G and k_L are the corresponding gas- and liquid-phase mass-transfer coefficients

a_e is the effective interfacial area for mass transfer provided by the packing

It is simple to perceive that Equation 5.10 will be accurate if the correlations used to predict the basic parameters are accurate.

When assuming binary or pseudo-binary systems; low concentrations of the solute, straight equilibrium, and operation lines (generally true in CC-SCFF); and constant molar overflow throughout the column, Equations 5.9 and 5.10 can be solved using conventional methods. The number of stages can be estimated using graphical methods such as those of McCabe–Thiele or Ponchon–Savarit as described by Brunner (1994, 1998, 2009) or using numerical integration of traditional expressions for the Number of Transfer Units (NTU) (Sinnott, 1999). For multicomponent separations, a more complex approach, based on the Stefan–Maxwell equation to estimate multicomponent diffusivities, is necessary. Studies reporting multicomponent diffusion using the Stefan–Maxwell approach under high pressure can be found in the work of Unsulu and Sunol (2004a,b). Recent improvements to the HETP model are also reported by Hanley and Chen (2012).

In general, the mathematical description of CC-SCFF considers only axial dispersion and the mass fluxes in terms of overall mass-transfer coefficients for the solutes into the SC-CO_2 (k_{i,CO_2}) and of the SC-CO_2 into the liquid phase ($k_{i,CO_2,L}$, nonnegligible in nonaqueous systems). Due to back-mixing effects, a correction factor dependent on the densities of both phases (ρ_{CO_2}, ρ_L) and the corresponding superficial velocities (U_{CO_2}, U_L) is usually introduced in the expression for the mass fluxes, Equations 5.11 and 5.12, to take into account the buoyancy effects produced by the slight density difference between the SC-CO_2 and the liquid phase (Ruivo et al., 2004):

$$J_i = k_{i,CO_2} \cdot \rho_{CO_2} \sqrt{\frac{\rho_{CO_2}}{\rho_L - \rho_{CO_2}} \cdot \frac{U_{CO_2}}{U_L}} \cdot (y_i^* - y_i), \tag{5.11}$$

$$J_{CO_2} = k_{CO_2,L} \cdot \rho_L \sqrt{\frac{\rho_{CO_2}}{\rho_L - \rho_{CO_2}} \cdot \frac{u_{CO_2}}{u_L}} \cdot (x_{CO_2}^* - x_{CO_2}). \tag{5.12}$$

From Equations 5.10 through 5.12, it is seen that phase equilibrium mole fractions (y_i^*, $x_{CO_2}^*$) are essential in the description of the process. Comprehensive reviews on correlations for estimating basic parameters, that is, mass-transfer coefficients of both phases and effective interfacial area for different types of packing can be found in the literature (de Haan and de Graauw, 1991; Oliveira et al., 2011; Wang et al., 2005). Several mass-transfer and hydrodynamics studies, along with CC-SCFF process modeling and simulation, can be found in literature, mainly for nonaqueous systems and very few applications related to aqueous systems. Hydrodynamics of CC packed columns consists of studying dry and wet pressure drops, liquid holdups, and flooding points. The work of Stockfleth and Brunner (1999, 2001a,b) demonstrated that the hydrodynamic behavior of CC columns at high pressure could be described with models developed for normal pressure operation if specific peculiarities, such as the real mixture properties of miscible systems under the relevant conditions of high-pressure operation, are appropriately addressed. Becker and Heydrich (2004) contributed greatly to clarify the fluid dynamic axial dispersion, the separation efficiency, and hydrodynamics of two types of packing for the (CO_2 + ethanol + water) mixture.

Brunner (1998, 2005, 2009) described a methodology to develop industrial CC-SCFF processes. The methodology involved (1) measurements of phase equilibrium; (2) simplified determination of the NTU and HTU or HETP using

McCabe–Thiele or Ponchon–Savarit methods for quasi-binary and quasi-ternary systems based on pseudo-components; (3) separation experiments carried out on laboratory columns to ensure the accuracy of separation factors estimated from (1) and to determine the dependence of the separation factors on concentration; (4) determination of hydrodynamic characteristics such as dry and wet pressure drops, flooding points, and liquid holdups; (5) multicomponent simulation including different conditions, methods of product recovery, and gas cycles; (6) pilot-scale experiments to produce samples for customer studies and product development, using different feed sources; (7) simulation experiments to optimize column operation and plant design; and (8) economical evaluation of the process to conclude the development stage of the design.

Very few studies focusing on extraction of phenolic compounds by CC-SCFF can be found in the scientific literature. Señoráns et al. (2001) optimized process conditions for the isolation of antioxidants compounds from orange juice by CC-SCFF in a 1.8-m column packed with glass beads. They found that the solvent-to-feed ratio was the parameter that had the most important effect on selectivity toward benzoic acid, hesperidin, narirutin, and naringin. Later, Simó et al. (2002) studied these antioxidants and concluded that CC-SCFF allowed for the improvement of the selectivity of the extraction process, yielding a better recovery of benzoic acid in the extract while increasing the content of flavanones in the raffinate. Persson et al. (2002) fractionated low-molecular-weight aromatic compounds from a model aqueous solution and a real sample (a lignocellulosic dilute-acid hydrolysate of spruce) using a 1.1-m column filled with Sulzer packing. Depending on the process conditions, they obtained extraction efficiencies ranging from 2% to 100% for the model solution and from 11% to 98% for the hydrolysate.

5.5.3 Chromatographic Purification

The application of SCF Chromatography (SCFC) for separations is covered in Chapter 7 focusing on analytical aspects in the book of Clifford (1999) and in Chapter 9 focusing on the scale-up for preparative applications in the book of Brunner (1994). A book section focusing on the analytical aspects of SCFC (Hirata, 2002) and a book chapter focusing on preparative SCFC applications for the food and nutraceutical industry (Montañés et al., 2015) can also be consulted. Relevant aspects of the last subject are covered as part of a special issue (volume 1250) of the *Journal of Chromatography A* devoted to SCFC and chromatography (Miller and Sebastian, 2012; Rajendran, 2012).

In chromatographic separations, an adsorbent and a mobile phase (SC-CO_2 with a polar cosolvent in the case of SCFC) are selected that accentuate the differences between migration velocities on the components of a complex solute mixture or extract, so that they are eluted from the packed bed column (emerge from its exit end) at different times. In elution chromatography, the chromatogram signals changes in the concentration of each component of the mixture as a function of process time that reaches component-dependent peak values for the corresponding average retention times. In analytic separations, the complete separation (resolution) of the bands representing each component of the mixture is favored. On the other hand, in preparative separations, some band overlapping is allowed that is compatible with the required purity of each component to increase the productivity of the system. To achieve this, the amount and/or concentration of the sample injected to the column is increased, the diameter of adsorbent particles (d_p) in the packed bed is increased, the superficial velocity of the mobile phase (U) is increased, and/or the column is shortened. The production capacity of the column is increased by increasing its diameter (D), but it is necessary to take into account the increase in axial dispersion and worsening in resolution associated with the increase in d_p, U, and/or D to reach the required purity of the main component in each band.

Specialized literature contains several examples on the use of SCFC for the preparative separation of plant phenolics. Ramírez et al. (2004) separated antioxidant compounds in rosemary extract (carnosic acid and carsonol) using a 0.5-mm-diameter and 25-cm-length column packed with porous silica particles (d_p = 10 μm) coated with SE-54 (5% phenyl, 95% methyl silicone) and CW20M [poly(ethylene glycol)] using pure SC-CO_2 at 100°C and different pressure programs that finished at 37.5 MPa. Subsequently, Ramírez et al. (2005) compared the performance of bare octadecylsilica (ODS) and silica particles or coated with 3% CW20M in a packed column (D = 0.5 mm, L = 25 cm, d_p = 10 μm) for the fractionation of rosemary extract with pure SC-CO_2 using a program with an increase in both system temperature (40°C–180°C at 3°C/min) and pressure (12–34 MPa at 0.4 MPa/min) and found the need to coat the adsorbent particles. These works were expanded by Ramírez et al. (2007), who developed an improved procedure to pack a 1-mm-diameter and 25-cm-length column with porous silica particles (d_p = 10 μm) coated with 3% SE-54 and effected separation at 80°C and 18 MPa using SC-CO_2 modified with 10% ethanol. The elution conditions were previously selected by Ramírez et al. (2006) who used a commercial preparative LC-Diol column (D = 1 mm, L = 25 cm, d_p = 5 μm) to isolate antioxidants and antimicrobials from equivalent SC-CO_2 rosemary extract fractions whose composition and functional (antioxidant and antimicrobial) properties were measured as part of the work. More recently, Fernández-Ponce et al. (2014) optimized the elution conditions in a pilot-scale SCFC unit for two typical phenolics (GA, methyl gallate) that ended up being 40°C (the lowest tested temperature) and 40 MPa (the highest tested pressure) using SC-CO_2 modified with methanol containing 0.5% formic acid (a gradient of increasing concentration that went from 5% to 50% w/w). They subsequently applied the selected gradient to increase the concentration of mangiferin in a mango leaf extract by a factor of five as compared to the original extract.

5.5.4 Production of Powders

The production of powders using SC-CO_2 was covered by Fages et al. (2004) as part of a special issue (volume 141, issue 3) of *Powder Technology* devoted to particle formation in pharmaceutical applications. It was also covered in several reviews in a special issue (volume 7, issue 5) of *Current Opinion in Solid State and Materials Science* on SCF application in materials sciences (Hakuta et al., 2003; Knez and Weidner, 2003; Shariati and Peters, 2003; Weidner et al., 2003), as well as two reviews (Meterc et al., 2009; Weidner, 2009) on pharmaceutical and food applications of SCFs to particle formation and encapsulation in a special issue (volume 47, issue 3, 20th anniversary) of *The Journal of Supercritical Fluids* devoted to future directions in SCFs science and technology. It was the single subject of a special issue (volume 60, issue 3) of *Advanced Drug Delivery Reviews* devoted on the application of SCF technologies to drug delivery. Additional reviews on the subject have been published by Tom and Debenedetti (1991), Reverchon (1999), Jung and Perrut (2001), Knez (2006), Silva and Meireles (2014), and Knez et al. (2015). This subject was also included on a chapter on nonextractive applications (high-pressure micronizing) in the book of Stahl et al. (1988) about the use of dense gases for extraction and refining of biologic materials, on a chapter on special applications (SCF nucleation and antisolvent recrystallization) in the book of McHugh and Krukonis (1994) about SCFE, on a book section on precipitation of solids with dense gases (Knez and Weidner, 2001), on a book section on crystallization using SCFs (Kitamura and Nagahama, 2002), and on a book chapter on the formation of food ingredient particles using SCFs (Rodríguez-Meizoso and Plaza, 2015).

In general, antioxidants are thermolabile compounds that degrade with the exposure to light and oxygen. One alternative to prevent their degradation and to improve viability, stability, and bioactivity is encapsulation, which helps also in improving their solubility in different media (Martín et al., 2010; Visentin et al., 2012). The use of SC-CO_2 technologies for encapsulation has not only the same advantages than its use in other processes such as moderate operational conditions and safety, but also the possibility of obtaining very narrow distributions of particle size (Fraile et al., 2014). Particle formation techniques using SC-CO_2 can be classified depending on the role of SC-CO_2 as a solvent, solute, antisolvent, or impregnating agent (Martín and Cocero, 2008).

Particles can be formed using SC-CO_2 as the solvent in the so-called Rapid Expansion of Supercritical Solutions (RESS) process. RESS is a method where the solute is dissolved in SC-CO_2 until saturation and then expanded through a nozzle to decrease solvent temperature and pressure with the result of a decrease in solubility and precipitation of the solute. The high supersaturation achieved in this depressurization step allows for the formation of very small particles with narrow particle size distribution (PSD) without increasing the temperature over 80°C (Martín and Cocero, 2008). RESS can be used to encapsulate benzoic acid (Türk, 1999). However, since this method is based on the solubility of the solute in SC-CO_2, it is only suitable for highly soluble compounds. The use of a cosolvent such as ethanol could help in overcoming this limitation, for the particle formation of small, homogeneous particles of curcumin (Sonkaew et al., 2012), anthocyanins (Santos and Meireles, 2013), and rutin (Santos and Meireles, 2013), among other solutes.

Several types of particle formation processes have been reported using SC-CO_2 as the antisolvent: semicontinuous Supercritical AntiSolvent (SAS), Solution-Enhanced Dispersion by Supercritical (SEDS) fluids, Supercritical Antisolvent Extraction (SAE), and Water Extraction and Particle Formation On-line (WEPO). For the SAS process, the solute is dissolved in a solvent such as ethanol or acetone and the solution is cocurrently sprayed with SC-CO_2 through a nozzle causing solute precipitation (Sosa et al., 2011). The main disadvantage of SAS is the use of organic solvents in the process (Martín and Cocero, 2008). Antioxidant particles produced using SAS include propolis (Catchpole et al., 2004; Chen et al., 2009; Wu et al., 2009), green tea polyphenols (Meterc et al., 2009; Sosa et al., 2011), 3,5-diprenyl-4-hydroxycinnamic acid (Wu et al., 2009), rosemary antioxidants (Visentín et al., 2011, 2012), bixin-rich extract (Santos and Meireles, 2013), quercetin (Fraile et al., 2014; Santos and Meireles, 2013), and curcuminoids (Osorio-Tobón et al., 2016).

For the SEDS process, the organic solution is mixed with the SC-CO_2 and then sprayed for precipitation through a capillary tubing (Franceschi et al., 2009). Grape seed extract has been encapsulated using this technique (Boschetto et al., 2013). SAE is based on the SC-CO_2 extraction of a polar solvent such as ethanol from the organic solution. Following extraction, some compounds remain soluble in the ethanol–CO_2 mixture, while others precipitate. SAE has been used to precipitate mangiferin, isomangiferin, quercetin 3-*O*-galactoside, quercetin 3-*O*-glucoside, quercetin 3-*O*-xyloside, quercetin 3-*O*-arabinoside, quercetin, and kaempferol (Meneses et al., 2015), polyphenols and anthocyanins from lyophilized grape residues (Floris et al., 2010), and catechin, EC, GA, and resveratrol from grape vinification waste (Marqués et al., 2013).

The WEPO process considers the production of an aqueous mixture by extraction (rosemary leaves) or homogenization of a biologic tissue (fresh onion) prior to particle formation by the SC-CO_2 antisolvent process (Rodríguez-Meizoso et al., 2009, 2012). WEPO is faster than performing extraction and particle formation separately, thus reducing degradation of bioactives. Antioxidants encapsulated using WEPO include phenolic diterpenes from rosemary leaves (Rodríguez-Meizoso et al., 2012) and mixtures of quercetin and quercetin glucosides from fresh onion (Andersson et al., 2012).

There are several names for alternative processes to form particles using SC-CO_2 as the solute. They can be called Particles from Gas-Saturated Solutions (PGSS), Concentrated Powder Form (CPF), and Supercritical Assisted Atomization (SAA). In the PGSS process, SC-CO_2 is dissolved in a melted solid and expanded through a nozzle. Due to the depressurization, CO_2 is vaporized and the associated cooling effect precipitates the solute. This process can be applied only for systems with low antisolvent effect of CO_2 (Martín and Cocero, 2008). PGSS has being used to form particles of green tea polyphenols (Meterc et al., 2008, 2009). In the CPF process, a liquid extract is mixed with SC-CO_2 and then rapidly expanded to precipitate powder particles (Vatai et al., 2008). CPF has been used to encapsulate anthocyanin concentrates from grape residues (Vatai et al., 2008). In the SAA process, SC-CO_2 is dissolved in a liquid solution containing the active component, and then the mixture is atomized through a nozzle. In this process, CO_2 is claimed to act as a cosolute and pneumatic agent (Reverchon, 2002). Atomization can be assisted by a Hydrodynamic Cavitation Mixer (HCM) that can form cavitation bubbles and cavities within the liquid phase, causing localized pressure drops that allow for the formation of spherical microparticles with narrow PSD (Cai et al., 2008). SAA with a HCM has been used to produce microparticles of ginkgo flavonoids (Miao et al., 2010).

The process of particle formation by Supercritical Solvent Impregnation (SSI) consists of the impregnation of a porous solid matrix with a solution of a solute dissolved in SC-CO_2, which is followed by CO_2 removal by depressurization (Rodríguez-Meizoso and Plaza, 2015). There is no information available on the use of SSI to encapsulate antioxidants.

5.6 CONCLUDING REMARKS

SC-CO_2 is not the best solvent for high-molecular-weight and polar polyphenol molecules that are the focal interest of this book. However, this chapter shows that SC-CO_2 can be supplemented with polar solvents such as ethanol, water, and the mixture of the two, or else compliment processes using conventional solvents to produce new polyphenol products with improved functional properties. Indeed, SC-CO_2 can remove prooxidant essential oils responsible for the characteristic aroma of the substrate, prooxidant triglycerides and minor lipids, and questionable metabolites such as alkaloids from the plant material prior to further processing using conventional solvents, thus increasing the concentration of polyphenols and other high-molecular-weight and polar antioxidants. In stating this possibility, this chapter raises as many queries as it answers.

All process possibilities using SC-CO_2 analyzed in this chapter are derived from phase equilibrium considerations. There is a lot of information on binary (CO_2 + phenolic) systems, and this chapter proposes the use of the Ziegler–Eckert framework to correlate solubility data, but this will demand additional information on physical properties (vapor pressure and molar volume and their dependence on temperature) that could be estimated using group contribution methods (Coutsikos et al., 2003). However, more efforts will be required to incorporate the effect of polar cosolvents on solubility and, even more challenging, to expand the use of this phase equilibrium information for real multicomponent systems that interact with solid matrices.

This chapter highlights the need of additional research efforts to get robust mathematical models describing the SC-CO_2 extraction of solid substrates. These models should explicitly consider the use of ethanol, water, and the mixture of the two as polar modifiers and be used to simulate industrial multivessel SCFE plants. To the best of the authors' knowledge, mathematical simulation will be the best alternative to optimize the process from an economic standpoint and to explore the commercial application of the technology. Then, pilot plant experiments will be required to validate results of simulation studies. Because of the more limited previous research efforts, extensive work should be carried out to explore the commercial application of the other technologies reviewed in this chapter that can be applied to process conventional liquid extracts, namely, CC fractionation in packed columns, preparative elusion chromatography in packed columns of adsorbent particles, and particle formation in spraying processes.

To help the reader in exploring these alternatives, this chapter cites many books, book chapters and sections, and peer-reviewed review articles on the SC-CO_2 processing of polyphenols and other antioxidants; the technologies involved for the batch extraction of solid substrates in packed beds, CC fractionation of liquid mixtures in packed columns, isolation of individual compounds and fractions by elusion chromatography in adsorbent packed columns, synthesis of particles with improved functionality using CO_2 as the cosolute or antisolvent, and the integration of several unit operations as required for the SC-CO_2 processing of polyphenols; and fundamental aspects related to phase equilibrium, transport phenomena, and mathematical modeling.

This chapter used rosemary phenolic diterpenes, grape seed coat phenolics, and tea catechins as subjects in case studies, but there are many additional examples of polyphenol sources including spices and herbs; fruits and vegetables; seeds, beans, and grains; tree barks and leaves; and propolis (many of which have been claimed to be used as medicinal plants), as well as by-products of their processing. The chapter includes selected examples of these alternative substrates being processed through several SC-CO_2-based technologies.

NOTATIONS

a_e	Effective interfacial area for mass transfer provided by the packing
c_f^*	Solute concentration in the SC-CO_2 phase near the surface of the particle
c_s	Solute concentration in the solid phase
d_p	Mean particle size
D	Diameter of the extraction vessel
D_e	Effective diffusivity of the solute in the solid
D_L	Axial dispersion coefficient
D_{12}	Binary diffusivity of solute in SC-CO_2
E	Solubility enhancement
F_M	Microstructural factor
f	Factor dependent of the solubility parameters
G_m	Molar flow rates per unit cross-sectional area of the gas phase
H	Height of the extraction vessel
L_m	Molar flow rates per unit cross-sectional area of the liquid phase
k_G	Gas-phase mass-transfer coefficient
k_L	Liquid-phase mass-transfer coefficient
k_{i,CO_2}	Overall mass-transfer coefficient for the solutes into the SC-CO_2
$k_{CO_2,L}$	Overall mass-transfer coefficient of the SC-CO_2 into the liquid phase
k_f	External mass-transfer coefficient
N	Number of theoretical stages
P	Pressure
P_c	Critical pressure
P_2^v	Vapor/sublimation pressure of the solute
Q	Flow rate of CO_2
Q_{VE}	Volumetric flow rate of expanded CO_2
Re	Dimensionless Reynolds number
Sc	Dimensionless Schmidt number
Sh	Dimensionless Sherwood number
T_c	Critical temperature
T_r	Reduced temperature
U, U_{CO_2}	Superficial velocity of CO_2
U_L	Superficial velocity of liquid phase
ν_2	Solute-dependent intercept in the linear relationship between $\log(E)$ and f
v_2^L	Molar volume of the solute
V_{c2}	Critical volume of a representative solute

V_E Volume of the extraction vessel
y_2 Molar fraction of the solute in a saturated SC-CO_2 phase
y_i^*, $x_{CO_2}^*$ Phase equilibrium mole fractions
Z Height of packing
Δ Ratio of the solubility parameters of SC-CO_2 to the solute

GREEK LETTERS

δ_1 Solubility parameter of SC-CO_2
δ_2 Solubility parameter of solute
λ Stripping factor
η_1 Solvent-dependent slope in the linear relationship between log(E) and f
ρ, ρ_{CO_2} CO_2 density
ρ_L Liquid-phase density

ACRONYMS

CC Countercurrent
CIAL Insituto de Investigacion en Ciencias de Alimentación
CPF Concentrated powder form
CSIC Consejo Superior de Investigaciones Científicas
EC Epicatechin
ECg EC gallate
EGC Epigallocatechin
EGCg EGC gallate
EtOH Ethanol
FID Flame ionization detector
GA Gallic acid
HCM Hydrodynamic cavitation mixer
HTU Height of a transfer unit
GC Gas chromatograph
GMP Good manufacturing practice
GRAS Generally recognized as safe
HETP Height equivalent to a theoretical plate
MeOH Methanol
MW Molecular weight of a representative solute
NTU Number of transfer units
PGSS Particles from gas saturated solutions
PSD Particle size distribution
SAA Supercritical assisted atomization
SAE Supercritical antisolvent extraction
SAS Supercritical antisolvent
SCFE Supercritical fluid extraction
SCFF Supercritical fluid fractionation
SCFs Supercritical fluids

SC-CO_2	Supercritical carbon dioxide
SEDS	Solution-enhanced dispersion by supercritical
SSI	Supercritical solvent impregnation
UAM	Universidad Autónoma de Madrid
WEPO	Water Extraction and Particle Formation On-line

ACKNOWLEDGMENTS

Inaki Abadie and Joaquín Amenabar helped with the data analysis in Section 5.3, and Arturo Bejarano wrote the first draft of Section 5.5.3. Their help is truly appreciated by the authors. The financial support of Grant Anillo en Ciencia y Tecnología (ACT1105 Project) is also acknowledged.

REFERENCES

Abbott, A.P., Corr, S., Durling, N.E., and Hope, E.G. 2002. Solubility of substituted aromatic hydrocarbons in supercritical difluoromethane. *Journal of Chemical and Engineering Data* 47:900–905.

Agostini, F., Bertussi, R.A., Agostini, G., Atti dos Santos, A.C., Rossato, M., and Vanderlinde, R. 2012. Supercritical extraction from vinification residues: Fatty acids, α-tocopherol, and phenolic compounds in the oil seeds from different varieties of grape. *The Scientific World Journal* 2012:ID 790486.

Aguilera, J.M. and Stanley, D.W. 1999. *Microstructural Principles of Food Processing and Engineering*, 2nd ed. Frederick, MD: Aspen.

Ahluwalia, S., Shivhare, U.S., and Basu, S. 2013. Supercritical CO_2 extraction of compounds with antioxidant activity from fruits and vegetables waste—A review. *Focusing Modern Food Industry* 2:43–62.

Aizpurua-Olaizola, O., Ormazabal, M., Vallejo, A., Olivares, M., Navarro, P., Etxebarria, N., and Usobiaga, A. 2015. Optimization of supercritical fluid consecutive extractions of fatty acids and polyphenols from *Vitis vinifera* grape wastes. *Journal of Food Science* 80:E101–E107.

Akgerman, A., Roop, R.K., Hess, R.K., and Yeo, S.D. 1991. Environmental application of supercritical extraction. In *Supercritical Fluid Technology—Reviews on Modern Theory and Application*, eds. T.J. Bruno and J.F. Ely, p. 479. Boca Raton, FL: CRC Press.

Al-Jabari, M. 2002. Kinetic models of supercritical fluid extraction. *Journal of Separation Science* 25:477–489.

Andersson, J.M., Lindahl, S., Turner, C., and Rodríguez-Meizoso, I. 2012. Pressurised hot water extraction with on-line particle formation by supercritical fluid technology. *Food Chemistry* 134:1724–1731.

Arce, L., Lista, A.G., Ríos, A., and Valcárcel, M. 2001. Screening of polyphenols in grape marc by on-line supercritical fluid extraction–flow through sensor. *Analytical Letters* 34:1461–1476.

Ashraf-Khorassani, M. and Taylor, L.T. 2004. Sequential fractionation of grape seeds into oils, polyphenols, and procyanidins via a single system employing CO_2-based fluids. *Journal of Agricultural and Food Chemistry* 52:2440–2444.

Babovic, N., Djilas, S., Jadranin, M., Vajs, V., Ivanovic, J., Petrovic, S., and Zizovic, I. 2010. Supercritical carbon dioxide extraction of antioxidant fractions from selected Lamiaceae herbs and their antioxidant capacity. *Innovative Food Science and Emerging Technologies* 11:98–107.

Becker, O. and Heydrich, G. 2004. Separation efficiency and axial mixing in packed high pressure extraction columns. In *Supercritical Fluids as Solvents and Reaction Media*, ed. G. Brunner, pp. 465–488. Hamburg, Germany: Elsevier Science and Technology Books.

Bejarano, A., Simões, P.C., and del Valle, J.M. 2016. Fractionation technologies for liquid mixtures using dense carbon dioxide. *The Journal of Supercritical Fluids* 107:321–348.

Berna, A., Cháfer, A., and Montón, J. 2001. High-pressure solubility data of the system resveratrol (3) + ethanol (2) + CO_2 (1). *The Journal of Supercritical Fluids* 19:133–139.

Berna, A., Cháfer, A., Montón, J.B., and Subirats, S. 2001. High-pressure solubility data of system ethanol (1) + catechin (2) + CO_2 (3). *The Journal of Supercritical Fluids* 20:157–162.

Bicchi, C., Binello, A., and Rubiolo, P. 2000. Determination of phenolic diterpene antioxidants in rosemary (*Rosmarinus officinalis* L.) with different methods of extraction and analysis. *Phytochemical Analysis* 11:236–242.

Bleve, M., Ciurlia, L., Erroi, E., Lionetto, G., Longo, L., Rescio, L., Schettino, T., and Vasapollo, G. 2008. An innovative method for the purification of anthocyanins from grape skin extracts by using liquid and sub-critical carbon dioxide. *Separation and Purification Technology* 64:192–197.

Boschetto, D.L., Dalmolin, I., de Cesaro, A.M., Rigo, A.A., Ferreira, S.R.S., Meireles, M.A.A., Batista, E.A.C., and Oliveira, J.V. 2013. Phase behavior and process parameters effect on grape seed extract encapsulation by SEDS technique. *Industrial Crops and Products* 50:352–360.

Brogle, H. 1982. CO_2 as a solvent: Its properties and applications. *Chemistry and Industry* 12:385–390.

Brunner, G. 1994. *Gas Extraction. An Introduction to Fundamentals of Supercritical Fluids and the Application to Separation Processes*. New York: Springer.

Brunner, G. 1998. Industrial process development countercurrent multistage gas extraction (SFE) processes. *The Journal of Supercritical Fluids* 13:283–301.

Brunner, G. 2005. Supercritical fluids: Technology and application to food processing. *Journal of Food Engineering* 67:21–33.

Brunner, G. 2009. Countercurrent separations. *The Journal of Supercritical Fluids* 47:574–582.

Brunner, G. 2010. Applications of supercritical fluids. *Annual Review of Chemical and Biomolecular Engineering* 1:321–342.

Brunner G. and Peter, S. 1981. The state of the art of extraction with compressed gases. *Chemie Ingenieur Technik* 53:529–542.

Cai, M.Q., Guan, Y.X., Yao, S.J., and Zhu, Z.Q. 2008. Supercritical fluid assisted atomization introduced by hydrodynamic cavitation mixer (SAA-HCM) for micronization of levofloxacin hydrochloride. *The Journal of Supercritical Fluids* 43:524–534.

Capuzzo, A., Maffei, M.E., and Occhipinti, A. 2013. Supercritical fluid extraction of plant flavors and fragrances. *Molecules* 18:7194–7238.

Carvalho R.N. Jr., Moura, L.S., Rosa, P.T.V., and Meireles, M.A.A. 2005. Supercritical fluid extraction from rosemary (*Rosmarinus officinalis*): Kinetic data, extract's global yield, composition, and antioxidant activity. *The Journal of Supercritical Fluids* 35:197–204.

Casas, L., Mantell, C., Rodríguez, M., Martínez de la Ossa, E.J., Roldán, A., De Ory, I., Caro, I., and Blandino, A. 2010. Extraction of resveratrol from the pomace of Palomino fino grapes by supercritical carbon dioxide. *Journal of Food Engineering* 96:304–308.

Catchpole, O.J., Grey, J.B., Mitchell, K.A., and Lan, J.S. 2004. Supercritical antisolvent fractionation of propolis tincture. *The Journal of Supercritical Fluids* 29:97–106.

Catchpole, O.J. and King, M.B. 1994. Measurement and correlation of binary diffusion coefficients in near critical fluids. *Industrial and Engineering Chemistry Research* 33:1828–1837.

Cavero, S., Jaime, L., Martín-Álvarez, P.J., Señoráns, F.J., Reglero, G., and Ibáñez, E. 2005. *In vitro* antioxidant analysis of supercritical fluid extracts from rosemary (*Rosmarinus officinalis* L.). *European Food Research and Technology* 221:478–486.

Cháfer, A., Berna, A., Montón, J., and Muñoz, R. 2002. High-pressure solubility data of system ethanol (1) + epicatechin (2) + CO_2 (3). *The Journal of Supercritical Fluids* 24:103–109.

Cháfer, A., Fornari, T., Berna, A., Ibáñez, E., and Reglero, G. 2005. Solubility of solid carnosic acid in supercritical CO_2 with ethanol as a co-solvent. *The Journal of Supercritical Fluids* 34:323–329.

Cháfer, A., Fornari, T., Berna, A., and Stateva, R.P. 2004. Solubility of quercetin in supercritical CO_2 + ethanol as a modifier: Measurements and thermodynamic modelling. *The Journal of Supercritical Fluids* 32:89–96.

Cháfer, A., Fornari, T., Stateva, R.P., Berna, A., and García-Reverter, J. 2007. Solubility of the natural antioxidant gallic acid in supercritical CO_2 + ethanol as a cosolvent. *Journal of Chemical and Engineering Data* 52:116–121.

Cháfer, A., Pascual-Martí, M.C., Salvador, A., and Berna, A. 2005. Supercritical fluid extraction and HPLC determination of relevant polyphenolic compounds in grape skin. *Journal of Separation Science* 28:2050–2056.

Chang, C.H., Chyau, C.C., Hsieh, C.L., Wu, Y.Y., Ker, Y.B., Tsen, H.Y., and Peng, R.Y. 2008. Relevance of phenolic diterpene constituents to antioxidant activity of supercritical CO_2 extract from the leaves of rosemary. *Natural Product Research* 22:76–90.

Chang, C.M.J., Chiu, K.L., Chen, Y.L., and Chang, C.Y. 2000. Separation of catechins from green tea using carbon dioxide extraction. *Food Chemistry* 68:109–113.

Chang, C.M.J., Chiu, K.L., Chen, Y.L., and Yang, P.W. 2001. Effect of ethanol content on carbon dioxide extraction of polyphenols from tea. *Journal of Food Composition and Analysis* 14:75–82.

Chen, C.R., Shen, C.T., Wu, J.J., Yang, H.L., Hsu, S.L., and Chang, C.M.J. 2009. Precipitation of sub-micron particles of 3,5-diprenyl-4-hydroxycinnamic acid in Brazilian propolis from supercritical carbon dioxide anti-solvent solutions. *The Journal of Supercritical Fluids* 50:176–182.

Chen, J.W. and Tsai, F.N. 1995. Solubilities of methoxybenzoic acid isomers in supercritical carbon dioxide. *Fluid Phase Equilibria* 107:189–200.

Choi, E.S., Noh, M.J., and Yoo, K.P. 1998. Solubilities of *o*-, *m*- and *p*-coumaric acid isomers in carbon dioxide at 308.15–323.15 K and 8.5–25 MPa. *Journal of Chemical and Engineering Data* 43:6–8.

Christov, M. and Dohrn, R. 2002. High-pressure fluid-phase equilibria: Experimental methods and systems investigated (1994–1999). *Fluid Phase Equilibria* 202:153–218.

Clifford, T. 1999. *Fundamentals of Supercritical Fluids*. Oxford, UK: Oxford University Press.

Conde, E., Moure, A., Domínguez, H., and Parajó, J.C. 2010. Extraction of natural antioxidants from plant foods. In *Separation, Extraction and Concentration Processes in the Food, Beverage and Nutraceutical Industries*, ed. S.S.H. Rizvi, pp. 506–594. Oxford, UK: Woodhead Publishing.

Cortesi, A., Kikic, I., Alessi, P., Turtoi, G., and Garnier, S. 1999. Effect of chemical structure on the solubility of antioxidants in supercritical carbon dioxide: Experimental data and correlation. *The Journal of Supercritical Fluids* 14:139–144.

Coutsikos, P., Magoulas, K., and Tassios, D. 1995. Solubilities of phenols in supercritical carbon dioxide. *Journal of Chemical and Engineering Data* 40:953–958.

Coutsikos, P., Voutsas, E., Magoulas, K., and Tassios, D.P. 2003. Prediction of vapor pressures of solid organic compounds with a group-contribution method. *Fluid Phase Equilibria* 207:263–281.

Da Porto, C., Decorti, D., and Natolino, A. 2014a. Water and ethanol as co-solvent in supercritical fluid extraction of proanthocyanidins from grape marc: A comparison and a proposal. *The Journal of Supercritical Fluids* 87:1–8.

Da Porto, C., Natolino, A., and Decorti, D. 2014b. Extraction of proanthocyanidins from grape marc by supercritical fluid extraction using CO_2 as solvent and ethanol–water mixture as co-solvent. *The Journal of Supercritical Fluids* 87:59–64.

Da Porto, C., Natolino, A., and Decorti, D. 2015. The combined extraction of polyphenols from grape marc: Ultrasound assisted extraction followed by supercritical CO_2 extraction of ultrasound-raffinate. *LWT: Food Science Technology* 61:98–104.

de Haan, A.B., and de Graauw, J. 1991. Mass transfer in supercritical extraction columns with structured packings for hydrocarbon processing. *Industrial and Engineering Chemistry Research* 30:2463–2470.

de Melo, M.M.R., Silvestre, A.J.D., and Silva, C. 2014. Supercritical fluid extraction of vegetable matrices: Applications, trends and future perspectives of a convincing green technology. *The Journal of Supercritical Fluids* 92:115–176.

del Valle, J.M. 2015. Extraction of natural compounds using supercritical CO_2: Going from the laboratory to the industrial application. *The Journal of Supercritical Fluids* 96:180–199.

del Valle, J.M. and Aguilera, J.M. 1989. Effects of substrate densification and CO_2 conditions on supercritical extraction of mushroom oleoresins. *Journal of Food Science* 54:135–141.

del Valle, J.M. and de la Fuente, J.C. 2006. Supercritical CO_2 extraction of oilseeds: Review of kinetic and equilibrium models. *Critical Reviews in Food Science and Nutrition* 46:131–160.

del Valle, J.M., de la Fuente, J.C., and Uquiche, E.L. 2012a. A refined equation for predicting the solubility of vegetable oils in high-pressure CO_2. *The Journal of Supercritical Fluids* 67:60–70.

del Valle, J.M., de la Fuente, J.C., Uquiche, E., Zetzl, C., and Brunner, G. 2011. Mass transfer and equilibrium parameters on high-pressure CO_2 extraction of plant essential oils. In *Food Engineering Interfaces*, eds. J.M. Aguilera, G.V. Barbosa-Cánovas, R. Simpson, J. Welti-Chanes, and D. Bermúdez-Aguirre, pp. 393–472. New York: Springer.

del Valle, J.M., Germain, J.C., Uquiche, E., Zetzl, C., and Brunner, G. 2006. Microstructural effects on internal mass transfer of lipids in prepressed and flaked vegetable substrates. *The Journal of Supercritical Fluids* 37:178–190.

del Valle, J.M., Glatzel, V., and Martínez, J.L. 2012b. Supercritical CO_2 extraction of allicin from garlic flakes: Screening and kinetic studies. *Food Research International* 45:216–224.

del Valle, J.M., Jiménez, M., and de la Fuente, J.C. 2003a. Extraction kinetics of pre-pelletized Jalapeño peppers with supercritical CO_2. *The Journal of Supercritical Fluids* 25:33–44.

del Valle, J.M., Jiménez, M., Napolitano, P., Zetzl, C., and Brunner, G. 2003b. Supercritical carbon dioxide extraction of pelletized Jalapeño peppers. *Journal of the Science of Food and Agriculture* 83:550–556.

del Valle, J.M. and Urrego, F. 2012. Free solute content and solute-matrix interactions affect apparent solubility and apparent solute content in supercritical CO_2 extractions. A hypothesis paper. *The Journal of Supercritical Fluids* 66:157–175.

Díaz-Reinoso, B., Moure, A., Domínguez, H., and Parajo, J.C. 2008. Antioxidants extraction by supercritical fluids. In *Supercritical Fluid Extraction of Nutraceutical and Bioactive Compounds*, ed. J.L. Martínez, pp. 275–303. Boca Raton, FL: CRC Press.

Díaz-Reinoso, B., Moure, A.S., Domínguez, H., and Parajo, J.C. 2006. Supercritical CO_2 extraction and purification of compounds with antioxidant activity. *Journal of Agricultural and Food Chemistry* 54:2441–2469.

Dobbs, J.M., Wong, J.M., and Johnston, K.P. 1986. Nonpolar co-solvents for solubility enhancement in supercritical fluid carbon dioxide. *Journal of Chemical and Engineering Data* 31:303–308.

Dobbs, J.M., Wong, J.M., Lahiere, R.J., and Johnston, K.P. 1987. Modification of supercritical fluid phase behavior using polar cosolvents. *Industrial and Engineering Chemistry Research* 26:56–65.

Dohrn, R. and Brunner, G. 1995. High-pressure fluid-phase equilibria: Experimental methods and systems investigated (1988–1993). *Fluid Phase Equilibria* 106:213–282.

Dohrn, R., Peper, S., and Fonseca, G.J.M.S. 2010. High-pressure fluid-phase equilibria: Experimental methods and systems investigated (2000–2004). *Fluid Phase Equilibria* 288:1–54.

Dunford, N.T., King, J.W., and List, G.R. 2003. Supercritical fluid extraction in food engineering. In *Extraction Optimization in Food Engineering*, eds. C. Tzia and G. Liadakis, pp. 57–93. New York: Marcel Dekker.

Edi-Soetaredjo, F., Ismadji, S., and Ju, Y.H. 2013a. Measurement and modeling of epicatechin solubility in supercritical carbon dioxide fluid. *Fluid Phase Equilibria* 340:7–10.

Edi-Soetaredjo, F., Ismadji, S., Liauw, M.Y., Angkawijaya, A.E., and Ju, Y.H. 2013b. Catechin sublimation pressure and solubility in supercritical carbon dioxide. *Fluid Phase Equilibria* 358:220–225.

Eggers, R. 1996. Supercritical fluid extraction (SFE) of oilseeds/lipids in natural products. In *Supercritical Fluid Technology in Oil and Lipid Chemistry*, eds. J.W. King and G.R. List, pp. 35–64. Champaign, IL: AOAC Press.

Eggers, R., Sievers, U., and Stein, W. 1985. High pressure extraction of oil seed. *Journal of the American Oil Chemists' Society* 62:1222–1230.

Esquivel, M.M., Ribeiro, M.A., and Bernardo-Gil, M.G. 1999. Supercritical extraction of savory oil: Study of antioxidant activity and extract characterization. *The Journal of Supercritical Fluids* 14:129–138.

Fages, J., Lochard, H., Letourneau, J.J., Sauceau, M., and Rodier, E. 2004. Particle generation for pharmaceutical applications using supercritical fluid technology. *Powder Technology* 141:219–226.

Fahmy, T.M., Paulaitis, M.E., Johnson, D.M., and McNally, M.P.E. 1993. Modifier effects in the supercritical fluid extraction of solutes from clay, soil, and plant materials. *Analytical Chemistry* 65:1462–1469.

Farías-Campomanes, A.M., Rostagno, M.A., and Meireles, M.A.A. 2013. Production of polyphenol extracts from grape bagasse using supercritical fluids: Yield, extract composition and economic evaluation. *The Journal of Supercritical Fluids* 77:70–78.

Fernández-Ponce, M.T., Casas, L., Mantell, C., and Martínez de la Ossa, E. 2014. Fractionation of Mangifera indica Linn polyphenols by reverse phase supercritical fluid chromatography (RP-SFC) at pilot plant scale. *The Journal of Supercritical Fluids* 95:444–456.

Fiori, L., De Faveri, D., Casazza, A.A., and Perego, P. 2009. Grape by-products: Extraction of polyphenolic compounds using supercritical CO_2 and liquid organic solvent—A preliminary investigation. *CyTA—Journal of Food Science* 7:163–171.

Floris, T., Filippino, G., Scrugli, S., Pinna, M.B., Argiolas, F., Argiolas, A., Murru, M., and Reverchon, E. 2010. Antioxidant compounds recovery from grape residues by a supercritical antisolvent assisted process. *The Journal of Supercritical Fluids* 54:165–170.

Fonseca, J.M.S., Dohrn, R., and Peper, S. 2011. High-pressure fluid-phase equilibria: Experimental methods and systems investigated (2005–2008). *Fluid Phase Equilibria* 300:1–69.

Fornari, R.E., Alessi, P., and Kikic, I. 1990. High pressure fluid phase equilibria: Experimental methods and systems investigated (1978–1987). *Fluid Phase Equilibria* 57:1–33.

Foster, N.R., Gurdial, G.S., Yun, J.S.L., Liong, K.K., Tilly, K.D., Ting, S.S.T., Singh, H., and Lee, J.H. 1991. Significance of the crossover pressure in solid-supercritical fluid phase equilibria. *Industrial and Engineering Chemistry Research* 30:1955–1964.

Fraile, M., Buratto, R., Gómez, B., Martín, Á., and Cocero, M.J. 2014. Enhanced delivery of quercetin by encapsulation in poloxamers by supercritical antisolvent process. *Industrial and Engineering Chemistry Research* 53:4318–4327.

Franceschi, E., De Cesaro, A.M., Ferreira, S.R.S., and Oliveira, J.V. 2009. Precipitation of beta-carotene microparticles from SEDS technique using supercritical CO_2. *Journal of Food Engineering* 95:656–663.

Galia, A., Argentino, A., Scialdone, O., and Filardo, G. 2002. A new simple static method for the determination of solubilities of condensed compounds in supercritical fluids. *The Journal of Supercritical Fluids* 24:7–17.

Germain, J.C., del Valle, J.M., and de la Fuente, J.C. 2005. Natural convection retards supercritical CO_2 extraction of essential oils and lipids from vegetable substrates. *Industrial and Engineering Chemistry Research* 44:2879–2886.

Ghafoor, K., Al-Juhaimi, F.Y., and Choi, Y.H. 2012. Supercritical fluid extraction of phenolic compounds and antioxidants from grape (*Vitis labrusca* B.) seeds. *Plant Foods Human Nutrition* 67:407–414.

Ghafoor, K., Park, J., and Choi, Y.H. 2010. Optimization of supercritical fluid extraction of bioactive compounds from grape (*Vitis labrusca* B.) peel by using response surface methodology. *Innovative Food Science and Emerging Technologies* 11:485–490.

Ghiasvand, A.R., Hosseini, M., Sharghi, H., Yamini, Y., and Shamsipur, M. 1999. Solubilities of some hydroxyxanthone derivatives in supercritical carbon dioxide. *Journal of Chemical and Engineering Data* 44:1135–1138.

Ghoreishi, S.M. and Heidari, E. 2012. Extraction of epigallocatechin gallate from green tea via modified supercritical CO_2: Experimental, modeling and optimization. *The Journal of Supercritical Fluids* 72:36–45.

Ghoreishi, S.M. and Heidari, E. 2013. Extraction of epigallocatechin-3-gallate from green tea via supercritical fluid technology: Neural network modeling and response surface optimization. *The Journal of Supercritical Fluids* 74:128–136.

Gopalan, B., Goto, M., Kodama, A., and Hirose, T. 2000. Supercritical carbon dioxide extraction of turmeric (*Curcuma longa*). *Journal of Agricultural and Food Chemistry* 48:2189–2192.

Goto, M. 2002. Extraction and separation of natural products. In *Supercritical Fluids: Molecular Interactions, Physical Properties, and New Applications*, eds. Y. Arai, T. Sako, and Y. Takebayashi, pp. 239–259 and 278–280. Berlin, Germany: Springer.

Goto, M., Roy, B.C., and Hirose, T. 1996. Shrinking-core leaching model for supercritical-fluid extraction. *The Journal of Supercritical Fluids* 9:128–133.

Goto, M., Sato, M., and Hirose, T. 1993. Extraction of peppermint oil by supercritical carbon dioxide. *Journal of Chemical Engineering of Japan* 26:401–407.

Gupta, R.B. and Shim, J.J. 2007. *Solubility in Supercritical Carbon Dioxide*. Boca Raton, FL: CRC Press.

Gurdial, G.S. and Foster, N.R. 1991. Solubility of o-hydroxybenzoic acid in supercritical carbon dioxide. *Industrial and Engineering Chemistry Research* 30:575–580.

Gurdial, G.S., Macnaughton, S.J., Tomasko, D.L., and Foster, N.R. 1993. Influence of chemical modifiers on the solubility of o-and m-hydroxybenzoic acid in supercritical carbon dioxide. *Industrial and Engineering Chemistry Research* 32:1488–1497.

Gutiérrez, J.E., Bejarano, A., and de la Fuente, J.C. 2010. Measurement and modeling of high-pressure (vapour+ liquid) equilibria of (CO_2 + alcohol) binary systems. *The Journal of Chemical Thermodynamics* 42:591–596.

Hakuta, Y., Hayashi, H., and Arai, K. 2003. Fine particle formation using supercritical fluids. *Current Opinion in Solid State and Materials Science* 7:341–351.

Hallberg, W., Jekat, H., Schutz, E., Stork, K., and Vollbrecht, H.R. 1989. Method of extraction hop pellets with liquified carbon dioxide. U.S. Patent 4,828,867.

Hanley, B. and Chen, C.C. 2012. New mass-transfer correlations for packed towers. *AIChE Journal* 58:132–152.

Hatami, T., Cavalcanti, R.N., Takeuchi, T.M., and Meireles, M.A.A. 2012. Supercritical fluid extraction of bioactive compounds from macela (*Achyrocline satureioides*) flowers: Kinetic, experiments and modeling. *Journal of Supercritical Fluids* 65:71–77.

Hawthorne, S.B. and Miller, D.J. 1994. Direct comparison of Soxhlet and low- and high-temperature supercritical CO_2 extraction efficiencies of organics from environmental solids. *Analytical Chemistry* 66:4005–4012.

Heidari, E. and Ghoreishi, S.M. 2013. Prediction of supercritical extraction recovery of EGCG using hybrid of Adaptive Neuro-Fuzzy Inference System and mathematical model. *The Journal of Supercritical Fluids* 82:158–167.

Herrero, M., Cifuentes, A., and Ibáñez, E. 2006. Sub- and supercritical fluid extraction of functional ingredients from different natural sources: Plants, food-by-products, algae and microalgae: A review. *Food Chemistry* 98:136–148.

Herrero, M., Mendiola, J.A., Cifuentes, A., and Ibáñez, E. 2010. Supercritical fluid extraction: Recent advances and applications. *Journal of Chromatography A* 1217:2495–2511.

Herzi, N., Camy, S., Bouajila, J., Destrac, P., Romdhane, M., and Condoret, J.S. 2013. Supercritical CO_2 extraction of *Tetraclinis articulata*: Chemical composition, antioxidant activity and mathematical modeling. *The Journal of Supercritical Fluids* 82:72–82.

Hirata, H. 2002. Separation and analysis by supercritical fluid chromatography. In *Supercritical Fluids: Molecular Interactions, Physical Properties, and New Applications*, eds. Y. Arai, T. Sako, and Y. Takebayashi, pp. 260–276 and 280. Berlin, Germany: Springer.

Huang, K.J., Wu, J.J., Chiu, Y.H., Lai, C.Y., and Chang, C.M.J. 2007. Designed polar cosolvent-modified supercritical CO_2 removing caffeine from and retaining catechins in green tea powder using response surface methodology. *Journal of Agricultural Food Chemistry* 55:9014–9020.

Huang, Z., Shi, X.H., and Jiang, W.J. 2012. Theoretical models for supercritical fluid extraction. *Journal of Chromatography A* 1250:2–26.

Ibáñez, E., Oca, A., de Murga, G., López-Sebastián, S., Tabera, J., and Reglero, G. 1999. Supercritical fluid extraction and fractionation of different preprocessed rosemary plants. *Journal of Agricultural and Food Chemistry* 47:1400–1404.

Ivanovic, J., Djilas, S., Jadranin, M., Vajs, V., Babovic, N., Petrovic, S., and Zizovic, I. 2009. Supercritical carbon dioxide extraction of antioxidants from rosemary (*Rosmarinus officinalis* L.) and sage (*Salvia officinalis* L.). *Journal of Serbian Chemical Society* 74:717–732.

Iwai, Y., Yamamoto, H., Tanaka, Y., and Arai, Y. 1990. Solubilities of 2,5- and 2,6-xylenols in supercritical carbon dioxide. *Journal of Chemical and Engineering Data* 35:174–176.

Joshi, D.K. and Prausnitz, J.M. 1984. Supercritical fluid extraction with mixed solvents. *AIChE Journal* 30:522–525.

Jung, J. and Perrut, M. 2001. Particle design using supercritical fluids: Literature and patent survey. *The Journal of Supercritical Fluids* 20:179–219.

Ke, J., Mao, C., Zhong, M., Han, B., and Yan, H. 1996. Solubilities of salicylic acid in supercritical carbon dioxide with ethanol cosolvent. *The Journal of Supercritical Fluids* 9:82–87.

Kim, W.J., Kim, J.D., Kim, J.H., Oh, S.G., and Lee, Y.W. 2008. Selective caffeine removal from green tea using supercritical carbon dioxide extraction. *Journal of Food Engineering* 89:303–309.

King, J.W. 2014. Modern supercritical fluid technology for food applications. *Annual Review of Food Science and Technology* 5:215–238.

King, J.W. and France, J.E. 1992. Basic principles of analytical supercritical fluid extraction. In *Analysis with Supercritical Fluids: Extraction and Chromatography*, ed. B. Wenclawiak, pp. 32–60. Berlin, Germany: Springer.

King, J.W. and Srinivas, K. 2009. Multiple unit processing using sub- and supercritical fluids. *The Journal of Supercritical Fluids* 47:598–610.

King, J.W. and Srinivas, K. 2015. Development of multiple unit-fluid processes and bio-refineries using critical fluids. In *High Pressure Fluid Technology for Green Food Processing*, eds. T. Fornari and R.P. Stateva, pp. 455–478. Cham, Switzerland: Springer.

King, M.B. and Catchpole, O. 1993. Physico-chemical data required for the design of near-critical fluid extraction process. In *Extraction of Natural Products Using Near-Critical Solvents*, eds. M.B. King and T.R. Bott, pp. 184–231. London, UK: Blackie Academic & Professional.

Kitamura, M. and Nagahama, K. 2002. Crystallization using a supercritical fluid. In *Supercritical Fluids: Molecular Interactions, Physical Properties, and New Applications*, eds. Y. Arai, T. Sako, and Y. Takebayashi, pp. 299–319 and 343. Berlin, Germany: Springer.

Knez, Ž. 2006. Particle formation using supercritical fluids—A short review. *Chemical Industry and Chemical Engineering Quarterly* 12:141–146.

Knez, Ž., Knez Hrnčič, M., and Škerget, M. 2015. Particle formation and product formulation using supercritical fluids. *Annual Review of Chemical and Biomolecular Engineering* 6:379–407.

Knez, Ž., Škerget, M., and Knez-Hrnčič, M. 2010. Principles of supercritical fluid extraction and applications in the food, beverage and nutraceutical industries. In *Separation, Extraction and Concentration Processes in the Food, Beverage and Nutraceutical Industries*, ed. S.S.H. Rizvi, pp. 3–38. Oxford, UK: Woodhead Publishing.

Knez, Ž. and Weidner, E. 2001. Precipitation of solids with dense gases. In *High Pressure Process Technology: Fundamentals and Applications*, eds. A. Bertucco and G. Vetter, pp. 587–611. Amsterdam, the Netherlands: Elsevier Science.

Knez, Ž. and Weidner, E. 2003. Particles formation and particle design using supercritical fluids. *Current Opinion in Solid State and Materials Science* 7:353–361.

Krukonis, V.J. and Kurnik, R.T. 1985. Solubility of solid aromatic isomers in carbon dioxide. *Journal of Chemical and Engineering Data* 30:247–249.

Kuo, C.F., Su, J.D., Chiu, C.H., Peng, C.C., Chang, C.H., Sung, T.Y., Huang, S.H., Lee, W.C., and Chyau, C.C. 2011. Anti-inflammatory effects of supercritical carbon dioxide extract and its isolated carnosic acid from *Rosmarinus officinalis* leaves. *Journal of Agricultural and Food Chemistry* 59:3674–3685.

Kurnik, R.T., Holla, S.J., and Reid, R.C. 1981. Solubility of solids in supercritical carbon dioxide and ethylene. *Journal of Chemical and Engineering Data* 26:47–51.

Kurnik, R.T. and Reid, R.C. 1982. Solubility of solid mixtures in supercritical fluids. *Fluid Phase Equilibria* 8:93–105.

Lack, E., Gamse, T., and Marr, R. 2001. Extraction from solids. In *High Pressure Process Technology: Fundamentals and Applications*, eds. A. Bertucco and G. Vetter, pp. 378–395. Amsterdam, the Netherlands: Elsevier Science.

Lack, E. and Seidlitz, H. 1993. Commercial scale decaffeination of coffee and tea using supercritical CO_2. In *Extraction of Natural Products Using Near-Critical Solvents*, eds. M.B. King and T.R. Bott, pp. 34–49, London, UK: Blackie Academic & Professional.

Langenfeld, J.J., Hawthorne, S.B., Miller, D.J., and Pawliszynl, J. 1993. Effects of temperature and pressure on supercritical fluid extraction efficiencies of polycyclic aromatic hydrocarbons and polychlorinated biphenyls. *Analytical Chemistry* 65:338–334.

Laurent, A. 2001a. Pressure distillation. In *High Pressure Process Technology: Fundamentals and Applications*, eds. A. Bertucco and G. Vetter, pp. 352–368, 376, and 377. Amsterdam, the Netherlands: Elsevier Science.

Laurent, A. 2001b. Packed towers: Random and structured packings. In *High Pressure Process Technology: Fundamentals and Applications*, eds. A. Bertucco and G. Vetter, pp. 368–371, 376, and 377. Amsterdam, the Netherlands: Elsevier Science.

Leal, P.F., Braga, M.E.M., Sato, D.N., Carvalho, J.E., Marques, M.O.M., and Meireles, M.A.A. 2003. Functional properties of spice extracts obtained via supercritical fluid extraction. *Journal of Agricultural and Food Chemistry* 51:2520–2525.

Le Floch, F., Tena, M. T., Rıos, A., and Valcarcel, M. 1998. Supercritical fluid extraction of phenol compounds from olive leaves. *Talanta* 46:1123–1130.

Lee, S., Park, M.K., Kim, K.H., and Kim, Y.S. 2007. Effect of supercritical carbon dioxide decaffeination on volatile components of green teas. *Journal of Food Science* 72:S497–S502.

Leeke, G.A., Santos, R., and King, M.B. 2001. Vapor-liquid equilibria for the carbon dioxide + carvacrol system at elevated pressures. *Journal of Chemical and Engineering Data* 46:541–545.

Lemmon, E.W., Huber, M.L., and McLinden, M.O. 2010. NIST Standard Reference Database 23: Mini-Reference Fluid Thermodynamic and Transport Properties (REFPROP), Version 9.0. National Institute of Standards and Technology, Standard Reference Data Program, Gaithersburg, MD.

Ligor, M., Kornyšova, O., Maruška, A., and Buszewski, B. 2008. Determination of flavonoids in tea and Rooibos extracts by TLC and HPLC. *Journal of Planar Chromatography* 21:355–360.

Louli, V., Ragoussis, N., and Magoulas, K. 2004. Recovery of phenolic antioxidants from wine industry by-products. *Bioresource Technology* 92:201–208.

Lucien, F.P. and Foster, N.R. 1996. Influence of matrix composition on the solubility of hydroxybenzoic acid isomers in supercritical carbon dioxide. *Industrial and Engineering Chemistry Research* 35:4686–4699.

Luengthanaphol, S., Mongkholkhajornsilp, D., Douglas, S., Douglas, P.L., Pengsopa, L.I., and Pongamphai, S. 2004. Extraction of antioxidants from sweet Thai tamarind seed coat : preliminary experiments. *Journal of Food Engineering* 63:247–252.

Mantell, C., Rodríguez, M., and Martínez de la Ossa, E. 2003. A screening analysis of the high-pressure extraction of anthocyanins from red grape pomace with carbon dioxide and cosolvent. *Engineering Life Sciences* 3:38–42.

Maran, J.P., Manikandan, S., Priya, B., and Gurumoorthi, P. 2013. Box-Behnken design based multi-response analysis and optimization of supercritical carbon dioxide extraction of bioactive flavonoid compounds from tea (*Camellia sinensis* L.) leaves. *Journal of Food Science and Technology* 52:92–104.

Maróstica, M.R. Jr., Leite, A.V., and Dragano, N.R.V. 2010. Supercritical fluid extraction and stabilization of phenolic compounds from natural sources—A review. *The Open Chemical Engineering Journal* 4:51–60.

Marqués, J.L., Della Porta, G., Reverchon, E., Renuncio, J.A.R., and Mainar, A.M. 2013. Supercritical antisolvent extraction of antioxidants from grape seeds after vinification. *The Journal of Supercritical Fluids* 82:238–243.

Martín, A. and Cocero, M.J. 2008. Micronization processes with supercritical fluids: Fundamentals and mechanisms. *Advanced Drug Delivery Reviews* 60:339–350.

Martín, Á., Varona, S., Navarrete, A., and Cocero, M.J. 2010. Encapsulation and co-precipitation processes with supercritical fluids: Applications with essential oils. *Open Chemical Engineering Journal* 4:31–41.

Martínez, J.L. and Vance, S.W. 2008. Supercritical extraction plants: Equipment, process, and cost. In *Supercritical Fluid Extraction of Nutraceutical and Bioactive Compounds*, ed. J.L. Martínez, pp. 25–49. Boca Raton, FL: CRC Press.

McHugh, M. and Krukonis, V. 1994. *Supercritical Fluid Extraction*, 2nd ed. Boston, MA: Butterworth-Heinemann.

Meneses, M.A., Caputo, G., Scognamiglio, M., Reverchon, E., and Adami, R. 2015. Antioxidant phenolic compounds recovery from *Mangifera indica* L. by-products by supercritical antisolvent extraction. *Journal of Food Engineering* 163:45–53.

Meterc, D., Cocero, M.J., and Weidner, E. 2009. Comparing PGSS and GAS process for drying of green tea extracts. *Proceedings of the Ninth International Symposium on Supercritical Fluids*, Arcachon, France.

Meterc, D., Petermann, M., and Weidner, E. 2008. Drying of aqueous green tea extracts using a supercritical fluid spray process. *The Journal of Supercritical Fluids* 45:253–259.

Miao, S.F., Yu, J.P., Du, Z., Guan, Y.X., Yao, S.J., and Zhu, Z.Q. 2010. Supercritical fluid extraction and micronization of ginkgo flavonoids from *Ginkgo biloba* leaves. *Industrial and Engineering Chemistry Research* 49:5461–5466.

Miller, L. and Sebastian, I. 2012.Evaluation of injection conditions for preparative supercritical fluid chromatography. *Journal of Chromatography A* 1250:256–263.

Montañés, F., Catchpole, O.J., Tallon, S.J., Rose, P.M., and Moreno, T. 2015. Advances in analytical and preparative supercritical fluid chromatography. In *High Pressure Fluid Technology for Green Food Processing*, eds. T. Fornari and R.P. Stateva, pp. 217–268. Cham, Switzerland: Springer.

Moraes, M.N., Zabot, G.L., Prado, J.M., and Meireles, M.A.A. 2013. Obtaining antioxidants from botanic matrices applying novel extraction techniques. *Food Public Health* 3:195–214.

Mori, Y., Shimizu, T., Iwai, Y., and Arai, Y. 1992. Solubilities of 3,4-xylenol and naphthalene + 2,5-xylenol in supercritical carbon dioxide at 35°C. *Journal of Chemical and Engineering Data* 37:317–319.

Moyler, D.A. 1993. Extraction of flavours and fragrances with compressed CO_2. In *Extraction of Natural Products Using Near-Critical Solvents*, eds. M.B. King and T.R. Bott, pp. 140–183. London, UK: Blackie Academic & Professional.

Mukhopadyay, M. 2000. *Natural Extracts Using Supercritical Carbon Dioxide*. Boca Raton, FL: CRC Press.

Mukhopadhyay M. and De, S.K. 1995. Fluid phase behavior of close molecular weight fine chemicals with supercritical carbon dioxide. *Journal of Chemical and Engineering Data* 40:909–913.

Murga, R., Ruiz, R., Beltrán, S., and Cabezas, J.L. 2000. Extraction of natural complex phenols and tannins from grape seeds by supercritical mixtures of carbon dioxide and alcohol. *Journal of Agricultural and Food Chemistry* 48:3408–3412.

Murga, R., Sanz, M.T., Beltrán, S., and Cabezas, J.L. 2002. Solubility of some phenolic compounds contained in grape seeds, in supercritical carbon dioxide. *The Journal of Supercritical Fluids* 23:113–121.

Murga, R., Sanz, M.T., Beltrán, S., and Cabezas, J.L. 2003. Solubility of three hydroxycinnamic acids in supercritical carbon dioxide. *The Journal of Supercritical Fluids* 27:239–245.

Murga, R., Sanz, M.T., Beltrán, S., and Cabezas, J.L. 2004. Solubility of syringic and vanillic acids in supercritical carbon dioxide. *Journal of Chemical and Engineering Data* 49:779–782.

Nagahama, K. 2002. Principles for supercritical fluid separation. In *Supercritical Fluids: Molecular Interactions, Physical Properties, and New Applications*, eds. Y. Arai, T. Sako, and Y. Takebayashi, pp. 207–224, 276, and 277. Berlin, Germany: Springer.

Nakada, M., Imai, M., and Suzuki, I. 2009. Impact of ethanol addition on the solubility of various soybean isoflavones in supercritical carbon dioxide and the effect of glycoside chain in isoflavones. *Journal of Food Engineering* 95:564–571.

Oliveira, D.A., Salvador, A.A., Smânia A. Jr., Smânia, E.F.A., Maraschin, M., and Ferreira, S.R.S. 2013. Antimicrobial activity and composition profile of grape (*Vitis vinifera*) pomace extracts obtained by supercritical fluids. *Journal of Biotechnology* 164:423–432.

Oliveira, E.L.G.G., Silvestre, A.J.D.D., and Silva, C.M. 2011. Review of kinetic models for supercritical fluid extraction. *Chemical Engineering Research and Design* 89:1104–1117.

Oman, M., Skerget, M., and Knez, Z. 2013. Applications of supercritical fluid extraction for the separation of nutraceuticals and other phytochemicals from plant material. *Macedonian Journal of Chemistry Chemical Engineering* 32:183–226.

Omar, J., Alonso, I., Garaikoetxea, A., and Etxebarria, N. 2013. Optimization of focused ultrasound extraction and supercritical fluid extraction of volatile compounds and antioxidants from aromatic plants. *Food Analytical Methods* 6:1611–1620.

Oreopoulou, V. 2003. Extraction of natural antioxidants. In *Extraction Optimization in Food Engineering*, eds. Tzia, C. and G. Liadakis, pp. 329–346. New York: Marcel Dekker.

Osorio-Tobón, J.F., Carvalho, P.I.N., Rostagno, M.A., Petenate, A.J., and Meireles, M.A.A. 2016. Precipitation of curcuminoids from an ethanolic turmeric extract using a supercritical antisolvent process. *The Journal of Supercritical Fluids* 108:26–34.

Palenzuela, B., Arce, L., Macho, A., Muñoz, E., Ríos, A., and Valcárcel, M. 2004. Bioguided extraction of polyphenols from grape marc by using an alternative supercritical-fluid extraction method based on a liquid solvent trap. *Analytical and Bioanalytical Chemistry* 378:202–2027.

Palenzuela, B., Rodríguez-Amaro, R., Ríos, A., and Valcarcel, M. 2002. Screening of polyphenols in grape marc by on-line supercritical fluid extraction–amperometric detection with a PVC-graphite composite electrode. *Electroanalysis* 14:1427–1432.

Palma, M. and Taylor, L.T. 1999. Extraction of polyphenolic compounds from grape seeds with near critical carbon dioxide. *Journal of Chromatography A* 849:117–124.

Palma, M., Taylor, L.T., Varela, R.M., Cutler, S.J., and Cutler, H.G. 1999. Fractional extraction of compounds from grape seeds by supercritical fluid extraction and analysis for antimicrobial and agrochemical activities. *Journal of Agricultural Food Chemistry* 47:5044–5048.

Parajó, J.C., Domínguez, H., Moure, A., and Díaz-Reinoso, B. 2009. Obtaining antioxidants by supercritical fluid extraction. In *Extracting Bioactive Compounds for Food Products: Theory and Applications*, ed. M.A.A. Meireles, s. 6.2, pp. 288–327. Boca Raton, FL: CRC Press.

Park, H.S., Choi, H.K., Lee, S.J., Park, K.W., Choi, S.G., and Kim, K.H. 2007a. Effect of mass transfer on the removal of caffeine from green tea by supercritical carbon dioxide. *The Journal of Supercritical Fluids* 42:205–211.

Park, H.S., Lee, H.J., Shin, M.H., Lee, K.W., Lee, H.J., Kim, Y.S., Kim, K.O., and Kim, K.H. 2007b. Effects of cosolvents on the decaffeination of green tea by supercritical carbon dioxide. *Food Chemistry* 105:1011–1117.

Pascual-Martí, M.C., Salvador, A., Chafer, A., and Berna, A. 2001. Supercritical fluid extraction of resveratrol from grape skin of *Vitis vinifera* and determination by HPLC. *Talanta* 54:735–740.

Pereira, P.G. and Meireles, M.A.A. 2010. Supercritical fluid extraction of bioactive compounds: Fundamentals, applications and economic perspectives. *Food and Bioprocess Technology* 3:340–372.

Persson, P., Barisic, Z., Cohen, A., Thörneby, L., and Gorton, L. 2002. Countercurrent supercritical fluid extraction of phenolic compounds from aqueous matrices. *Analytica Chimica Acta* 460:1–12.

Perva-Uzunalic A., Skerget, M., Knez, Z., Weinreich, B., Otto, F., and Grüner, S. 2006. Extraction of active ingredients from green tea (*Camellia sinensis*): Extraction efficiency of major catechins and caffeine. *Food Chemistry* 96:597–605.

Pilavtepe, M., Yucel, M., Helvaci, S.S., Demircioglu, M., and Yesil-Celiktas, O. 2012. Optimization and mathematical modeling of mass transfer between *Zostera marina* residues and supercritical CO_2 modified with ethanol. *The Journal of Supercritical Fluids* 68:87–93.

Pinelo, M., Ruiz-Rodríguez, A., Sineiro, J., Señoráns, F.J., Reglero, G., and Núñez, M.J. 2007. Supercritical fluid and solid–liquid extraction of phenolic antioxidants from grape pomace: A comparative study. *European Food Research and Technology* 226:199–205.

Puiggené, J., Larrayoz, M.A., and Recasens, F. 1997. Free liquid-to-supercritical fluid mass transfer in packed beds. *Chemical Engineering Science* 52:195–212.

Rahimi, E., Prado, J.M., Zahedi, G., and Meireles, M.A.A. 2011. Chamomile extraction with supercritical carbon dioxide: Mathematical modeling and optimization. *The Journal of Supercritical Fluids* 56:80–88.

Rajendran, A. 2012. Design of preparative-supercritical fluid chromatography. *Journal of Chromatography A* 1250:227–249.

Ramírez, P., Fornari, T., Señoráns, F.J., Ibáñez, E., and Reglero, G. 2005. Isolation of phenolic antioxidant compounds by SFC. *The Journal of Supercritical Fluids* 35:128–132.

Ramírez, P., García-Risco, M.R., Santoyo, S., Señoráns, F.J., Ibáñez, E., and Reglero, G. 2006. Isolation of functional ingredients from rosemary by preparative-supercritical fluid chromatography (Prep-SFC). *Journal of Pharmaceutical and Biomedical Analysis* 41:1606–1613.

Ramírez, P., Santoyo, S., García-Risco, M.R., Señoráns, F.J., Ibáñez, E., and Reglero, G. 2007. Use of specially designed columns for antioxidants and antimicrobials enrichment by preparative supercritical fluid chromatography. *Journal of Chromatography A* 1143:234–242.

Ramírez, P., Señoráns, F.J., Ibáñez, E., and Reglero, G. 2004. Separation of rosemary antioxidant compounds by supercritical fluid chromatography on coated packed capillary columns. *Journal of Chromatography A* 1057:241–245.

Reverchon, E. 1999. Supercritical antisolvent precipitation of micro- and nano-particles. *Journal of Supercritical Fluids* 15:1–21.

Reverchon, E. 2002. Supercritical-assisted atomization to produce micro- and/or nanoparticles of controlled size and distribution. *Industrial and Engineering Chemistry Research* 41:2405–2411.

Reverchon, E. and Marrone, C. 2001. Modeling and simulation of the supercritical CO_2 extraction of vegetable oils. *Journal of Supercritical Fluids* 19:161–175.

Reverchon, E. and Senatore, F. 1992. Isolation of rosemary oil: Comparison between hydrodistillation and supercritical CO_2 extraction. *Flavour and Fragrance Journal* 7:227–230.

Ribeiro, M.A., Bernardo-Gil, M.G., and Esquivel, M.M. 2001. *Melissa officinalis* L.: Study of antioxidant activity in supercritical residues. *Journal of Supercritical Fluids* 21:51–60.

Rodríguez-Meizoso, I., Castro-Puyana, M., Börjesson, P., Mendiola, J.A., Turner, C., and Ibáñez, E. 2012. Life cycle assessment of green pilot-scale extraction processes to obtain potent antioxidants from rosemary leaves. *Journal of Supercritical Fluids* 72:205–212.

Rodríguez-Meizoso, I. and Plaza, M. 2015. Particle formation of food ingredients by supercritical fluid technology. In *High Pressure Fluid Technology for Green Food Processing*, eds. T. Fornari and R.P. Stateva, pp. 155–183. Cham, Switzerland: Springer.

Rodríguez-Meizoso, I., Turner, C., Ibáñez, E., Cifuentes, A., Mendiola, J.A., Reglero, G., and Señoráns, F.J. 2009. Dispositivo y procedimiento para la extracción y secado en línea de extractos complejos. January 21, available at: http://digital.csic.es/handle/10261/40685 (accessed November 25, 2015).

Rombaut, N., Savoire, R., Thomasset, B., Bélliard, T., Castello, J., Van Hecke, É., and Lanoisellé, J.L. 2014. Grape seed oil extraction: Interest of supercritical fluid extraction and gas-assisted mechanical extraction for enhancing polyphenol co-extraction in oil. *Comptes Rendus Chimie* 17:284–292.

Rosa, P.T.V. and Meireles, M.A.A. 2009. Fundamentals of supercritical extraction from solid matrices. In *Extracting Bioactive Compounds for Food Products: Theory and Applications*, ed. M.A.A. Meireles, S. 6.1, pp. 272–288. Boca Raton, FL: CRC Press.

Ruivo, R.M., Paiva, A., Mota, J.P.B., and Simões, P.C. 2004. Dynamic model of a counter-current packed column operating at high pressure conditions. *Journal of Supercritical Fluids* 32:183–192.

Saldaña, M.D., Zetzl, C., Mohamed, R.S., and Brunner, G. 2002. Decaffeination of guaraná seeds in a microextraction column using water-saturated CO_2. *Journal of Supercritical Fluids* 22:119–127.

Sánchez-Camargo, A.D.P., Valdés, A., Sullini, G., García-Cañas, V., Cifuentes, A., Ibáñez, E., and Herrero, M. 2014. Two-step sequential supercritical fluid extracts from rosemary with enhanced anti-proliferative activity. *Journal of Functional Foods* 11:293–303.

Sanders, N. 1993. Food legislation and the scope for increased use of near-critical fluid extraction operations in the food, flavouring and pharmaceutical industries. In *Extraction of Natural Products Using Near-Critical Solvents*, eds. M.B. King and T.R. Bott, pp. 34–49. London, UK: Blackie Academic & Professional.

Santos, D. and Meireles, M.A.A. 2013. Micronization and encapsulation of functional pigments using supercritical carbon dioxide. *Journal of Food Process Engineering* 36:36–49.

Schmitt, W.J. and Reid, R.C. 1986. Solubility of monofunctional organic solids in chemically diverse supercritical fluids. *Journal of Chemical and Engineering Data* 31:204–212.

Schwarz, K., Ternes, W., and Schmauderer, E. 1992. Antioxidative constituents of *Rosmarinus officinalis* and *Salvia officinalis* III. Stability of phenolic diterpenes of rosemary extracts under thermal stress as required for technological process. *Zeitschrift für Lebensmittel-Untersuchung und Forschung* 195:104–107.

Señoráns, F.J., Ibáñez, E., Cavero, S., Tabera, J., and Reglero, G. 2000. Liquid chromatographic-mass spectrometric analysis of supercritical-fluid extracts of rosemary plants. *Journal of Chromatography A* 870:491–499.

Señoráns, F.J., Ruiz-Rodríguez, A., Cavero, S., Cifuentes, A., Ibáñez, E., and Reglero, G. 2001. Isolation of antioxidant compounds from orange juice by using countercurrent supercritical fluid extraction (CC-SFE). *Journal of Agricultural and Food Chemistry* 49:6039–6044.

Shariati, A. and Peters, C.J. 2003. Recent developments in particle design using supercritical fluids. *Current Opinion in Solid State and Materials Science* 7:371–383.

Shi, L.E., Zhang, Z.L., Xing, L.Y., Yang, D.D., Guo, Y.P., Guo, X.F., Zhao, L.M., and Tang, Z.X. 2011. Antioxidants extraction by supercritical CO_2. *Journal of Medicinal Plants Research* 5:300–308.

Sielfeld, C., del Valle, J.M., and Sastre, F. Effect of pelletization on supercritical CO_2 extraction of rosemary antioxidants. *Journal of Supercritical Fluids* (in press), doi: 10.1016/j.fluid.2016.04.010.

Silva, E.K. and Meireles, M.A.A. 2014. Encapsulation of food compounds using supercritical technologies: Applications of supercritical carbon dioxide as an antisolvent. *Food and Public Health* 4:247–258.

Simó, C., Ibáñez, E., Señoráns, F.J., Barbas, C., Reglero, G., and Cifuentes, A. 2002. Analysis of antioxidants from orange juice obtained by countercurrent supercritical fluid extraction, using micellar electrokinetic chromatography and reverse-phase liquid chromatography. *Journal of Agricultural and Food Chemistry* 50:6648–6652.

Sinnott, R.K. 1999. Separation columns (distillation, absorption and extraction). In *Chemical Engineering*, 3rd ed., eds. J.M. Coulson and J.F. Richardson, pp. 592–603. Oxford, UK: Butterworth-Heinemann.

Škerget, M., Knez, Ž., and Knez-Hrnčič, M. 2011. Solubility of solids in sub- and supercritical fluids: A review. *Journal of Chemical and Engineering Data* 56:694–719.

Song, Q., Zhu, J., Wan, J., and Cao, X. 2010. Measurement and modeling of epigallocatechin gallate solubility in supercritical carbon dioxide fluid with ethanol cosolvent. *Journal of Chemical and Engineering Data* 55:3946–3951.

Sonkaew, P., Sane, A., and Suppakul, P. 2012. Antioxidant activities of curcumin and ascorbyl dipalmitate nanoparticles and their activities after incorporation into cellulose-based packaging films. *Journal of Agricultural and Food Chemistry* 60:5388–5399.

Sosa, M.V., Rodríguez-Rojo, S., Mattea, F., Cismondi, M., and Cocero, M.J. 2011. Green tea encapsulation by means of high pressure antisolvent coprecipitation. *Journal of Supercritical Fluids* 56:304–311.

Souza-Machado, B.A., Pereira, C.G., Nunes, S.B., Padilha, F.F., and Umsza-Guez, M.A. 2013. Supercritical fluid extraction using CO_2: Main applications and future perspectives. *Separation Science and Technology* 48:2471–2760.

Sovová, H. 2001. Solubility of ferulic acid in supercritical carbon dioxide with ethanol as cosolvent. *Journal of Chemical and Engineering Data* 46:1255–1257.

Sovová, H. and Stateva, R.P. 2011.Supercritical fluid extraction of vegetable materials. *Reviews in Chemical Engineering* 27:79–156.

Stahl, E., Quirin, K.W., and Gerard, D. 1988. *Dense Gases for Extraction and Refining.* Berlin, Germany: Springer.

Stanley-Wood, N.G. 1990. Size enlargement. In *Principles of Powder Technology*, ed. M.J. Rhodes, pp. 193–226. Chichester, UK: John Wiley & Sons.

Stassi, A., Bettini, R., Gazzaniga, A., Giordano, F., and Schiraldi, A. 2000. Assessment of solubility of ketoprofen and vanillic acid in supercritical CO_2 under dynamic conditions. *Journal of Chemical and Engineering Data* 45:161–165.

Stockfleth, R. and Brunner, G. 1999. Hydrodynamics of a packed countercurrent column for the gas extraction. *Industrial and Engineering Chemistry Research* 38:4000–4006.

Stockfleth, R. and Brunner, G. 2001a. Holdup, pressure drop, and flooding in packed countercurrent columns for the gas extraction. *Industrial and Engineering Chemistry Research* 40:347–356.

Stockfleth, R. and Brunner, G. 2001b. Film thickness, flow regimes, and flooding in countercurrent annular flow of a falling film at high pressures. *Industrial and Engineering Chemistry Research* 40:6014–6020.

Sun, Q.L., Hua, S., Ye, J.H., Lu, J.L., Zheng, X.Q., and Liang, Y.R. 2010. Decaffeination of green tea by supercritical carbon dioxide. *Journal of Medicinal Plants Research* 4:1161–1168.

Sun, Y., Li, S., and Quan, C. 2005. Solubility of ferulic acid and tetramethylpyrazine in supercritical carbon dioxide. *Journal of Chemical and Engineering Data* 50:1125–1128.

Sunarso, J. and Ismadji, S. 2009. Decontamination of hazardous substances from solid matrices and liquids using supercritical fluids extraction: A review. *Journal of Hazardous Materials* 161:1–20.

Tabera, J., Guinda, Á., Ruiz-Rodríguez, A., Señoráns, F.J., Ibáñez, E., Albi, T., and Reglero, G. 2004. Countercurrent supercritical fluid extraction and fractionation of high-added-value compounds from a hexane extract of olive leaves. *Journal of Agricultural and Food Chemistry* 52:4774–4779.

Tan, C.S., Liang, S.K., and Liou, D.C. 1988. Fluid-solid mass transfer in a supercritical fluid extractor. *Chemical Engineering Journal* 38:17–22.

Tena, M.T., Valcárcel, M., Hidalgo, P.J., and Ubera, J.L. 1997. Supercritical fluid extraction of natural antioxidants from rosemary: Comparison with liquid solvent sonication. *Analytical Chemistry* 69:521–526.

Tom, J.W. and Debenedetti, P.G. 1991. Particle formation with supercritical fluids—A review. *Journal of Aerosol Science* 22:555–584.

Tsai, K.L. and Tsai, F.N. 1995. Solubilities of methylbenzoic acid isomers in supercritical carbon dioxide. *Journal of Chemical and Engineering Data* 40:264–266.

Türk, M. 1999. Formation of small organic particles by RESS: Experimental and theoretical investigations. *Journal of Supercritical Fluids* 15:79–89.

Uchiyama, H., Mishima, K., Oka, S., Ezawa, M., Ide, M., Takai, T., and Park, P.W. 1997. Solubilities of flavone and 3-hydroxyflavone in supercritical carbon dioxide. *Journal of Chemical and Engineering Data* 42:570–573.

Unlusu, B. and Sunol, A.K. 2004a. Modeling of equilibration times at high pressure for multicomponent vapor-liquid diffusional processes. *Fluid Phase Equilibria* 226:15–25.

Unlusu, B. and Sunol, A.K. 2004b. Multicomponent interphase diffusion of carbon dioxide-methanol-water under near-critical conditions. *Chemical Engineering Science* 59:1923–1929.

Uquiche, E., del Valle, J.M., and Ortiz, J. 2004. Supercritical carbon dioxide extraction of red pepper (*Capsicum annuun* L.) oleoresin. *Journal of Food Engineering* 65:55–66.

Uquiche, E., del Valle, J.M., and Ihl, M. 2005. Microstructure-extractability relationships in the extraction of prepelletized Jalapeño peppers with supercritical carbon dioxide. *Journal of Food Science* 70:E379–E386.

Urrego, F.A., Núñez, G.A., Donaire, Y.D., and del Valle, J.M. 2015. Equilibrium partition of rapeseed oil between supercritical CO_2 and prepressed rapeseed. *Journal of Supercritical Fluids* 102:80–91.

van Leer, R.A. and Paulaitis, M.E. 1980. Solubilities of phenol and chlorinated phenols in supercritical carbon dioxide. *Journal of Chemical and Engineering Data* 25:257–259.

Vatai, T., Škerget, M., Knez, Ž., Kareth, S., Wehowski, M., and Weidner, E. 2008. Extraction and formulation of anthocyanin-concentrates from grape residues. *Journal of Supercritical Fluids* 45:32–36.

Vicente, G., García-Risco, M.R., Fornari, T., and Reglero, G. 2012. Supercritical fractionation of rosemary extracts to improve the antioxidant activity. *Chemical Engineering and Technology* 35:176–182.

Vicente, G., Molina, S., González-Vallinas, M., García-Risco, M.R., Fornari, T., Reglero, G., and Ramírez de Molina, A. 2013. Supercritical rosemary extracts, their antioxidant activity and effect on hepatic tumor progression. *Journal of Supercritical Fluids* 79:101–108.

Visentín, A., Cismondi, M., and Maestri, D. 2011. Supercritical CO_2 fractionation of rosemary ethanolic oleoresins as a method to improve carnosic acid recovery. *Innovative Food Science and Emerging Technologies* 12:142–145.

Visentin, A., Rodríguez-Rojo, S., Navarrete, A., Maestri, D., and Cocero, M.J. 2012. Precipitation and encapsulation of rosemary antioxidants by supercritical antisolvent process. *Journal of Food Engineering* 109:9–15.

Wakao, N. and Kaguei, S. 1982. *Heat and Mass Transfer in Packed Beds*. New York: Gordon and Breach.

Wang, G.Q., Yuan, X.G., and Yu, K.T. 2005. Review of mass-transfer correlations for packed columns, *Industrial and Engineering Chemistry Research* 44:8715–8729.

Weidner, E. 2009. High pressure micronization for food applications. *Journal of Supercritical Fluids* 47:556–565.

Weidner, E., Petermann, M., and Knez, Ž. 2003. Multifunctional composites by high-pressure spray processes. *Current Opinion in Solid State and Materials Science* 7:385–390.

Williams, M.A. 1997. Preparing of oil-bearing materials for extraction. In *Technology and Solvents for Extraction Oilseeds and Nonpretroleum Oils*, eds. P.J. Wan and P.J. Wakelyn, pp. 121–136. Champaign, IL: AOCS Press.

Wu, J.J., Shen, C.T., Jong, T.T., Young, C.C., Yang, H.L., Hsu, S.L., Chang, C.M.J., and Shieh, C.J. 2009. Supercritical carbon dioxide anti-solvent process for purification of micronized propolis particulates and associated anti-cancer activity. *Separation and Purification Technology* 70:190–198.

Yesil-Celiktas, O., Bedir, E., and Sukan, F. 2007. In vitro antioxidant activities of *Rosmarinus officinalis* extracts treated with supercritical carbon dioxide. *Food Chemistry* 101:1457–1464.

Yilmaz, E.E., Özvural, E.B., and Vural, H. 2011. Extraction and identification of proanthocyanidins from grape seed (*Vitis vinifera*) using supercritical carbon dioxide. *Journal of Supercritical Fluids* 55:924–928.

Ziger, D.H. and Eckert, C.A. 1983. Correlation and prediction of solid-supercritical fluid phase equilibria. *Industrial and Engineering Chemistry Process Design and Development* 22:582–588.

6 Analysis and Characterization of Polyphenol Extracts

Sonia Guadalupe Sáyago-Ayerdi, Gilberto Mercado-Mercado, Sara Ramos-Romero, Josep Lluís Torres, and Jara Pérez-Jiménez

CONTENTS

6.1 INTRODUCTION

When identifying a natural product or a derived extract as a potential source of polyphenols, it is necessary to carry out a comprehensive analysis of their polyphenol content and profile. However, this may become a difficult task for a nonexpert in the topic because of the wide diversity that polyphenols exhibit in different relevant aspects. Polyphenols differ in polarity (from the very hydrophilic simplest phenolic acids to the lipophilic alkylresorcinols), molecular weight (MW) (from small phenolic acids to polymeric hydrolyzable tannins), and molecular structure, such as isomers (e.g., *cis*-resveratrol and *trans*-resveratrol). In addition, they present different forms of associations with food matrices (they may appear free, associated with weaker bonds or strongly linked with covalent bonds). Therefore, this chapter is a practical guide for those who are not familiar with the field. We will cover topics ranging from the simplest spectrophotometric methodologies to the chromatography analysis, especially liquid chromatography (LC), which allows for a more detailed characterization, and ending with the more sophisticated mass spectrometry (MS) techniques. Proanthocyanidins (PAs), a particularly complex class of polyphenols, will be used as a case study of polyphenol analysis. Finally, the particular case of the analysis of the commonly ignored nonextractable polyphenols (NEPPs) will be briefly explored.

6.2 SPECTROPHOTOMETRY TECHNIQUES

The most common and easiest methods for analyzing polyphenol profiles in a sample are spectrophotometric methods, including the Folin–Ciocalteau (F-C) assay for the analysis of total polyphenols, pH differential method for anthocyanins, and butanol–HCl or Porter assay for the analysis of PAs (see Section 6.5).

6.2.1 The Folin–Ciocalteau Assay

The Folin assay was originally developed in 1927 for the measurement of tyrosine and tryptophan but it evolved later, modifying the reagent to determine total polyphenols. The F-C assay is based on electron transfer reactions between a redox reagent (molybdotungstophosphate heteropolyanion, $PMoW_{11}O_{40}{}^{4-}$) and phenolic compounds, which form a blue chromophore complex under alkaline conditions (Singleton et al., 1999). F-C is widely used and there are published data for many different samples (Pérez-Jiménez et al., 2010; Vinson et al., 2001), it can be easily automatized for routine analyses, and it is a useful tool for a quick screening of samples before performing further analysis.

However, it should be noted that many nonphenolic reducing substances may react with this polyanion (iron II, bisulfites, proteins, carbohydrates as fructose, amino acids, nucleotides, thiols, unsaturated fatty acids, vitamins, amines, nitrites, sulfide, cyanide, aldehydes, and ketones), leading to an overestimation of content of polyphenols (Everette et al., 2010; Naczk and Shahidi, 2004).

The most common reference compound used in this method is gallic acid, available at low cost in a highly purified form and endowed with an average reactivity. Therefore, the total phenolic content is conventionally expressed as gallic acid equivalent (mg/L GAE). In some cases, it is preferable to use other standards that are closer to the phenolic composition of the sample, such as catechin or quercetin, for flavonoid-rich samples; in the case of seaweeds, with a specific profile of phenolic compounds, it is common to use phloroglucinol as the standard (Díaz-Rubio et al., 2009).

6.2.2 pH Differential Method for Anthocyanidins

Anthocyanidins are natural water-soluble pigments found in vascular plant vacuoles, where they are usually present as different glycosides. They reversibly change color with pH, as shown in Figure 6.1: at low pH (1–3), anthocyanidins appear as red flavylium cations; at pH 4–5, simultaneous hydration and deprotonation of the

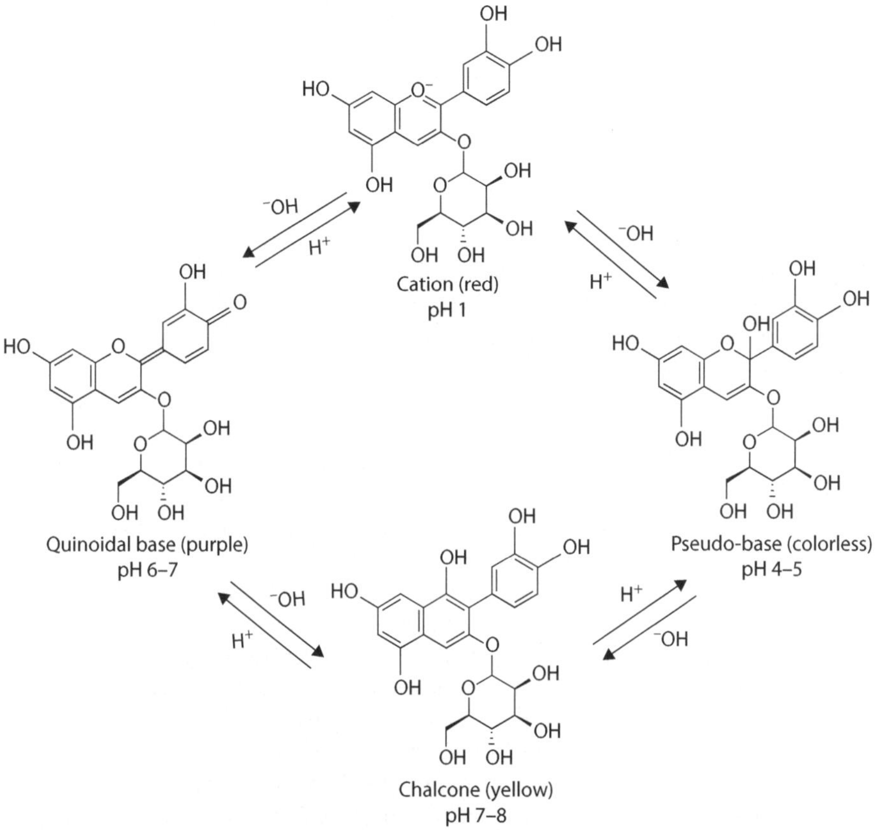

FIGURE 6.1 Cyanidin 3-*O*-glucoside transformation depending on pH.

cation takes place, leading to a colorless carbinol pseudo-base; at pH 6–7, anthocyanidins appear as a blue quinoidal base; and at pH 7–8, they are found as yellow chalcones. These modifications provide an easy and convenient method for measuring total pigment concentration, that is, the pH differential method for total anthocyanidins quantification. This method, developed several decades ago (Sondheimer and Kertesz, 1948; Swain and Hillis, 1959), evaluates the change in absorbance at the maximum wavelength at pH 1 and 3.5. It is a methodology widely applied to many matrices such as berries or teas (Fredes et al., 2014; Kerio et al., 2012). In addition, it is useful not only for comparing different samples but also for measuring the evolution in anthocyanidin content, for instance, to assess different storage methods (Symaladevi et al., 2011).

This technique was validated by an interlaboratory study (Lee et al., 2005) in which 11 different organizations (including commercial laboratories, academic institutions, and government laboratories) participated. These organizations analyzed total anthocyanidins in seven samples; the relative standard deviation of the repeatability varied from 1.06% to 4.16% and the relative standard deviation of the reproducibility ranged from 2.69% to 10.12%, which are acceptable ranges. In addition, a high correlation was found between the total anthocyanidin content obtained by the pH differential method and the more accurate high-performance liquid chromatography (HPLC) analysis (Lee et al., 2008). However, when both techniques were used for evaluating the degradation kinetics of individual anthocyanidins, similar results were obtained for cyanidin 3-glucoside, but the spectrophotometric method originated in an overestimation of the results in the case of cyanidin 3-(6′-malonyl)-glucoside (Cao et al., 2009); this different response of individual anthocyanidins to the pH differential method should be considered when applying it. Recently, coupling of the pH differential method with HPLC was suggested, allowing for the proper identification of individual anthocyanidins (Wang, 2014).

6.3 CHROMATOGRAPHY TECHNIQUES

Chromatography is needed for obtaining more detailed information on polyphenol profiles than is provided by spectrophotometric methods. To date, the most widely used techniques are based on HPLC coupled with photodiode array (PDA) detection, often combined with MS, as described in Section 6.2.

As general concepts, separation in LC is based on the differential partitioning of the compounds of interest between two nonmiscible phases: the stationary phase and the mobile phase. This partitioning is determined by the interactions that the compounds exhibit with both phases and may be of different types, that is, hydrogen bonding, dipole–dipole interactions, coulomb interactions, charged transfer or π–π interactions, or hydrophobic interactions (Di-Stefano et al., 2012). The mobile phase is an aqueous or organic medium, while the stationary phase is a solid, porous, or surface-active material in a small particle form or a viscous liquid immobilized in these particles. They exhibit different polarities, and the two existing combinations result in reversed-phase HPLC (RP-HPLC) when the mobile phase is polar and the stationary phase is nonpolar, while they result in normal-phase HPLC (NP-HPLC) when there is the opposite combination. In HPLC systems, a pump controls the flow rate and generates sufficient pressure to drive the mobile phase through the column.

Different factors should be considered when designing an HPLC method for polyphenol analysis: sample purification, mobile phase, stationary phase (column type), and detectors.

Regarding sample purification, a single general aqueous-organic extraction may be adequate in some cases (Vergara-Salinas et al., 2012), but in others, it may be necessary to carry out successive extractions with different solvents or conditions in order to selectively isolate the different polyphenol classes (Touriño et al., 2008). In addition, it is important to make sure that the conditions used for polyphenol extraction will not affect the HPLC analysis; for instance, extreme pH, needed to extract some polyphenol classes, may damage some columns or the MS equipment in case they are coupled with the chromatography separation (Pérez-Jiménez and Saura-Calixto, 2015).

In the mobile phase, the RP-HPLC phase described earlier is the most common mode for the separation of mixtures of polyphenols; the most polar compounds will have a higher affinity with the mobile phase and will elute first, while the less polar ones will be more likely to be retained in the stationary phase. In order to ensure that these retained compounds are also eluted, polyphenol analysis by HPLC usually includes a binary solvent system with a slightly acidified aqueous phase and an organic modifier. Then, a binary gradient is programmed including, in the mobile phase, increasing percentages of the organic solvent over time. An acidic mobile phase suppresses ionization, thereby increasing retention and decreasing the peak broadening that is caused by the formation of the deprotonated form, especially of phenolic acids (Gonçalves et al., 2012). These acidified solvents are either water with a small percentage of formic acid, acetic acid, or trifluoroacetic acid or salts such as ammonium acetate and ammonium formate (Cacciola et al., 2015). The two primary organic solvents used for polyphenol analysis are methanol and acetonitrile; generally, acetonitrile gives sharper peak shapes, although methanol is often preferred because of its less hazardous properties (Robbins and Bean, 2004). Nevertheless, the choice of mobile phases is also influenced by pressure and selectivity considerations, affecting resolution and retention times. For example, the addition of formic or trifluoroacetic acid to the mobile phase only affects chromatographic resolution, whereas the addition of acetic acid also results in a considerable decrease in the retention times of flavonoid *O*-glycosides (Becker et al., 2013; Biesaga and Pyszynska, 2013).

When selecting the stationary phase, the most common HPLC columns used in polyphenol analysis are octyl (C8) or octadecyl (C18) bonded silica columns (Klejdus et al., 2001). In this kind of column, polyphenol subclasses generally elute in the following order: hydroxybenzoic acids, flavan-3-ols, hydroxycinnamic acids, coumarins, flavanones, dihydrochalcones, flavonols, and flavones (Whelan et al., 2005). Within the same subclass, an increase in the number of hydroxyl groups leads to a reduction in retention time, while acylation, prenylation, or methylation has the opposite effect. The specific position of glycosylation can also have a significant impact on the retention time, with flavone *C*-glycosides generally eluting at lower retention times than the corresponding *O*-glycosides (Kelm et al., 2005; Whelan et al., 2005). HPLC analysis of polyphenols is mostly performed at ambient temperature, but placing the column in an oven with moderately higher temperatures (30°C–40°C) has also been recommended, since it leads to a reduction in the analysis time and more reproducible elution times (Whelan et al., 2005).

The most common detector for polyphenol analysis is the PDA (also called the diode array detector [DAD]) that allows for the recording of the UV-VIS spectra of each chromatographic peak. Multiple conjugate double bonds make polyphenols strong chromophores. Therefore, chromatographic peaks detected at specific wavelengths may then be attributed to a polyphenol subclass, since they exhibit characteristic UV-VIS spectra. Thus, hydroxybenzoic acids exhibit a maximum absorbance of about 280 nm and hydroxycinnamic acids exhibit absorbance at about 320 nm (Khoddami et al., 2013; Robbins, 2003). The UV spectra of most flavonoids exhibit two major absorption peaks in the region of 240–400 nm, commonly referred to as band I (300–380 nm) and band II (240–280 nm). Band I is associated with the absorption due to the B-ring cinnamoyl system and band II with the absorption involving the A-ring benzoyl system. Flavones present intense band I (339–350 nm) and band II (261–279 nm); flavonol-3-*O*-glucosides are characterized by a lower intensity of band I (347–370 nm) compared to that of band II (324–336 nm), and flavan-3-ols exhibit an intense band II (269–279 nm) and do not present band I absorption due to the lack of conjugation between the A-ring and B-ring (Rijke et al., 2006; Whelan et al., 2005). Flavanones have a distinctive UV spectrum, since the A-ring and B-ring are not conjugated: band I is not defined, but the benzoyl group shows an unresolved band at approximately 330 nm (Rijke et al., 2006; Robbins et al., 2004).

These different absorbance spectra allow the detection, in the same analysis, of polyphenols belonging to different classes, just as with simultaneous monitoring of the appropriate wavelengths. If the peak has the same retention time as a standard compound, it may be identified and quantified from a calibration curve of the standard. However, if no commercial standards are available for all the compounds in the sample, it is possible to carry out a semiquantitative analysis, expressing all the polyphenols belonging to a certain class as equivalents to a particular polyphenol of that class, that is, flavones as apigenin equivalent or hydroxycinnamic acids as caffeic acid equivalents (Vergara-Salinas et al., 2012). Overall, RP-HPLC coupled with PDA detection is the most suitable approach to carry out a quantitative analysis of all polyphenols present in a sample, without the drawbacks of spectrophotometric methods and the complexity associated with MS techniques. However, some methods have been developed using other detectors, such as fluorescence detector* (Piñeiro et al., 2006) or multichannel electrochemical coulometric detection†; this second one was used to analyze 32 phenolic compounds in samples of different nature, exhibiting at least one order of magnitude lower sensitivity than single-channel coulometric detection (Jandera et al., 2005).

Even though a general HPLC method may be used for the simultaneous analysis of all polyphenol classes, when a sample has a polyphenol profile particularly rich in one class, it may be more interesting to design a class-specific HPLC method. Therefore, particular methods have been suggested for the separation and quantitation of phenolic acids (Mattila et al., 2006), stilbenes (Piñeiro et al., 2006), capsaicinoids (Al-Othman et al., 2011), etc. Although most polyphenol analyses are carried out using LC, gas chromatography (GC) is the most common technique for the analysis of the specific class of alkylresorcinols (Prinsen et al., 2014).

Some new developments have appeared in chromatography techniques during the last years. Thus, ultrahigh-performance LC (UHPLC), using <2 mm packing materials and working with pressures of up to 1300 bar, is an important advancement for fast analysis of polyphenols. In particular, UHPLC-MS is from 3 to 7 times faster, has quantification limits that are 50% lower, and is less susceptible to matrix effects than HPLC (Churchwell et al., 2005; Ortega et al., 2010). For instance, 19 phenolic compounds belonging to different classes could be separated in 1.9 min by using UHPLC (Klejdus et al., 2008). Finally, the most complex polyphenol classes may be analyzed using two-dimensional HPLC. In this case, the separation in each

* Based on the principle that some compounds fluoresce when exposed to UV light.

† The compounds, after elution from the column, are subjected to an electrochemical reaction, generating an electric current that is detected.

dimension is based on a different kind of chromatography; using this approach, polymers of seaweed phlorotannins of up to 17 units could be separated by combining RP-HPLC in one dimension and hydrophilic interaction LC chromatography (HILIC) in the second one (Montero et al., 2014).

6.4 MASS SPECTROMETRY TECHNIQUES

Analysis of complex samples, such as ingredients, foods, and biological samples, have been widely improved by means of MS. Nowadays, this is considered one of the most reliable techniques for identification of phenolic compounds in their native forms, even in the absence of commercial standards and in the presence of interfering compounds that commonly remain in plant extracts after clean-up procedures.

Mass spectrometers are based on detection of gas-phase ions, either positive or negative, separated according to their mass-to-charge ratio (*m/z*). Mass spectrometers include three main elements: an ion source, a mass analyzer, and a detector. Selecting the first two elements is crucial to characterize the polyphenol profile of a sample effectively. Their main characteristics are described in the following, including examples of their applications to the analysis of polyphenols in different foods and extracts.

It should be kept in mind that when analyzing food polyphenols, MS is commonly coupled with a previous separation technique, usually chromatography. GC has been used widely in the past for the analysis of volatile phenols (Kivilompolo et al., 2007). Moreover, GC-MS presents some advantages when compared to HPLC-MS, such as its large linear range, its high stability, and the possibility of direct extrapolation of instrumental conditions from literature (Kivilompolo et al., 2007). However, since GC-MS requires more steps in the sample preparation than LC-MS and it can be only directly applied to volatile compounds, requiring derivatization for the rest (Kivilompolo et al., 2007), it is nowadays being substituted by LC-MS. Another separation technique usually coupled with MS is capillary electrophoresis (CE). CE is a fast, efficient, and low-cost technique for the separation of phenolic compounds that has been employed together with MS to analyze polyphenols in buckwheat (Verardo et al., 2011) and in almond skin extracts (Arraez-Roman et al., 2010).

6.4.1 Ionization Sources of MS

The two most common ionization techniques—electron ionization or impact (EI)* and chemical ionization (CI)†—require sample vaporization, but numerous biological compounds are nonvolatile and/or thermally labile. In the specific case of polyphenols, temperature affects their quantitative and qualitative extract profiles (Vergara-Salinas et al., 2012). Even though derivatization could resolve these limitations, the interpretation of the fragment pattern of this technique is very difficult.

* The sample, previously vaporized, is bombed by accelerated electrons that, given their high kinetic energy and their low mass, confer a high energy to the molecules of the sample and cause their ionization.

† The sample, previously vaporized, is ionized after collision with the ions produced by bombing a reactive gas with electrons.

Thus, the application of these ionization techniques for the analysis of polyphenols has been limited to certain specific applications such as the isoflavone profile (Burdette and Marcus, 2013).

Consequently, the newest ionization methods, such as electrospray ionization (ESI)* and matrix-assisted laser desorption/ionization (MALDI),† are the most commonly used for the analysis of phenolic compounds (Monagas et al., 2010). ESI produces cationic or anionic species with little fragmentation, and it has been widely used to characterize polyphenols in foods. MALDI offers several advantages over ESI for the analysis of complex polyphenols, such as greater tolerance for impurities, detection of predominantly single-charged molecular ions, possibility of reanalyzing the same sample without consuming it, and especially, its adequacy for the analysis of polyphenols of higher MW than those analyzed by ESI (Monagas et al., 2010). Up until now, MALDI is only applicable for qualitative analysis. Most of the examples of polyphenol analysis described as follows were obtained using ESI as the ionization source, although with different mass analyzers.

6.4.2 Mass Analyzers

The mass analyzer is the device that physically separates ions based on their *m/z* values by using electrical and/or magnetic fields. Currently, the most used mass analyzers for polyphenols are ion trap (IT), quadrupole (Q), time of flight (TOF), and orbitrap. These analyzers differ in many characteristics (see Table 6.1), and their selection depends on the nature of the experiment being performed (El-Aneed et al., 2009). Most of them may be used for either quantitative or qualitative analysis (excepting TOF coupled with MALDI as the ionization source, which is only suitable for qualitative analysis), but the triple quadrupole (QqQ) is still the best option for quantitative analysis. Moreover, many of them, especially when they present in hybrid forms (see Table 6.1), can include tandem MS (MS/MS or MS^2), that is, a combination of MS analyzers that allows further fragmentation of the original MS fragments that may be used to identify the exact structure of the original compound.

Ion trap analyzer. IT analyzers are based on the capture of charged particles in a vacuum by a combination of the oscillating potential generated by a magnetic field and the voltage generated by an electrostatic field. Due to their ability to perform MS^n experiments (successive MS^2 experiments, where the fragment of each experiment is further fragmented), the IT and the linear IT (LIT) analyzers are particularly useful to determine compound structures and to distinguish between isomers with identical fragmentation patterns for the first generation products but different for the subsequent ones (Lin et al., 2011). In this context, IT has been used for the determination of

* The sample, dissolved in a proper solvent, passes through a metal capillary at atmospheric pressure and where a high voltage is applied. This treatment originates an aerosol constituted by droplets with high charge. Solvent evaporation increases the charge density and the repulsions, producing gas-phase ions.

† The sample, dissolved or not, is mixed with a matrix that allows for the formation of a crystal in the MALDI sample plate. After solvent evaporation, the crystal is irradiated with a laser that allows for the desorption of gas-phase ions.

TABLE 6.1
Characteristics of the Most Common Mass Analyzers Used for Polyphenol Analysis

	IT	QqQ	TOF	Orbitrap
Mass accuracy (ppm)[a]	50–100	100	5–50	2–5
Mass-resolving power[b]	10^3–10^4	10^2–10^4	10^3–10^4	1.5×10^5
Sensitivity	Picomole	Attomole to femtomole	Femtomole	Femtomole
Mass range[c]	Up to 2×10^4	4×10^3	Up to 10^5	6×10^3
Precision[d]	0.2%–5%	0.1%–5%	0.1%–1%	Not available
Linear dynamic range[e]	10^2–10^5	10^7	10^2–10^4	10^3–10^4
Abundance sensitivity[f]	10^3	10^4–10^6	Up to 10^6	—
Efficiency[g]	<1%–95%	<1%–95%	1%–100%	—
Scan rate	Moderate	Moderate	Fast	Moderate–fast
Vacuum (Torr)	10^{-3}	10^{-4}–10^{-5}	10^{-7}	10^{-5}
Cost	Low–moderate	Low–high	Moderate–high	High
MS^n capabilities	MS^n [h]	Up to MS^2	Up to MS^2 [i] (MS^3 for Qq-TOF)	MS (MS^n [h] for QLIT–orbitrap)
Possible hybrid forms	QIT	QqQ–LIT	QTOF	QLIT–orbitrap

Sources: Adapted from Gentili, A. et al., *TrAC Trends Anal. Chem.*, 27, 888, 2008; Han, X. et al., *Curr. Opin. Chem. Biol.*, 12, 483, 2008.

Abbreviations: IT, ion trap; Q, quadrupole; QqQ, triple quadrupole; LIT, linear ion trap; TOF, time of flight; MS, mass spectrometry.

[a] Mass accuracy = Ratio of the *m/z* measurement error/true *m/z* (ppm = parts per million).

[b] Mass-resolving power (*M/DMx*) is based on a single well-resolved peak. *M* = Mass associated with the apex of the peak, *DMx* = Full width of the peak at half-maximum (FWHM).

[c] Mass range = Range of *m/z* amenable to analysis.

[d] Precision = Reproducibility with which ion abundances can be determined.

[e] Linear dynamic range = Range over which ion signal is linear with analyte concentration.

[f] Abundance sensitivity = Inverse of the ratio obtained/signal level associated with a large peak by the signal level of the background at one mass-to-charge unit lower or higher. It is particularly interesting in the measurement of isotope ratios.

[g] Efficiency = Transmission of an analyzer × its duty cycle. Duty cycle is the fraction of the ions of interest formed in the ionization step that are subjected to mass analysis. The efficiency of the analyzer can depend greatly on the scan type.

[h] Up to 13.

[i] Fragmentation achievable by postsource decay.

flavonoids in food supplements, including structurally related compounds (e.g., diosmin and hesperidin), obtaining relative standard deviations lower than 10% for the complete analytical procedure and limits of detection in the range of 2.5–120 $\mu g\ L^{-1}$ (Llorent-Martínez et al., 2014). Similarly, LIT–orbitrap equipment was used for a comprehensive analysis of the phenolic profile of beer, identifying a total of 47 compounds, with 7 of them reported for the first time for this beverage (Quifer-Rada et al., 2015).

Quadrupole analyzer. The analysis of phenolic compounds is often based on the use of a single Q analyzer coupled with LC. In the Q analyzer, an electric field is created by four parallel metal rods; specific voltages are applied in order to filter the ions depending on their *m/z*. The main limitation of Q analyzers is the low selectivity that could induce false positives when the study is performed in real samples, because the presence of several compounds with the same *m/z* is quite common in complex matrices and many times, there are no commercial standards allowing for the identification of each signal. In order to improve this limitation, Q analyzers can be coupled with a tandem mass spectrometer. Thus, the QqQ analyzer is an excellent detector in a multiple reaction monitoring (MRM)* acquisition mode due to its superior selectivity and sensitivity (highest signal/noise ratios) (Gentili et al., 2008). QqQ can also provide MS^2 acquisition modes, such as product–ion scan (PIS),† neutral-loss scan (NLS),‡ or precursor–ion scan (PrIS),§ which allow for the confirmation of identities. A limitation of QqQ is its resolution, which does not allow for the determination of the molecular formula of unknown compounds. The hybrid QIT presents the same level of resolution as a QqQ but it provides other benefits, such as a higher sensitivity in full-scan mode and a superior identification power in MS^n experiments (Gentili et al., 2008). Thus, a study reported that MALDI–QIT–TOF for the analysis of phenolic compounds in *Rosa chinensis* flowers allowed for the identification of two ellagitannins (rugosins B and C) that had not been identified by LC–ESI–Q (Cai et al., 2005). Other hybrid mass spectrometers with different analysis characteristics are QqQLIT and QTOF.

TOF analyzer. The TOF mass analyzers are based on an electric field in which the ions are forced to follow a linear trajectory and are discriminated by the time they take to reach the end of a tube. They provide accurate mass measurements of the ions (up to four decimals), allowing for the obtainment of exact molecular formulas and structural information of the analytes. TOF analyzers present the highest mass range and the ability to acquire the entire spectrum from a single laser pulse event (see Table 6.1). TOF combined with the MALDI ionization technique (MALDI–TOF) and without a previous chromatographic separation allows for the detection of high mass compounds with precision. In this sense, MALDI–TOF is a good candidate to analyze complex polyphenols (Reed et al., 2005) such as PAs (Monagas et al., 2010) and hydrolyzable tannins (Sáyago-Ayerdi et al., 2013). In addition, this technique recently provided the first data on oxidative oligomerization of flavanols during black tea production (Yassin et al., 2014).

Orbitrap. The orbitrap analyzer is based on the use of a quadro-logarithmic field for ion separation. This type of MS analyzer provides the highest resolution based on full width at half maximum (FWHM) and a mass accuracy of 2–5 ppm,

---

* A selected ion in Q1 is fragmented by collision-induced dissociation (CID) in Q2, and then only a specific fragment ion (transition ion) is detected in Q3.

† A selected ion in the Q1 is fragmented by CID in Q2, and then all charged fragments are scanned in Q3.

‡ Q1 and Q3 are synchronized, and any molecule passing through Q1 and losing in Q2 a neutral fragment previously established will be then transmitted to Q3 and detected.

§ A selected charged fragment obtained by CID in Q2 is detected in Q3, and all ions that passed through Q1 and originated those charged fragments are scanned.

although its cost is high (see Table 6.1). Orbitrap coupled with UHPLC was used for the identification of phenolic compounds belonging to different families (hydroxybenzoic acids, flavanols, and flavonols) in tablets and capsules obtained from green tea (López-Gutiérrez et al., 2015).

6.4.3 Approaches for the Identification of Polyphenols

Although the MS equipment has particular pros and cons, one of the most interesting aspects of applying MS techniques to a sample is the possibility of characterizing its composition when lacking commercial standards. This is especially relevant in the case of polyphenols since, given the wide variety of structures that they may present, it is difficult to have all the existing commercial standards in a laboratory, for instance, when carrying out the screening of several medicinal plants with different polyphenol profiles. Moreover, for an important percentage of the polyphenols that have been described in plants or foods, no commercial standards are available yet, making the use of techniques such as MS indispensable. A general approach may be proposed for the identification of polyphenols in a sample using MS.

The first step would be to perform a full-scan MS, in order to detect all the ions present in a sample. These ions will be then cross-checked with one of the existing databases on food polyphenols, such as Phenol-Explorer (Neveu et al., 2010), as well as with the previous literature on those samples, since databases are based on quantitative information and they do not include any on polyphenols that have been detected only qualitatively. Depending on the mass accuracy of the equipment, this first identification will be more or less tentative. For instance, if an IT is used (mass accuracy of 50–100 ppm), this will have to be necessarily combined with at least MS^2 experiments. Nevertheless, with a TOF or an orbitrap (mass accuracy of 5–50 and 2–5 ppm, respectively), this identification may already be quite exact. For instance, using HPLC-Q-TOF equipment, 17 polyphenols were identified in a pressurized hot water extract from thyme, including the detection of seven compounds for the first time in the genus *Thymus*. One of them was caffeic acid 4-*O*-glucoside; the monocharged ion had a theoretical *m/z* of 341.0873 and the observed signal was 341.0862, so this identity could be assigned with an error of 3.1 ppm (Vergara-Salinas et al., 2012).

For the analysis of the most complex polyphenols, usually the equipment does not have enough mass accuracy or there are doubts about the particular structure of the polyphenols. In these cases, MS^2 experiments (sometimes followed by successive MS^3, MS^4, etc., experiments on selected fragments) are needed to elucidate the identity of the original compounds from the obtained fragments. Figure 6.2 shows an example of the fragments found when carrying out the MS^2 experiment of hexa-*O*-galloyl glucose, a hydrolyzable tannin found in mango peel (Berardini et al., 2004; Sáyago-Ayerdi et al., 2013). The compound detected in the full-scan MS as a monocharged anion had an *m/z* of 1091.7736, corresponding to six galloyl substituents attached to a central glucose core. The fragment with an *m/z* of 304.2086 corresponded to the loss of two gallic acid molecules, which produced tetra-*O*-galloyl glucose as another fragment (*m/z* 787.5650). At the same time, it is possible that the original molecule loses a fragment of an *m/z* of 170.1195, corresponding to the loss of

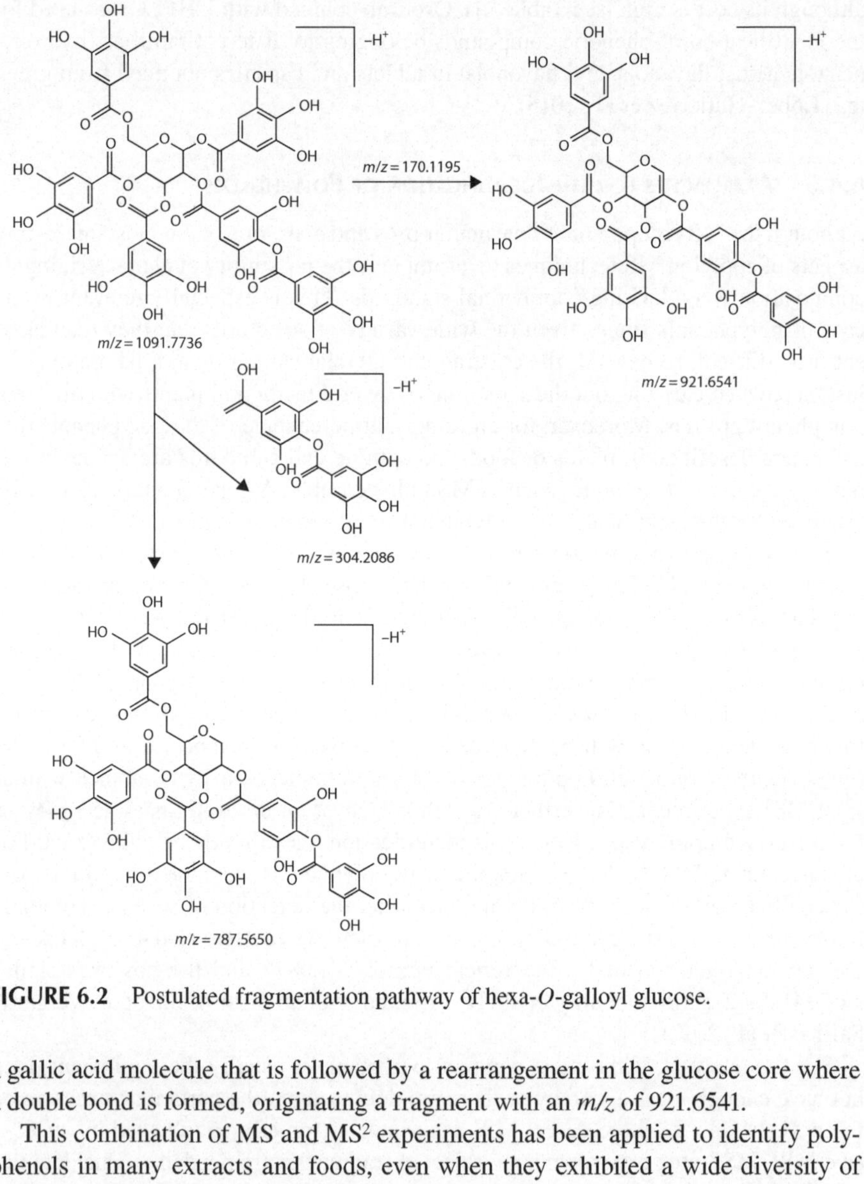

FIGURE 6.2 Postulated fragmentation pathway of hexa-*O*-galloyl glucose.

a gallic acid molecule that is followed by a rearrangement in the glucose core where a double bond is formed, originating a fragment with an *m/z* of 921.6541.

This combination of MS and MS^2 experiments has been applied to identify polyphenols in many extracts and foods, even when they exhibited a wide diversity of structures, such as in a matrix derived from grapes that contained more than one hundred polyphenols, including hydroxybenzoic acids, hydroxycinnamic acids, flavanols, flavonols, and anthocyanins (Touriño et al., 2008).

6.5 PROANTHOCYANIDINS AS A CASE STUDY

PAs are oligomers and polymers made of flavanol units (Figure 6.3). Flavanols typically establish bonds between position 4 and either position 6 or 8 of a second unit (B-type PA). A-type PAs show a second covalent bond (oxygen bridge)

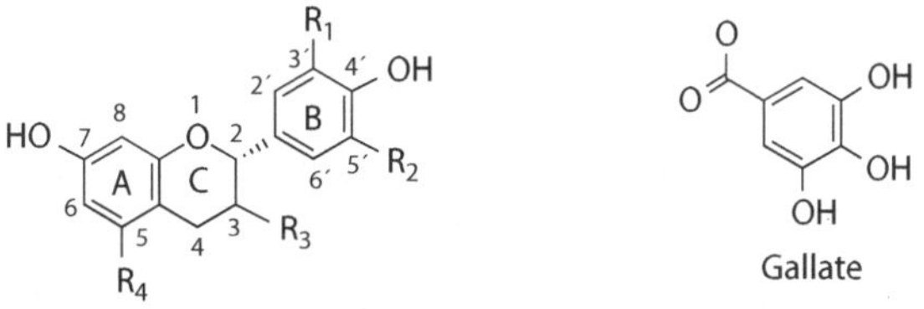

(epi)afzelechin: $R_1 = H$, $R_2 = H$, $R_3 = OH$, $R_4 = OH$
(epi)catechin: $R_1 = H$, $R_2 = OH$, $R_3 = OH$, $R_4 = OH$
(epi)gallocatechin: $R_1 = OH$, $R_2 = OH$, $R_3 = OH$, $R_4 = OH$
(epi)catechingallate: $R_1 = H$, $R_2 = OH$, $R_3 =$ gallate, $R_4 = OH$
epigallocatechigallate: $R_1 = OH$, $R_2 = OH$, $R_3 =$ gallate, $R_4 = OH$
(2R, 3R configuration)
(epi)fisetinidol: $R_1 = H$, $R_2 = OH$, $R_3 = OH$, $R_4 = H$
(epi)guibourtinidol: $R_1 = H$, $R_2 = H$, $R_3 = OH$, $R_4 = H$

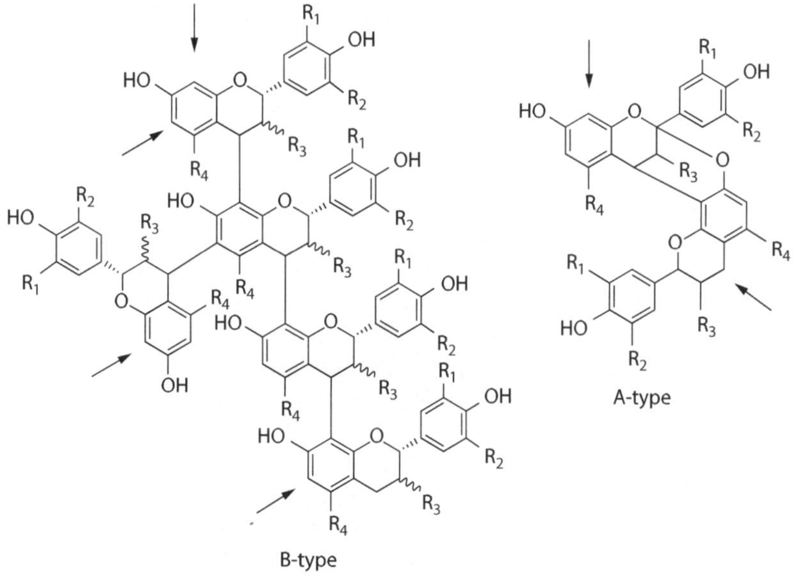

Arrows indicate possible polymerization positions

FIGURE 6.3 Some representative flavanol units and oligomeric proanthocyanidins.

between positions 2 and the hydroxyl group at position 5 or 7 of a second unit. PAs are not limited to these structures; a variety of other less common flavanol units and links between units occur naturally in plants (Ferreira and Slade, 2002). PAs are particularly difficult to analyze by LC because of their amphiphilic (both hydrophilic and hydrophobic) nature. Monomers are more hydrophobic, less water soluble than dimers and trimers; polymers are the most water-soluble PAs (Haslam, 1998). Paradoxically, their elution order in RP-HPLC is negatively related to their hydrophobic character; the relatively poor water-soluble (epi)catechin elutes before its oligomers and the fully water-soluble polymers (Torres and

Bobet, 2001). Moreover, oligomers larger than trimers generate broad peaks that make their separation nearly impossible (Rohr et al., 2000). The reason is that oligomers are flexible enough to interact with the environment through both the hydrophilic hydroxyl groups and the hydrophobic aromatic scaffold with a capacity to expose any of the two regions or both, as suggested by studies on oil-in-water emulsions (Torres et al., 2002). This dual behavior results in broad peaks in systems of hydrophobic stationary phases eluted with polar solvents. On the other hand, normal-phase and hydrophilic interaction chromatography have been successfully applied to the separation of PA up to tetradecamers (Kelm et al., 2006), although larger PA still coeluted as an unresolved hump (Tarascou et al., 2010). PAs have also been separated into fractions of different MWs at semipreparative scale by the so-called size-exclusion chromatography on stationary phases such as Sephadex LH-20 and Toyopearl TSK HW-50 eluted with aqueous alcohol or aqueous acetone solvents (Labarbe et al., 1999; Oki et al., 2002; Pérez-Ilzarbe et al., 1992). This type of chromatography is called size exclusion by analogy with the size-exclusion chromatography of proteins carried out on similar stationary phases. Nevertheless, PAs elute in fractions of ascending MW probably due to a combination of pore permeation and absorption by hydrophobic interactions.

Given the problems described earlier for the direct analysis of PA, an alternative approach for the estimation of the monomeric composition and overall degree of polymerization of B-type PA is the use of RP-LC after a depolymerization step under acidic conditions in the presence of nucleophiles (Figure 6.4). Treatment with acids (e.g., HCl in butanol or methanol) breaks the C–C bonds between subunits* (Porter et al., 1986) and presumably generates an electrophilic position. This position is rapidly attacked by nucleophiles such as thiols (Thompson et al., 1972; Torres and Bobet, 2001) or phloroglucinol (Foo and Porter, 1978; Matsuo et al., 1984) to yield a mixture of conjugates (extension units) and flavanols (terminal units). These mixtures are best analyzed by RP-HPLC (Kennedy and Jones, 2001; Matthews et al., 1997; Rigaud et al., 1991; Torres and Selga, 2003), and the mean degree of polymerization (mDP) may be obtained after quantification of the terminal and the extension units. The most widely used method for this approach involves toluene-α-thiol (benzylmercaptane, benzylthioether) (Guyot et al., 2001; Rigaud et al., 1991), although there are other user-friendly odorless alternatives, such as phloroglucinol (Kennedy and Jones, 2001) or cysteamine (Cya) (Torres and Lozano, 2001). In the case of Cya, as it contains an amino function, the derivatized extension units can be separated from the depolymerization mixtures by cation exchange (Torres and Selga, 2003) and CE may be used as an alternative method to RP-HPLC (Herrero-Martínez et al., 2003).

Depolymerization in the presence of nucleophiles has been applied to the analysis of PA in a variety of food products of plant origin, such as hop (mDP up to 24) (Kennedy and Taylor, 2003; Taylor et al., 2003), grape seeds (mDP up to 24), grape skins (mDP up to 39) (Kennedy and Taylor, 2003), or mountain birch leaves

* Acid cleavage is widely used as a first approach to the total PA content of plant extracts. The red color developed is due to the anthocyanidin generated by depolymerization and can be easily measured by colorimetry.

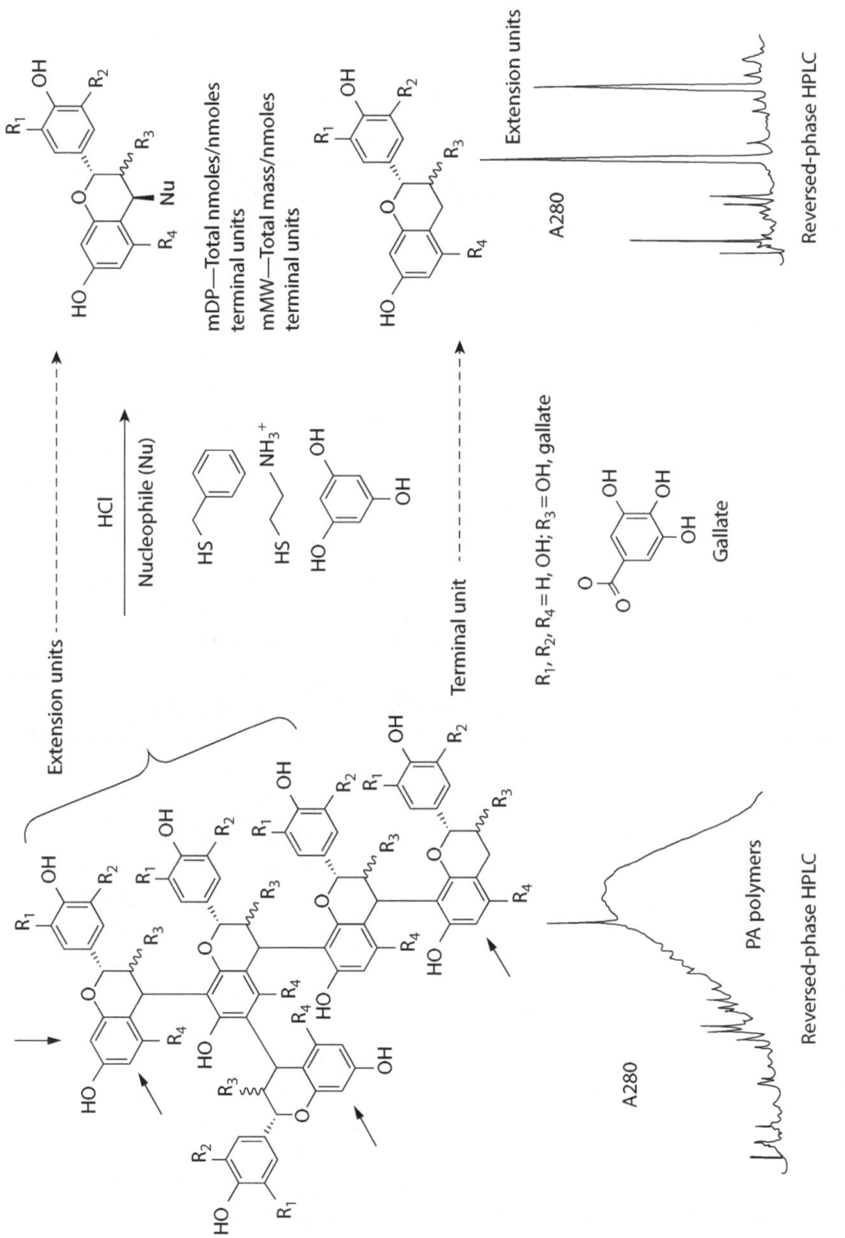

FIGURE 6.4 Depolymerization of proanthocyanidins.

(mDP up to 31) (Karonen et al., 2007) after fractionation by size exclusion on LH-20. Overall, the amount of PA in plant extracts and their degree of polymerization (DP) depends mainly on the extraction temperature and solvent, usually acetone, methanol, or ethanol under slightly acidic conditions (Scalbert, 1992). In this way, an mDP of up to 12 for grape PAs has been obtained after extraction with acetone/methanol/water followed by thioacidolysis with toluene-α-thiol (thiolysis) and up to 20 by direct thiolysis of the solid residues after extraction (Hellstrom and Matrila, 2008). Also, higher mDP was estimated for grape PA after size-exclusion chromatography (Labarbe et al., 1999) and for apple PA before and after fractionation of an aqueous acetone extract on reversed-phase HPLC (Guyot et al., 2001). Similar mDP for grape PA was obtained with phloroglucinol and benzylmercaptane (Kennedy and Jones, 2001).

Depolymerization and derivatization of PA provide valuable information about PA mean size and composition but give very limited insight into the structure of the individual oligomers and polymers in the mixtures. Moreover, analysis of A-type PA is not straightforward because the second intermolecular bond is insensitive to the cleavage conditions. MS is a powerful technique for structure elucidation of polymers such as proteins (Nesvizhskii, 2007), which is being increasingly applied to the study of both B-type and A-type PA (see Section 6.4). Incidentally, the partially depolymerized fragments of A-type-rich PA after thioacidolysis have been successfully analyzed by MS on a TOF analyzer after UPLC and ESI (Kimura et al., 2011). HPLC– or UPLC–ESI–MS with TOF, Q, or IT analyzers has been applied to the identification of PA from grape, wine, and pine bark through the examination of molecular ions and their fragmentation (Fulcrand et al., 1999; Karaonen et al., 2004; Monagas et al., 2003; Peng et al., 2001; Touriño et al., 2008). Structural information may be better obtained from MS^2, which provides valuable information about the subunit composition of the fragments and the parent PA. Typical MS^2 experiments include the MRM, NLS, PIS, and PrIS described earlier on a QqQ analyzer. HPLC–ESI–QqQ–MS/MS has been applied to the study of PA oligomers from sources such as grape (Touriño et al., 2008), hop (Callemien and Collin, 2008; Li and Deinzer, 2006), almond blanch water (Pérez-Jiménez and Torres, 2012), and the antiviral extract of *Chamaecrista nictitans* (Mateos-Martín et al., 2014). There is a limitation though as to the size of the oligomers (up to tetramers) that can be analyzed by HPLC–ESI–QqQ–MS/MS. Larger PA oligomers are best detected by MALDI–TOF. For instance, PAs from almond skin (Monagas et al., 2007), pine bark (Jerez et al., 2009), hop (Taylor et al., 2003), grape (Vivas et al., 2004), and other sources (Monagas et al., 2010) have been analyzed by MALDI–TOF. This technique requires the addition of a cationizing agent to the sample, which promotes ionization (Xiang et al., 2003). In this way, sodium and cesium salts have been successfully used and the choice of one or another depends on the individual constituents of the PA in the sample for instance, (catechins or delphinidins) along with their MWs (Monagas et al., 2010; Pérez-Jiménez and Torres, 2012).

Tandem experiments with two TOF analyzers (MALDI–TOF/TOF) are a step forward in the study of the structure of PA oligomers. Recently, a combination of MALDI–TOF and MALDI–TOF/TOF has revealed the subunit composition of individual PA oligomers from almond blanch water (Pérez-Jiménez and Torres, 2012) with a DP of up to 10 and from cinnamon (Mateos-Martín et al., 2012) with a DP of up to 11. The technique has also revealed that almond seed coats contain galloylated species (Pérez-Jiménez and Torres, 2012) and cinnamon PAs contain subunits of (epi)gallocatechin and (epi)catechin gallate apart from (epi)catechin and (epi)afzelechin (Mateos-Martín et al., 2012). The polymers include both A-type and B-type PAs. Coupling of MALDI–TOF/TOF with LC may be in the near future an even more powerful tool for PA structural analysis.

In summary, the total PA content is easily estimated by colorimetry after acid hydrolysis. For structural analysis, the two most used techniques are normal-phase HPLC analysis of the intact structures (with polymers coeluting in an unresolved peak), and depolymerization by thiolysis, followed by RP-HPLC or RP-HPLC-MS. Depolymerization and analysis of the subunits are fast, do not require sophisticated equipment beyond a liquid chromatograph, and are the best option to estimate the size of large PA polymers. Unless each individual PA is purified to homogeneity, a daunting task for oligomers with more than three subunits, thiolysis and HPLC will provide the mean DP of a mixture. MS/MS techniques are the most powerful tools to unravel the internal subunit structure of individual PA, although polymers larger than 10 units are hard to detect. MALDI–TOF/TOF is currently the best MS technique to provide comprehensive information about the structure of PAs. Previous solvent fractionation and/or chromatographic separation steps facilitate the analysis in all cases.

6.6 PARTICULAR CASE OF NONEXTRACTABLE POLYPHENOLS

All the strategies for the analysis of polyphenols described earlier are carried out in the supernatants of the extractions performed on vegetal materials. These extractions, as described in other chapters, may be classical treatments with aqueous-organic mixtures, extractions with pressurized solvents, microwave-assisted extraction, etc., which present different efficiency rates. However, what they all have in common is that they leave a fraction of polyphenols in the corresponding residues of extractions, the so-called NEPPs or macromolecular antioxidants. Due to the intrinsic problems associated with the analysis of NEPPs, these compounds have been commonly ignored—even when it has been known for decades that those polyphenols measured in the supernatants of the extractions, the extractable polyphenols (EPPs), were only a fraction of food polyphenols (Bravo et al., 1994). Indeed, this has caused EPPs to be commonly referred to as total polyphenols, even if that is not the case.

NEPPs comprise two kinds of structures: (1) nonextractable PAs (NEPAs), which are high-MW PAs presenting some interactions with food matrices; and (2) hydrolyzable polyphenols (HPPs), which are low-MW phenolic compounds of different classes (phenolic acids, flavonoids) as well as hydrolyzable tannins. NEPAs interact mostly with proteins and pectins through hydrogen bonds and hydrophobic interactions. HPPs may be encapsulated into the hydrophobic pockets of proteins also; all of them are strongly associated with proteins and polysaccharides through covalent bonds (Pérez-Jiménez and Torres, 2011). These specific characteristics of NEPPs explain why they have been traditionally ignored in polyphenol analysis. There are still technical problems to characterize compounds of such high

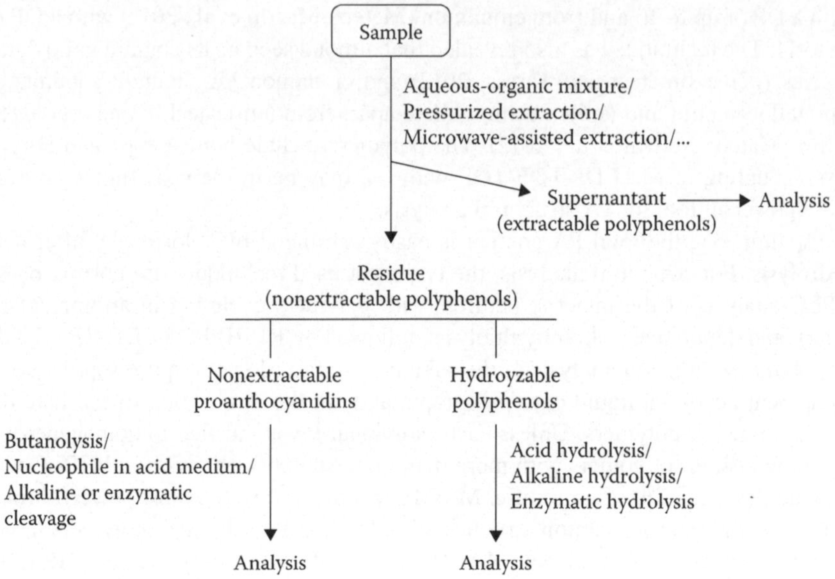

FIGURE 6.5 Analysis of nonextractable polyphenols.

MW as NEPAs, as well as to release HPPs from their association with food macromolecules without significantly altering its original structures. The analysis of NEPPs has been routinely carried out in the case of cereals only, where it has been widely known that most phenolic acids were associated with the cell wall. Nevertheless, recent data have shown that NEPPs are important contributors to the total polyphenol content in all plant foods and not only in cereals (Arranz and Saura-Calixto, 2010), emphasizing the need to include them in polyphenol analysis. Indeed, NEPPs have been defined as "the hidden face of food phenolic compounds" (Tarascou et al., 2010). Moreover, due to their arrival as intact molecules to the colon and their common association with dietary fiber, NEPPs may have relevant biologic activities in relation to gastrointestinal health and some chronic diseases (Pérez-Jiménez et al., 2013).

Although the analysis of NEPPs is much less developed than that of EPPs, some general strategies have already been suggested. As shown in Figure 6.5, a previous step for the analysis of NEPPs is the release of EPPs by some of the existing extraction techniques. Then, different replicates of the residue should be used for the analysis of the two fractions of NEPPs, HPPs and NEPAs, considering that in both cases a release step will be needed in order to obtain a liquid hydrolyzate where it will be possible to carry out the determination of these compounds.

6.6.1 Analysis of Hydrolyzable Polyphenols

HPPs are also called insoluble or bound polyphenols and have traditionally been identified in cereals, although they are present in all the food groups of vegetable origin. Different strategies may be used in order to release them from the food matrix. Alkaline hydrolysis, in order to cleave the ester bond between the carboxylic group of phenolic acids and the food matrix, was the traditional procedure. This procedure is very efficient for the release of ferulic acid, the most abundant phenolic compound in cereal products (Verma et al., 2009). However, a higher release of other phenolic compounds is achieved by acid hydrolysis (Arranz and Saura-Calixto, 2010). Hence, some authors have suggested a combination of both acid and alkaline hydrolysis (Verma et al., 2009). A particular way of carrying out acid hydrolysis is in a methanolic medium; this procedure was initially developed for the release of hydrolyzable tannins, but it is equally able to release HPPs belonging to other families of phenolic compounds (Arranz et al., 2009). Although different enzymatic hydrolysis processes have been tested, usually they are not as efficient as chemical hydrolysis. For instance, when different enzymes were used in spent barley grain, a by-product from the brewing process, enzymes were only able to release 70% of the ferulic acid and 8% of the *p*-coumaric acid previously released using chemical hydrolysis (Bartolomé et al., 1999).

After obtaining an HPP hydrolyzate by any of these procedures, the content and profile of these compounds may be estimated. The simplest approach is to carry out a Folin assay in order to know the total HPP content; the limitations of this method described earlier are applicable here also. Alternatively, a more detailed analysis of HP structures can be carried out using RP-HPLC, commonly with acetonitrile or methanol as organic phases. A more sophisticated characterization of these compounds may be performed by HPLC-MS techniques, allowing for the identification of individual compounds (Arranz and Saura-Calixto, 2010). Any of these analytical techniques may be applied in the supernatants of the extractions containing EPP also, therefore, allowing for the comparison between EPP and HPP content and profile. For instance, ferulic acid content as present in extractable and nonextractable forms in different cereal products is shown in Table 6.2.

An important aspect is that before applying any chromatographic technique, the pH of the hydrolyzate should be considered, since it may be necessary either to modify it or to purify HPP with solid-phase extraction or similar procedures (Pérez-Jiménez and Saura-Calixto, 2015); otherwise, the HPLC column or the ionization source in the MS equipment may be damaged. It is important to consider that the drastic conditions needed to release HPP from food matrices may cause some degradations. If that happens, the phenolic compounds identified in the hydrolyzates are not the same as those originally present in the sample (Pérez-Jiménez and Saura-Calixto, 2015). Future analytical developments should allow the characterizing HPPs as originally present in the food matrix, without needing to release them.

6.6.2 Analysis of NEPAs

In order to analyze NEPAs in the residue from the extractions, a previous treatment is also needed. In this case, in addition to releasing NEPAs from their potential associations with the food matrix, the goal is to partially break the longest NEPA chains, with several dozens of units, because they cannot be directly analyzed. The treatment chosen determines the analytical procedure applied in the derived NEPA

TABLE 6.2
Content of Extractable and Nonextractable Ferulic Acid (High-Performance Liquid Chromatography Analysis) in Different Cereal Samples (mg/100 g)

Sample	Extractable	Nonextractable	References
Barley	n.d.	22.2 ± 0.03	Yu et al. (2001)
Corn	0.17 ± 0.004	175.61 ± 1.76	Adom and Liu (2002)
Oat	0.13 ± 0.008	35.70 ± 3.14	Adom and Liu (2002)
Rice	0.13 ± 0.01	29.63 ± 1.68	Adom and Liu (2002)
Wheat	0.11 ± 0.004	332.87 ± 3.14	Adom and Liu (2002)
Wheat bran	1.15 ± 0.2	219.3 ± 5.1	Arranz and Saura-Calixto (2010)
Wheat bran	123.9 ± 10.6	292.2 ± 47.8	Verma et al. (2009)
Wheat flour	0.018 ± 0.007	0.20 ± 0.02	Arranz and Saura-Calixto (2010)
Wheat flour	0.5	5.7	Hatcher and Kruger (1997)

Abbreviations: HPLC, high-performance liquid chromatography; n.d., not detected.

concentrate, which will be one of those described in Section 6.4. Therefore, the following approaches exist:

- Acid-catalyzed depolymerization in the presence of a nucleophilic reagent, followed by RP-HPLC (Hellström et al., 2008). Any of the commonly used nucleophiles, that is, benzylmercaptan, benzylthioether, or phloroglucinol, may be used for the analysis of NEPAs by this approach.
- Alkaline or enzymatic cleavage, followed by direct-phase HPLC (DP-HPLC) (Pérez-Jiménez et al., 2009; White et al., 2010). With these approaches, NEPA chains are partially broken and the smaller structures may be directly analyzed by direct-phase LC. This is a good strategy if the fragments formed are below 10 units, the maximum length that may be properly characterized by DP-HPLC. As described for the release of HP, a higher efficiency in this reaction is achieved with chemical treatments than with enzymatic treatments (Pérez-Jiménez et al., 2009).
- Butanolysis followed by spectrophotometric determination. This method, which degrades the original structures into anthocyanins, allows the most comprehensive release of NEPA that has been achieved to date (Pérez-Jiménez et al., 2009). However, spectrophotometric determination presents general deficiencies in relation to chromatographic techniques and this method causes the highest degradation of the original NEPA chains as compared with the other methods (Pérez-Jiménez and Torres, 2011).

Any of these approaches may also be carried out in the supernatants of extraction, containing EPP (in case DP-HPLC is chosen, the previous cleavage step is not needed), allowing for the comparison of content, profile, or DP between extractable PA and NEPA. They can also be directly applied on the whole sample (Guyot et al., 2002), providing the value for total PA. Anyway, the strategies for the analysis of NEPA should still be improved, since all the current ones imply a cleavage of the original NEPA chains and/or transformations of the original constituents of the chains. In that way, there is a need to develop methods that are able to detect NEPAs as they are originally present in the food matrix, in order to establish their food content properly, their intakes, or their potential association with health effects.

6.7 CONCLUDING REMARKS

- Spectrophotometric techniques for polyphenol quantification, although quick and useful for the screening of a high number of samples, are also subjected to several interferences that should be considered when analyzing the results.
- RP-HPLC, coupled to DAD, is the most suitable approach to carry out a quantitative analysis of all polyphenols present in a sample, also if they belong to different classes.
- MS techniques may allow for the identification of the structures of phenolic compounds present in a sample, even when lacking commercial standards.
- A comprehensive analysis of certain polyphenol classes, such as PA, requires the combination of different analytical techniques that provide complementary information.
- The residues of polyphenols extractions contain significant amounts of polyphenols (NEPPs) that should be included in routine polyphenol analysis in order to have a complete determination of the polyphenol content in natural products.

REFERENCES

Adom, K.K. and Liu, R. 2002. Antioxidant activity of grains. *Journal of Agricultural and Food Chemistry* 50:6182–6187.

Al-Othman, Z.-A., Hadj-Ahmed, Y.-B., Habila, M.-A., and Ghafar, A.-A. 2011. Determination of capsaicin and dihydrocapsaicin in *Capsicum* fruit samples using high-performance liquid chromatography. *Molecules* 16:8919–8929.

Arraez-Roman, D., Fu, S., Sawalha, S.M.S., Segura-Carretero, A., and Fernandez-Gutierrez, A. 2010. HPLC/CE-ESI-TOF-MS methods for the characterization of polyphenols in almond-skin extracts. *Electrophoresis* 31:2289–2296.

Arranz, S. and Saura-Calixto, F. 2010. Analysis of polyphenols in cereals may be improved performing acid hydrolysis: A study in wheat flour and wheat bran and cereals of the diet. *Journal of Cereal Science* 51:313–318.

Arranz, S., Saura-Calixto, F., Shaha, T., and Kroon, P.A. 2009. High contents of nonextractable polyphenols in fruits suggest that polyphenol contents of plant foods have been underestimated. *Journal of Agricultural and Food Chemistry* 57:7298–7303.

Bartolomé, B. and Gómez-Cordovés, C. 1999. Barley spent grain: Release of hydroxycinnamic acids (ferulic acid and *p*-coumaric acids) by commercial enzyme preparations. *Journal of the Science of Food and Agriculture* 79:435–439.

Becker, C., Klaring, H.P., Kroh, L.W., and Krumbein, A. 2013. Temporary reduction of radiation does not permanently reduce flavonoid glycosides and phenolic acids in red lettuce. *Plant Physiology and Biochemistry* 72:154–160.

Berardini, N., Reinholdt, C., and Schieber, A. 2004. Characterization of gallotannins and benzophenone derivatives from mango (*Mangifera indica* L. cv. 'Tommy Atkins') peels, pulp and kernels by high-performance liquid chromatography/electrospray ionization mass spectrometry. *Rapid Communications in Mass Spectrometry* 18:2208–2216.

Biesaga, M. and Pyszynska, K. 2013. Stability of bioactive polyphenols from honey during different extraction methods. *Food Chemistry* 136:46–54.

Bravo, L., Abia, R., and Saura-Calixto, F. 1994. Polyphenols as dietary fiber associated compounds: Comparative study on in vivo and in vitro properties. *Journal of Agricultural and Food Chemistry* 42:1481–1487.

Burdette, C.Q. and Marcus, K. 2013. Determination of isoflavone content in soy, red clover, and kudzu dietary supplement materials by liquid chromatography-particle beam/electron ionization mass spectrometry. *Journal of AOAC International* 96:925–932.

Cacciola, F., Russo, M.F., Beccaria, M., Donato, P., Farnetti, S., Dugo, P., and Mondello, L. 2015. Comprehensive two-dimensional liquid chromatography methods for analysis of polyphenols in food sample. *Austin Chromatography* 2:1–6.

Cai, Y.Z., Xing, J., Sun, M., Zhan, Z.Q., and Corke, H. 2005. Phenolic antioxidants (hydrolyzable tannins, flavonols, and anthocyanins) identified by LC-ESI-MS and MALDI-QIT-TOF MS from *Rosa chinensis* flowers. *Journal of Agricultural and Food Chemistry* 53:9940–9948.

Callemien, D. and Collin, S. 2008. Use of RP-HPLC-ESI(–)-MS/MS to differentiate various proanthocyanidin isomers in lager beer extracts. *Journal of the American Society of Brewing Chemists* 66:109–115.

Cao, S., Liu, L., Pan, S., Lu, Q., and Xu, X. 2009. A comparison of two determination methods for studying degradation kinetics of the major anthocyanins from blood orange. *Journal of Agricultural and Food Chemistry* 57:245–249.

Churchwell, M.I., Twaddle, N.C., Meeker, L.R., and Doerge, D.R. 2005. Improving LC-MS sensitivity through increases in chromatographic performance: Comparisons of UPLC-ES/MS/MS to HPLC-ES/MS/MS. *Journal of Chromatography B—Analytical Technologies in the Biomedical and Life Sciences* 825:134–143.

Díaz-Rubio, M.E., Pérez-Jiménez, J., and Saura-Calixto, F. 2009. Dietary fiber and antioxidant capacity in *Fucus vesiculosus* products. *International Journal of Food Sciences and Nutrition* 60:23–34.

Di-Stefano, V., Avellones, G., Bongiorno, D., Cunsolo, V., Muccilli, V., Sforza, S., Dossena, A., Drahos, L., and Vekey, K. 2012. Applications of HPLC-MS for food analysis. *Journal of Chromatography A* 1259:74–85.

El-Aneed, A., Cohen, A., and Banoub, J. 2009. Mass spectrometry, review of the basics: Electrospray, MALDI, and commonly used mass analyzers. *Applied Spectroscopy Reviews* 44:210–230.

Everette, J.-D., Bryant, Q.M., Green, A.M., Abbey, Y.A., Wangila, G.M., and Walker, R.B. 2010. Thorough study of reactivity of various compound classes toward the Folin-Ciocalteu reagent. *Journal of Agricultural and Food Chemistry* 58:8139–8144.

Ferreira, D. and Slade, D. 2002. Oligomeric proanthocyanidins: Naturally occurring *O*-heterocycles. *Natural Products Report* 19:517–541.

Foo, L. and Porter, L.J. 1978. Prodelphinidin polymers: Definition of structural units. *Journal of the Chemical Society: Perkin Transactions* 10:1186–1190.

Fredes, C., Yousef, G.G., Robert, P.C., Grace, M.H., Lila, M.A.N., Gómez, M.F., Gebauer, M., and Montenegro, G. 2014. Anthocyanin profiling of wild maqui berries (*Aristotelia chilensis* [Mol.] Stuntz) from different geographical regions in Chile. *Journal of the Science of Food and Agriculture* 94:2639–2648.

Fulcrand, H., Remy, S., Souquet, J.M., Cheynier, V., and Montounet, M. 1999. Study of wine tannin oligomers by on-line liquid chromatography electrospray ionization mass spectrometry. *Journal of Agricultural and Food Chemistry* 47:1023–1028.

Gentili, A., Marchese, S., and Perret, D. 2008. MS techniques for analyzing phenols, their metabolites and transformation products of environmental interest. *TrAC: Trends in Analytical Chemistry* 27:888–903.

Gonçalves, J.L., Silva, C.L.U., Castilho, P.C., and Camara, J.S. 2012. An attractive, sensitive and high-throughput strategy based on microextraction by packed sorbent followed by UHPLC-PDA analysis for quantification of hydroxybenzoic and hydroxycinnamic acids in wines. *Microchemical Journal* 106:129–138.

Guyot, S., Bourvellec, C.L., Marnet, N., and Drilleau, J.F. 2002. Procyanidins are the most abundant polyphenols in dessert apples at maturity. *LWT: Food Science and Technology* 35:289–291.

Guyot, S., Marnet, N., and Drilleau, J.-F. 2001. Thiolysis-HPLC characterization of apple procyanidins covering a large range of polymerization states. *Journal of Agricultural and Food Chemistry* 49:14–20.

Han, X., Aslanian, A.S., and Yates, J.R.R. 2008. Mass spectrometry for proteomics. *Current Opinion in Chemical Biology* 12:483–490.

Haslam, E. 1998. Polyphenols-structure and biosynthesis. In *Practical Polyphenolics: From Structure to Molecular Recognition and Physiological Action*, ed. E. Haslam. Cambridge, UK: Cambridge University Press.

Hatcher, D.W. and Kruger, J.E. 1997. Simple phenolic acids in flours prepared from Canadian wheat: Relationship to ash content, color, and polyphenol oxidase activity. *Cereal Chemistry* 74:337–343.

Hellström Jarkko, J. and Mattila, P.H. 2008. HPLC determination of extractable and unextractable proanthocyanidins in plant materials. *Journal of Agricultural and Food Chemistry* 56:7617–7624.

Herrero-Martínez, J.M., Ràfols, C., Rosés, M., Torres, J.L., and Bosch, E. 2003. Mixed micellar electrokinetic capillary chromatography separation of depolymerized grape procyanidins. *Electrophoresis* 24:707–713.

Jandera, P., Skerikova, V., Rehova, L., Hajek, T., Baldrinova, L., Skopova, G., Kellner, V., and Horna, A. 2005. RP-HPLC analysis of phenolic compounds and flavonoids in beverages and plant extracts using a CoulArray detector. *Journal of Separation Science* 28:1005–1022.

Jerez, M., Sineiro, J., Guitián, E., and Núñez, M.J. 2009. Identification of polymeric procyanidins from pine bark by mass spectrometry. *Rapid Communications in Mass Spectrometry* 23:4013–4018.

Karaonen, M., Loponen, J.M., Ossipov, V.I., and Pihlaja, K. 2004. Analysis of procyanidins in pine bark with reversed-phase and normal-phase high-performance liquid chromatography-electrospray ionization mass spectrometry. *Analytica Chimica Acta* 522:105–112.

Karonen, M., Leikas, A., Loponen, J., Sinkkonen, J., Ossipov, V., Pihlaja, K. 2007. Reversed-phase HPLC-ESI/MS analysis of birch leaf proanthocyanidins after their acidic degradation in the presence of nucleophiles. *Phytochemical Analysis* 18: 378.

Kelm, M.A., Hammerstone, J.F., and Schmitz, H.H. 2005. Identification and quantification of flavanols and proanthocyanidins in foods: How good are the data? *Clinical and Developmental Immunology* 12:35–41.

Kelm, M.A., Johnson, P.C., Robbins, R.J., Hammerstone, J.F., and Schmitz, H.H. 2006. High-performance liquid chromatography separation and purification of cacao (*Theobroma cacao* L.) procyanidins according to degree of polymerization using a diol stationary phase. *Journal of Agricultural and Food Chemistry* 54:1571–1576.

Kennedy, J.A. and Jones, G.P. 2001. Analysis of proanthocyanidin cleavage products following acid-catalysis in the presence of excess phloroglucinol. *Journal of Agricultural and Food Chemistry* 49:1740–1746.

Kennedy, J.A. and Taylor, A.W. 2003. Analysis of proanthocyanidins by high-performance gel permeation chromatography. *Journal of Chromatography A* 995:99–107.

Kerio, L.C., Wachira, F.N.Y., Wanyoko, J.K., and Rotich, M.K. 2012. Characterization of anthocyanins in Kenyan teas: Extraction and identification. *Food Chemistry* 131:31–38.

Khoddami, A., Wilkes, M.A., and Roberts, T.H. 2013. Review techniques for analysis of plant phenolic compounds. *Molecules* 18:2328–2375.

Kimura, H., Ogawa, S., Akihiro, T., and Yokota, K. 2011. Structural analysis of A-type or B-type highly polymeric proanthocyanidins by thiolytic degradation and the implication in their inhibitory effects on pancreatic lipase. *Journal of Chromatography A* 1218:7704–7712.

Kivilompolo, M., Oburka, V., and Hyotylainen, T. 2007. Comparison of GC-MS and LC-MS methods for the analysis of antioxidant phenolic acids in herbs. *Analytical and Bioanalytical Chemistry* 388:881–887.

Klejdus, B., Třináctý, J., Hrdlička, P., and Kubáň, V. 2001. Comparison of solid-phase extraction procedures for isolation of plant isoflavones prior to liquid chromatographic determination. *Chemical Papers* 55:285–289.

Klejdus, B., Vacek, J., Lojková, L., Benšová, L., and Kubáň, V. 2008. Ultrahigh-pressure liquid chromatography of isoflavones and phenolic acids on different stationary phases. *Journal of Chromatography A* 1195:52–59.

Labarbe, B., Cheynier, V., Brossaud, F., Souquet, J.M., and Motounet, M. 1999. Quantitative fractionation of grape proanthocyanidins according to their degree of polymerization. *Journal of Agricultural and Food Chemistry* 47:2719–2723.

Lee, J., Durst, R.W., and Wrolstad, R.E. 2005. Determination of total monomeric anthocyanin pigment content of fruit juices, beverages, natural colorants, and wines by the pH differential method: Collaborative study. *Journal of AOAC International* 88:1269–1278.

Lee, J., Rennaker, C., and Wrolstad, R.E. 2008. Correlation of two anthocyanin quantification methods: HPLC and spectrophotometric methods. *Food Chemistry* 110:782–786.

Li, H. and Deinzer, M.L. 2006. Structural identification and distribution of proanthocyanidins in 13 different hops. *Journal of Agricultural and Food Chemistry* 54:4048–4056.

Lin, L.-Z., Sun, J., Chen, P., and Harnly, J. 2011. UHPLC-PDA-ESI/HRMS/MSn analysis of anthocyanins, flavonol glycosides, and hydroxycinnamic acid derivatives in red mustard greens (*Brassica juncea* Coss Variety). *Journal of Agricultural and Food Chemistry* 59:12059–12072.

Llorent-Martínez, E.J., Molina-García, L., Ruiz-Medina, A., and Ortega-Barrales, P. 2014. Quantitation of selected polyphenols in plant-based food supplements by liquid chromatography-ion trap mass spectrometry. *Food Analytical Methods* 7:2177–2183.

López-Gutiérrez, N., Romero-González, R., Plaza-Bolaños, P., Luis Martínez-Vidal, J., and Garrido-Frenich, A. 2015. Identification and quantification of phytochemicals in nutraceutical products from green tea by UHPLC-Orbitrap-MS. *Food Chemistry* 173:607–618.

Mateos-Martín, M.L., Fuget, E., Jiménez-Ardón, A., Herrero-Uriba, L., Tamayo-Castillo, G., and Torres, J.L. 2014. Identification of polyphenols from antiviral *Chamaecrista nictitans* extract using high-resolution LC–ESI–MS/MS. *Analytical Bioanalytical Chemistry* 406:5501–5506.

Mateos-Martín, M.L., Fuget, E., Quero, C., Pérez-Jiménez, J., and Torres, J.L. 2012. New identification of proanthocyanidins in cinnamon (*Cinnamomum zeylanicum* L.) using MALDI-TOF/TOF mass spectrometry. *Analytical and Bioanalytical Chemistry* 402:1327–1336.

Matsuo, T., Tamaru, K., and Itoo, S. 1984. Chemical degradation of condensed tannin with phloroglucinol in acidic solvents. *Agricultural and Biological Chemistry* 48:1199–1204.

Matthews, S., Mila, I., Scalbert, A., Pollet, B., Lapierre, C., Hervé du Penhoat, C.L.M., Rolando, C., and Donnelly, D.M.X. 1997. Method for estimation of proanthocyanidins based on their acid depolymerization in the presence of nucleophiles. *Journal of Agricultural and Food Chemistry* 45:1195–1201.

Mattila, P., Hellström, J., and Törrönen, R. 2006. Phenolic acids in berries, fruits and beverages. *Journal of Agricultural and Food Chemistry* 54:7193–7199.

Monagas, M., Garrido, I., Lebrón-Aguilar, R., Bartolomé, B., and Gómez-Cordovés, C. 2007. Almond (*Prunus dulcis* (Mill.) D.A. Webb) skins as a potential source of bioactive polyphenols. *Journal of Agricultural and Food Chemistry* 55:8498–8507.

Monagas, M., Gómez-Cordovés, C., Bartolomé, B., Laureano, O., and Ricardo da Silva, J.M. 2003. Monomeric, oligomeric, and polymeric flavan-3-ol composition of wines and grapes from *Vitis vinifera* L. cv. Graciano, Tempranillo, and Cabernet Sauvignon. *Journal of Agricultural and Food Chemistry* 51:6475–6481.

Monagas, M., Quintanilla-López, J.E., Gómez-Cordóvés, C., Bartolomé, B., Lebrón-Aguilar, R. 2010. MALDI-TOF MS analysis of plant proanthocyanidins. *Journal of Pharmaceutical and Biomedical Analysis* 51:358–372.

Montero, L., Herrero, M., Ibáñez, E., and Cifuentes, A. 2014. Separation and characterization of phlorotannins from brown algae *Cystoseira abies-marina* by comprehensive two-dimensional liquid chromatography. *Electrophoresis* 35:1644–1651.

Naczk, M. and Shahidi, F. 2004. Extraction and analysis of phenolics in food. *Journal of Chromatography A* 1054:95–111.

Nesvizhskii, A.I. 2007. Protein identification by tandem mass spectrometry and sequence database searching. In *Mass Spectrometry Data Analysis in Proteomics*, ed. R. Matthiesen, Vol. 367, pp. 87–119. Totowa, NJ: Humana Press.

Neveu, V., Pérez-Jiménez, J., Vos, F. et al. 2010. Phenol-Explorer: An online comprehensive database on polyphenol content in foods. *Database* 2010:bap024.

Oki, T., Masuda, M., Kobayashi, M., Nishiba, Y., Furuta, S., Suda, I., and Sato, T. 2002. Polymeric procyanidins as radical-scavenging components in red-hulled rice. *Journal of Agricultural and Food Chemistry* 50:7524–7529.

Ortega, N., Romero, M.-P., Macià, A., Reguant, J., Angles, N., Morello, J.R., and Motilva, M.-J. 2010. Comparative study of UPLC-MS/MS and HPLC-MS/MS to determine procyanidins and alkaloids in cocoa samples. *Journal of Food Composition and Analysis* 23:298–305.

Peng, Z., Hayasaka, Y., Iland, P.G., Sefton, M.A., Höj, P.B., and Waters, E.J. 2001. Quantitative analysis of polymeric procyanidins (tannins) from grape (*Vitis vinifera*) seeds by reverse phase high-performance liquid chromatography. *Journal of Agricultural and Food Chemistry* 49:26–31.

Pérez-Ilzarbe, F.J., Martínez, V., Hernández, T., and Estrella, I. 1992. Liquid chromatographic determination of apple pulp procyanidins. *Journal of Liquid Chromatography* 15:637–646.

Pérez-Jiménez, J., Arranz, S., and Saura-Calixto, F. 2009. Proanthocyanidin content in foods is largely underestimated in the literature data: An approach to the quantification of the missing proanthocyanidins. *Food Research International* 42:1381–1388.

Pérez-Jiménez, J., Díaz-Rubio, M.E., and Saura-Calixto, F. 2013. Non-extractable polyphenols, a major dietary antioxidant: Occurrence, metabolic fate and health effects. *Nutrition Research Reviews* 26:118–129.

Pérez-Jiménez, J., Neveu, V., Vos, F., and Scalbert, A. 2010. Identification of the 100 richest dietary sources of polyphenols—An application of the Phenol-Explorer database. *European Journal of Clinical Nutrition* 64:s112–s120.

Pérez-Jiménez, J. and Saura-Calixto, F. 2015. Macromolecular antioxidants or non-extractable polyphenols in fruit and vegetables: Intake in four European countries. *Food Research International* 74:315–323.

Pérez-Jiménez, J. and Torres, J.L. 2011. Analysis of non-extractable phenolic compounds in foods: The current state of the art. *Journal of Agricultural and Food Chemistry* 59:12713–12724.

Pérez-Jiménez, J. and Torres, J.L. 2012. Analysis of proanthocyanidins in almond blanch water by HPLC–ESI–QqQ–MS/MS and MALDI–TOF/TOF MS. *Food Research International* 49:798–806.

Piñeiro, Z., Palma, M., and Barroso, C.G.A. 2006. Determination of trans-resveratrol in grapes by pressurised liquid extraction and fast high-performance liquid chromatography. *Journal of Chromatography A* 1110:61–65.

Porter, L.J., Hrstich, L.N., and Chan, B.G. 1986. The conversion of procyanidins and prodelphinidins to cyanidin and delphinidin. *Phytochemistry* 25:223–230.

Prinsen, P., Gutiérrez, A., Faulds, C.B., and Carlos del Río, J. 2014. Comprehensive study of valuable lipophilic phytochemicals in wheat bran. *Journal of Agricultural and Food Chemistry* 62:1664–1673.

Quifer-Rada, P., Vallverdú-Queralt, A., Martínez-Huélamo, M., Chiva-Blanch, G., Jáuregui, O., Estruch, R., and Lamuela-Raventós, R. 2015. A comprehensive characterisation of beer polyphenols by high resolution mass spectrometry (LC-ESI-LTQ-Orbitrap-MS). *Food Chemistry* 169:336–343.

Reed, J.D., Krueger, C.G., and Vestling, M.M. 2005. MALDI-TOF mass spectrometry of oligomeric food polyphenols. *Phytochemistry* 66:2248–2263.

Rigaud, J., Pérez-Ilzarbe, F.J., Ricardo da Silva, J.M., and Cheynier, V. 1991. Micro method for the identification of proanthocyanidin using thiolysis monitored by high-performance liquid chromatography. *Journal of Chromatography A* 540:401–405.

Rijke, E., Out, P., Niessen, W.M.A., Ariese, F., Gooijer, C., and Brinkman, U.A.Th. 2006. Analytical separation and detection methods for flavonoids. *Journal of Chromatography A* 1112:31–63.

Robbins, R.J. 2003. Phenolic acids in foods: An overview of analytical methodology. *Food Chemistry* 51:2866–2887.

Robbins, R.J. and Bean, S.R. 2004. Development of a quantitative high-performance liquid chromatography-photodiode array detection measurement system for phenolic acids. *Journal of Chromatography* 1038:97–105.

Rohr, G.E., Meier, B., and Sticher, O. 2000. Analysis of procyanidins. *Studies in Natural Products Chemistry* 21:497–570.

Sáyago-Ayerdi, S.G., Moreno-Hernández, C.L., Montalvo-González, E., Garcia-Magaña, M.L., Mata-Montes de Oca, M., Torres, J.L., and Pérez-Jiménez, J. 2013. Mexican 'Ataulfo' mango (*Mangifera indica* L.) as a source of hydrolyzable tannins analysis by MALDI-TOF/TOF MS. *Food Research International* 51:188–194.

Scalbert, A. 1992. Quantitative methods for the estimation of tannins in plant tissues. In *Plant Polyphenols*, eds. R.W. Hemingway and P.E. Laks. New York: Plenum Press.

Singleton, V.L., Orthofer, R., and Lamuela-Raventós, R.M. 1999. Analysis of total phenols and other oxidation substrates and antioxidants by means of Folin-Ciocalteau reagent. *Methods of Enzymology* 299:152–178.

Sondheimer, E. and Kertesz, Z.I. 1948. Anthocyanin pigments: Colorimetric determination in strawberries and strawberry products. *Analytical Chemistry* 20:245–248.

Swain, T. and Hillis, W.E. 1959. The phenolic constituents of *Prunus domestica*. I. The quantitative analysis of phenolic constituents. *Journal of the Science of Food and Agriculture* 10:63–68.

Symaladevi, R.M., Sablani, S.S., Tang, J., Powers, J.R., and Swanson, B.G. 2011. Stability of anthocyanins in frozen and freeze-dried raspberries during long-term storage: In relation to glass transition. *Journal of Food Science* 76:E414–E421.

Tarascou, I., Souquet, J.M., Mazauric, J.P., Carrillo, S., Coq, S., Canon, F., Fulcrand, H., and Cheynier, V. 2010. The hidden face of food phenolic composition. *Archives of Biochemistry and Biophysics* 501:16–22.

Taylor, A.W., Barofsky, E.L., Kennedy, J.A., and Deinzer, M.L. 2003. Hop (*Humulus lupulus* L.) proanthocyanidins characterized by mass spectrometry, acid catalysis, and gel permeation chromatography. *Journal of Agricultural and Food Chemistry* 51:4104–4110.

Thompson, R.S., Jacques, D., Haslam, E., and Tanner, R.J.N. 1972. Plant proanthocyanidins. Part I. Introduction: The isolation, structure, and distribution in nature of plant procyanidins. *Journal of the Chemical Society: Perkin Transactions* 1:1387–1399.

Torres, J.L. and Bobet, R.A. 2001. New flavanol-derivatives from grape (*Vitis vinifera*) by-products. Antioxidant aminoethylthio-flavan-3-ol conjugates from a polymeric waste fraction used as a source of flavanols. *Journal of Agricultural and Food Chemistry* 49:4627–4634.

Torres, J.L. and Lozano, C. 2001. Chromatographic characterisation of proanthocyanidins after thiolysis with cysteamine. *Chromatographia* 54:523–526.

Torres, J.L. and Selga, A. 2003. Procyanidin size and composition by thiolysis with cysteamine hydrochloride and chromatography. *Chromatographia* 57:441–445.

Torres, J.L., Varela, B., García, M.T., Carrilla, J., Matito, C., Centelles, J.J., Cascane, M., Sort, X., and Bobet, R. 2002. Valorization of grape (*Vitis vinifera*) by-products: Antioxidant and biological properties of polyphenolic fractions differing in procyanidin composition and flavonol content. *Journal of Agricultural and Food Chemistry* 50:7548–7555.

Touriño, S., Fuguet, E., Jáuregui, O., Saura-Calixto, F., Cascante, M., and Torres, J.L. 2008. High-resolution liquid chromatography/electrospray ionization time-of-flight mass spectrometry combined with liquid chromatography/electrospray ionization tandem mass spectrometry to identify polyphenols from grape antioxidant dietary fiber. *Rapid Communications in Mass Spectrometry* 22:3489–3500.

Verardo, V., Gómez-Caravaca, A.M., Segura-Carretero, A., Caboni, M.F., and Fernández-Gutiérrez, A. 2011. Development of a CE-ESI-microTOF-MS method for a rapid identification of phenolic compounds in buckwheat. *Electrophoresis* 32:669–673.

Vergara-Salinas, J.R., Pérez-Jiménez, J., Torres, J.L., Agosin, E., and Pérez-Correa, J.R. 2012. Effects of temperature and time on polyphenolic content and antioxidant activity in the pressurized hot water extraction of deodorized thyme (*Thymus vulgaris*). *Journal of Agricultural and Food Chemistry* 60:10920–10929.

Verma, B., Hucl, P., and Chibbar, R.N. 2009. Phenolic acid composition and antioxidant capacity of acid and alkali hydrolysed wheat bran fractions. *Food Chemistry* 116:947–954.

Vinson, J.A., Xuehui, S., Zubik, L., and Bose, P. 2001. Phenol antioxidant quantity and quality in foods: Fruits. *Journal of Agricultural and Food Chemistry* 49:5315–5321.

Vivas, N., Nonier, M.F., Vivas de Gaulejac, N., Absalon, C., Bertrand, A.K., and Mirabel, M. 2004. Differentiation of proanthocyanidin tannins from seeds, skins and stems of grapes (*Vitis vinifera*) and heartwood of quebracho (*Schinopsis balansae*) by matrix-assisted laser desorption/ionization time-of-flight mass spectrometry and thioacidolysis/liquid chromatography/electrospray ionization mass spectrometry. *Analytica Chimica Acta* 513:247–256.

Wang, H. 2014. Rapid quantitative analysis of individual anthocyanin content based on high-performance liquid chromatography with diode array detection with the pH differential method. *Journal of Separation Science* 37:2535–244.

Whelan, T.J., Gray, M.J., Slonecker, P.J., Shalliker, R.A., and Wilson, M.A. 2005. Study of the selectivity of reversed-phase columns for the separation of plycarboxylic acids and polyphenols compounds. *Journal of Chromatography A* 1097:148–156.

White, B.L., Howard, L.R., and Prior, R.L. 2010. Release of bound procyanidins from cranberry pomace by alkaline hydrolysis. *Journal of Agricultural and Food Chemistry* 58:7572–7579.

Xiang, P., Lin, Y., Lin, P., Xiang, C., Yang, Z., and Lu, Z. 2003. Effect of cationization reagents on the matrix-assisted laser desorption/ionization time-of-flight mass spectrum of Chinese gallotannins. *Journal of Applied Polymer Science* 105:859–864.

Yassin, G.H., Koek, J.H., and Kuhnert, N. 2014. Identification of trimeric and tetrameric flavan-3-ol derivatives in the SII black tea thearubigin fraction of black tea using ESI-tandem and MALDI-TOF mass spectrometry. *Food Research International* 63:317–327.

Yu, J., Vassanthan, T., and Temelli, F. 2001. Analysis of phenolic acids in barley by high-performance liquid chromatography. *Journal of Agricultural and Food Chemistry* 49:4352–4358.

7 Purification of Polyphenolic Extracts from Natural Sources

Adsorption and Membrane Technologies

María P. Casas, María Luisa Soto, Beatriz Díaz-Reinoso, Elena M. Balboa, Andrés Moure, Enma Conde, Herminia Domínguez, and Juan Carlos Parajó

CONTENTS

7.1 INTRODUCTION

The increasing consumer demand for ingredients and additives derived from natural sources, and the growing market for functional foods and nutraceuticals have favored the search for bioactive compounds and fractions. Phenolic compounds are one of the most numerous and widespread natural structures that show biological effects, such as antioxidant, antimicrobial, antiallergenic, cardioprotective, anticancer, and anti-inflammatory (Del Río et al., 2013). Solvent extraction is the first step to obtaining target compounds from plant materials. When the concentration of phenolics in the extracts is low, further separation and concentration have to be carried out. In an industrial process, the separation stages (e.g., evaporation and distillation for solvent removal) often account for an important part of the cost of the product. However, the energy-associated costs and the degradation caused by exposure to high temperatures should be avoided. Adsorption–desorption and membranes are energy-efficient technologies for separation of thermolabile compounds.

This chapter focuses on the recovery, purification, and concentration of polyphenols by adsorption and membrane processes. These separation techniques are relatively simple and flexible, and represent promising alternatives to other more energy-consuming processes. After a brief summary of the fundamentals, some examples of recovery, concentration, and fractionation of phenolics from a variety of sources are presented.

7.2 ADSORPTION

7.2.1 Fundamentals

Adsorption can provide an efficient separation and concentration of the bioactive compounds found in plant extracts and allows for a relatively simple design and control of the operational parameters. Therefore, adsorption is a widely applied separation technology in industry (Yang, 2003) and is being increasingly used to recover phenolic compounds (Soto et al., 2011). Adsorption is driven by attractive forces between the adsorbate (solute present in the solution) and the adsorbent (solid phase) and can involve physical adsorption (physisorption) or chemical adsorption (chemisorption), depending on the nature of the bonding between adsorbate and adsorbent (Duong, 1998; Ruthven, 2005; Thomas and Crittenden, 1998; Toth, 2002).

Different adsorbents have been successfully proposed for the recovery and concentration of phenolic compounds from natural sources. Activated charcoals are widely used for the adsorption of phenolic compounds from aqueous solutions, and could be useful for the recovery of flavonoids (Aehle et al., 2004), although irreversible adsorption could limit the efficiency of the desorption process (Soto et al., 2008). Activated charcoals, composed of hydrophobic graphite layers with heterogeneous surfaces and hydrophilic functional groups, are usually produced from low-cost materials, such as coal and agricultural by-products (Paraskeva et al., 2008; Soto et al., 2008). Minerals and low-cost materials from renewable mineral, lignocellulosic, and polysaccharide materials are potential adsorbents for phenolics. However, the most studied adsorbents for selective recovery are resins. These hydrophilic or hydrophobic synthetic polymer adsorbents can be polystyrene–divinylbenzene copolymers, polymethacrylate, divinylbenzene–ethylvinylbenzene copolymers, and vinylpyridine. They are highly advantageous for their durability, chemical stability and high adsorption capacity, efficiency, selectivity, and ease of regeneration, with relatively low cost and limited toxicity.

7.2.2 Adsorption Kinetics

The kinetic analysis allows for the discovery of the contact time required to reach the equilibrium and modeling the system behavior for process optimization and scaling-up. The adsorption kinetics can be represented by a number of mathematical models, which are usually classified as adsorption reaction models and adsorption diffusion models. In general, the sorption rate is controlled by a series of resistances, according to the following steps:

1. Mass transfer from the fluid phase to the external surface of the sorbent particles (film diffusion resistance)
2. Pore diffusion in the fluid phase (pore diffusion resistance)
3. Adsorption reaction at the phase boundaries (surface reaction resistance)
4. Diffusion in the sorbed state (particle phase diffusion resistance or solid diffusion resistance) which can occur in a pore-surface layer

The rate of the overall process can be limited by one or several of these steps, depending on the characteristics of the adsorbent, adsorbate, and solution. In poorly mixed systems, the transport in the bulk phase may be of major importance, particularly if dilute adsorbates, small-sized adsorbent particles, and high-affinity adsorbent–adsorbate are involved. Adsorption is controlled by intraparticle mass transfer in systems with good mixing involving large-sized adsorbent particles, high adsorbate concentration, and low-affinity adsorbent–adsorbate (Soto et al., 2011). The first mechanism is related to the external mass-transfer resistance associated with diffusion through the laminar fluid film surrounding the particle. External film diffusion occurs by molecular diffusion through the boundary layer that surrounds the adsorbent particle and is driven by the concentration difference between the bulk fluid and the adsorbent surface. The external film diffusion is influential in the early adsorption stages due to the rapid decrease in the driving force. The second mechanism is related to the macropore mass-transfer resistance associated with diffusion through the solution inside the pores (pore diffusion) or diffusion along the adsorbent surface (surface diffusion). The third mechanism is micropore diffusion. When this process is fast, the overall uptake rate will be controlled by macropore diffusion. However, if micropore diffusion is slow, it is rather difficult to identify the limiting step.

In order to identify the main controlling mechanisms, several experimental parameters should be studied, including particle size, pH, agitation speed, temperature, and initial concentration (Ahmaruzzaman, 2008; Alhamed, 2009; Geng et al., 2009; Kammerer et al., 2010).

To establish whether the external film diffusion is the controlling step, a series of batch kinetic experiments at different agitation speeds should be performed. When the agitation speed increases, the sorbate transport rate from the bulk liquid to the liquid boundary layer surrounding the particles (and likewise, the external mass-transfer coefficients) becomes higher. External film diffusion is affected by the adsorbent particle size. In small particles, both the specific external surface area of the adsorbent and the initial uptake rate are high, reaching equilibrium in short times. In larger particles, intraparticle macropore diffusion is the limiting step, requiring long contact times to reach equilibrium.

The uptake rates are affected by the adsorbate concentration if the sorption process is controlled by either external or intraparticle mass transfer. When the external mass transfer is the main mechanism, high initial solute concentrations result in low external mass transfer and low initial uptake rates. However, when the intraparticle mass transfer is the limiting step, high initial concentrations result in steep uptake curves and high intraparticle transport rates. When the adsorption is reaction-controlled, the uptake rate depends only on the sorbate concentration (Ho et al., 2000). pH affects the functional groups on the sorbent surface and the functional properties of the sorbate. The ionic interactions between adsorbent surface and solute molecules are determined by pH. If the surface–sorbate ionic interaction is strong, the reaction rate is not the limiting step.

Distinguishing between the diffusion and reaction process is often difficult. The dependence on temperature is much stronger in chemically controlled systems than in diffusion-controlled ones. Therefore, the magnitude of the activation energy (Ea) indicates the most influential stage. In physical sorption, equilibrium is usually rapidly attained and easily reversible; energy requirements are small because weak forces are involved. Chemical sorption is specific and involves much stronger interactions.

The rate-controlling mechanism in the kinetic models is the adsorption reaction of the solute in the active sites of the sorbent surface. In diffusional models, intraparticle diffusion is the rate-controlling mechanism and can involve pore volume diffusion, surface diffusion, or a combination of both.

The intraparticle diffusion model is based on the theory proposed by Weber and Morris (1962). Different forms are applied to low surface coverages (initial adsorption stages) and long adsorption times (when the system approaches to equilibrium). The basic formula might correlate experimental data for a wide range of adsorption times, and the simplified (lumped) forms of this equation are suitable for limited ranges of surface coverages. According to this theory

$$q = k_d\sqrt{t} \tag{7.1}$$

where

k_d is the intraparticle diffusion rate constant (mg/(g·min$^{1/2}$))
q is the amount adsorbed at time t (mg/g)
t is the adsorption time (min)

A dimensionless equation based on the intraparticle diffusion model was proposed for assessing the initial adsorption stages, and four zones were identified: weak, intermediate, strong, and approaching equilibrium. During the initial adsorption, the intermediate and strong zones are the most important (Wu et al., 2009). Adsorption of single phenolic compound onto a polymeric adsorbent has been assessed using a parallel diffusion model (liquid-phase diffusion and surface diffusion) and surface adsorption from multicomponent solutions controlled by diffusion was studied with the effective diffusion coefficients (Mijangos et al., 2001). Intraparticle mass transfer in the absence of reaction has been modeled with a linear driving force to approximate the nonlinear pore- or surface-diffusion mechanisms. Simplified rate equations based on linear driving forces have been applied for interpretation of data obtained with model phenols (Otero et al., 2005).

Useful formulae applicable when the surface reaction step controls the overall sorption rate have been reviewed (Plazinski et al., 2009; Rudzinski and Plazinski, 2007). The pseudo first-order kinetic model considers a reversible equilibrium of molecules between a liquid and a solid phase. Pseudo first-order can assess different adsorption situations, including (1) systems close to equilibrium, (2) systems with time-independent solute concentration or linear equilibrium adsorption isotherms, and (3) special cases for more complex systems. The integral expression of the pseudo first-order rate equation of Lagergren is generally written as

$$\log\left(q_e - q_t\right) = \log q_e - \frac{K_{ads1}}{2.303} \cdot t \tag{7.2}$$

where

q_e and q_t (mg/g) are the sorption capacities at equilibrium and at time t, respectively
K_{ads1} (1/min) is the rate constant of pseudo first-order sorption

The integral form of the pseudo second-order kinetic model may be represented by

$$\frac{t}{q_t} = \frac{1}{K_{ads2}\, q_e^2} + \frac{1}{q_e} t \tag{7.3}$$

where

q_e and q_t (mg/g) are the sorption capacities at equilibrium and at time t, respectively
K_{ads2} is the rate constant of pseudo second-order sorption (g/(mg·min))

The Elovich equation is widely used and describes the different mechanisms that can occur, including bulk and surface diffusion, and can be expressed as follows:

$$q_t = \left(\frac{1}{b}\right)\ln\left(ab\right) + \left(\frac{1}{b}\right)\ln t \tag{7.4}$$

where

q_t (mg/g) is the sorption capacity at time, t
a (mg/(g·min)) is the initial sorption rate
b (g/mg) is the desorption constant

7.2.3 Adsorption Equilibrium

Equilibrium provides information related to adsorbate–adsorbent interactions, essential for practical operation because it defines the affinity or the capacity of the adsorbent. The isotherm shape provides qualitative information on the type of adsorbate–adsorbent interactions. Table 7.1 summarizes the literature information on isotherm models used to assess the adsorption of phenolic compounds.

The Langmuir isotherm is the most widely used equation and assumes that adsorption is a physical process of interaction between neutral particles and homogeneous sorbent surfaces, with uniform adsorption energies along the adsorbent surface and monomolecular coverage. This isotherm is described by the following equation:

$$q_e = \frac{q_m K_L C_e}{1 + K_L C_e} \tag{7.5}$$

where

q_e (mg/g) measures the amount of solute adsorbed at equilibrium
C_e (mg/L) is the equilibrium concentration in the solution
q_m (mg/g) is the maximum adsorption capacity
K_L (L/mg) is the constant related to the free energy of adsorption, which can be used to define the constant separation factor, R_L:

$$R_L = \frac{1}{1 + K_L C_0} \tag{7.6}$$

The isotherm is linear for $R_L = 1$; the process is irreversible for $R_L = 0$; unfavorable for values higher than 1; and favorable for values between 0 and 1.

The Freundlich isotherm is an empirical equation suited for nonideal systems that are highly heterogeneous. This model assumes that an infinite adsorption can occur and describes the equilibrium on heterogeneous sites with different adsorption energies. The corresponding equation is as follows:

$$q_e = K_F Ce^{1/n} \tag{7.7}$$

where

K_F provides an indication of the adsorption capacity of the adsorbent (depending on temperature, surface area of the adsorbent, and the relative attraction of the solutes in a mixture for the solid surface)
$1/n$ is the temperature-dependent adsorption exponent (lower than 1 when adsorption is favorable)

TABLE 7.1
Examples of Adsorption Isotherms

Model Name	Equation	Model Parameters	Other Variables
BET Brunauer–Emmett–Teller	$\frac{X}{m} = \frac{B_{\mathrm{BET}}\, C_e \left(\frac{X}{m}\right)}{\left(C_s - C_e\right)\left(1 + \left(B_{\mathrm{BET}} - 1\right)\frac{C_e}{C_s}\right)}$	B_{BET} = dimensionless constant	C_s = saturation adsorbate concentration (X/m) = monolayer adsorption capacity C_e = equilibrium adsorbate concentration X = adsorbed adsorbate m = adsorbent mass
Dubinin–Radushkevich–Kaganer	$n_{ads} = n_{m\mathrm{DRK}}\ \exp\left\{-\left[\frac{RT}{E_s}\ln\left(\frac{C^*}{C_e}\right)\right]^n\right\}$	$n_{m\mathrm{DRK}}$ = monolayer adsorption capacity	C^* = saturation concentration E_s = adsorption energy C_e = equilibrium solute concentration n_{ads} = adsorption capacity
Elovich	$\frac{q_e}{q_m} = K_E\, C_e \exp\left(-\frac{q_e}{q_m}\right)$	K_E = Elovich constant q_m = Elovich maximum adsorption capacity	q_e = adsorptive capacity at equilibrium C_e = equilibrium solute concentration
Freundlich	$q_e = K_F C_e^{1/n}$	K_F = rate constant $1/n$ = indicates the intensity of the adsorption	q_e = adsorptive capacity at equilibrium C_e = equilibrium solute concentration
Frumkin	$q_e = n_T \ln K_T + n_T \ln C_e$	n_T, K_T = constants	C_e = equilibrium solute concentration q_e = adsorptive capacity at equilibrium
Langmuir	$q_e = \frac{q_m\, b C_e}{1 + b C_e}$	q_m = maximum adsorption capacity b = model constant	q_e = adsorptive capacity at equilibrium C_e = equilibrium solute concentration
Myers	$C_{e,i} = \frac{q_{e,i}}{H_i}\ \exp(K_i q_{e,i}^{pi})$	H_i, K_i, pi = regression parameters	$C_{e,i}$ = equilibrium solute concentration $q_{e,i}$ = equilibrium surface loading
Nitta	$\frac{q}{q_m} = K_N C_e \left(1 - \frac{q}{q_m}\right)^a$	a = number of sites occupied by the adsorbate K_N = Nitta constant	C_e = equilibrium solute concentration q_m = maximum adsorption capacity

(Continued)

TABLE 7.1 (*Continued*)
Examples of Adsorption Isotherms

Model Name	Equation	Model Parameters	Other Variables
Redlich–Peterson	$q_e = \frac{K_R C_e}{1 + B_R C_e^{\beta}}$	K_R = rate constant B_R = adsorption constant β = constant parameter $0 < \beta < 1$	q_e = adsorptive capacity at equilibrium C_e = equilibrium solute concentration
Sips	$q_e = \frac{q_{ms} K_S C_e^{ms}}{1 + K_S C_e^{ms}}$	q_{ms} = maximum adsorption capacity K_S = equilibrium constant ms = model exponent	q_e = adsorptive capacity at equilibrium C_e = equilibrium solute concentration
Temkin	$\Theta = \frac{RT}{\Delta Q} \ln K_0 C_e$	K_0 = Temkin constant	Θ = fractional coverage R = universal gas constant T = temperature $\Delta Q = (-\Delta H)$ variation of adsorption energy C_e = equilibrium solute concentration
Toth	$q_m = q_0 \frac{(bC_e)}{\left(1 + (bC_e)^t\right)^{1/t}}$	b = model parameter	q_m = adsorptive capacity at equilibrium C_e = equilibrium solute concentration t = service time

7.2.4 Equipment

7.2.4.1 Adsorption in Stirred Vessels

For laboratory experiments, adsorption–desorption studies are usually performed in batch assays in stirred vessels, since this configuration allows flexibility for evaluating the influence of the most important factors on the adsorption kinetics and equilibrium. Stirring is achieved by mechanical means, but the integrity of the particles could be compromised. Semicontinuous mode, with continuous flow of the liquid phase, could also be employed (Figure 7.1a).

7.2.4.2 Adsorption in Packed Beds

Adsorption in continuous-flow fixed-bed columns is simple to operate and offers high yields and valuable data for scaling-up. Industrial operation is usually performed using two columns (when one is being regenerated after exhaustion, the other is operative) (Figure 7.1b). The main problems with fixed beds are channeling, the existence of dead zones, and the pumping cost needed to compensate for the pressure drop across the bed (Correa et al., 2007). Expanded and fluidized beds could offer a solution for these problems.

In packed-bed sorption systems, resin loading is performed until the concentration of the solution leaving the bed exceeds a fixed threshold, which indicates exhaustion of the resin, and is referred to as breakthrough. In bed sorption processes, breakthrough loading is commonly performed instead of equilibrium loading (at which much higher compound leakage takes place). Subsequently, the resin needs to be regenerated or substituted. The mass transfer, the equilibrium, and the mass balance are required to design the adsorption process (Correa et al., 2007). Breakthrough curves offer information on the adsorption equilibrium and dynamics; therefore, they are useful to design and operate fixed beds.

7.2.4.3 Breakthrough Curves

Breakthrough curves represent graphically the dynamic adsorption process and offer information about the interaction of adsorbate with adsorbent. The shape of breakthrough curves is highly dependent on the variables involved in the process

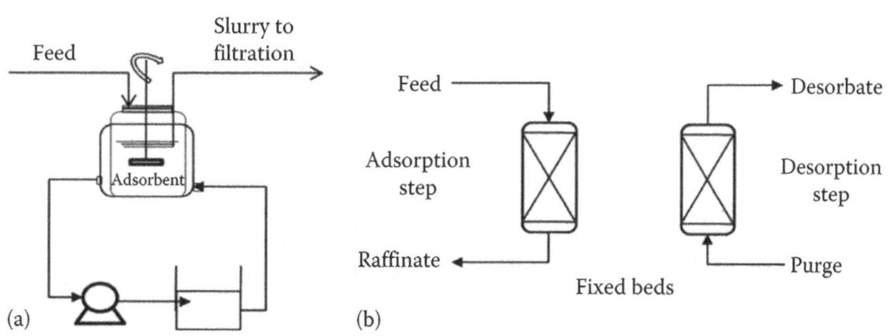

FIGURE 7.1 Schemes of the semicontinuous operation during adsorption, (a) contacting in a stirred tank and (b) in a fixed bed.

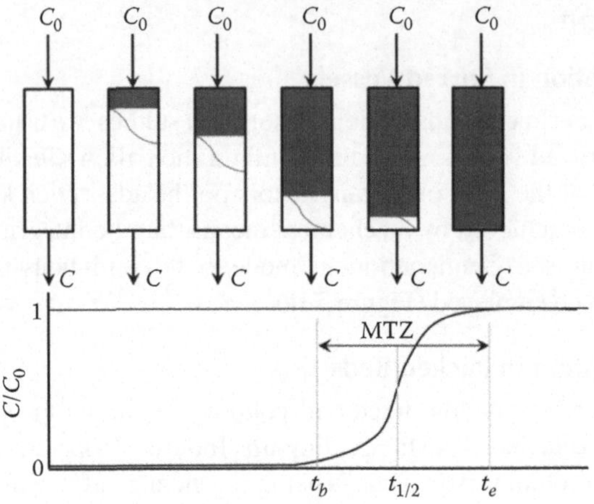

FIGURE 7.2 Breakthrough curves for fixed-bed adsorption, with the mass-transfer zone (MTZ).

(Hall et al., 1966; Negrea et al., 2011; Shafeeyan et al., 2014). The analysis of breakthrough curves serves two purposes (1) to assess whether or not the adsorbent is efficient for the required separation, and (2) to establish the break point (Knaebel, 1999).

The breakthrough curves are obtained by plotting the dimensionless concentration C_t/C_0 versus time or volume of the effluent, where C_t and C_0 are the outlet and inlet adsorbate concentrations, respectively (Arsuaga et al., 2014). The breakthrough time, t_b, is achieved when the C_t/C_0 ratio reaches the value 0.10. The saturation time (often called exhaustion time, t_e) is achieved when the ratio C_t/C_0 reaches the value 0.95 (Figure 7.2).

The mathematical models used to represent a dynamic adsorption process differ mainly in the form of the mass-transfer rate, the equilibrium isotherm, and the pressure drop along the bed. Some mathematical models for fixed-bed column adsorption are shown in Table 7.2. The Thomas model is applied to estimate the capacity of the adsorbent and to predict breakthrough curves based on the second-order reversible reaction kinetics and the Langmuir isotherm (Ghasemi et al., 2011; Han et al., 2008). Theoretically, it is suitable when external and internal diffusion resistances are extremely small (Aksu and Gönen, 2004). The Adams–Bohart model is useful for describing the initial part of the breakthrough curve. The Yoon–Nelson model does not require data concerning the characteristics of the adsorbate, the type of adsorbent, and the physical properties of the adsorption bed.

7.2.5 Examples

Data concerning the recovery and concentration of phenolic compounds from real effluents and solutions, most of them generated from waste, are summarized in Table 7.3. Elution can be performed successfully using alcoholic solutions. Ethanol is advantageous for this purpose based on environmental and toxicity reasons. Commercial resins are widely used for the recovery of natural phenolic compounds

TABLE 7.2
Mathematical Models Commonly Used to Fit Data from Breakthrough Curves

Model	Mathematical Equation	Parameters	Other Variables
Bed depth service time (BDST)	$\ln\left(\frac{C_0}{C_b}-1\right)=\ln(e^{K_a qH/v}-1)-K_aC_0t$	K_a = rate constant q = adsorption capacity	C_0 = initial solute concentration C_b = adsorbate concentration at breakthrough H = bed depth v = linear flow velocity t = service time
Bohart–Adams	$\ln\frac{C}{C_0}=k_{AB}C_0t-k_{AB}q_{AB}\frac{H}{v}$	k_{AB} = rate constant q_{AB} = adsorption capacity	C_0 = initial solute concentration C = effluent solute concentration H = bed depth v = linear flow velocity t = service time
Clark	$\ln\left[\left(\frac{C}{C_0}\right)^{n-1}-1\right]=\ln B-rt$ $B=\left(\frac{C_0^{n-1}}{C_b^{n-1}}\right)e^{r't}$ $r'=(n-1)\frac{K_T}{Q_a}\frac{dz}{dt}$	C_b = breakthrough concentration K_T = rate constant	C_0 = initial solute concentration C = effluent solute concentration t = service time Q_a = volumetric flow rate
Modified dose–response	$\ln\frac{C}{C_0-C}=a'\ln(C_0Qt)-a'\ln(qm)$	q = adsorption capacity a' = model parameter	C_0 = initial solute concentration C = effluent solute concentration m = bed mass t = service time Q = volumetric flow rate
Thomas	$\ln\left(\frac{C_t}{C_0}-1\right)=\frac{K_{Th}qm}{Q}-\frac{K_{Th}C_0}{Q}V$	K_{Th} = rate constant q = adsorption capacity	C_0 = initial solute concentration C = effluent solute concentration V = effluent volume Q = flow rate m = bed mass
Wolborska	$\ln\frac{C}{C_0}=\frac{\beta_aC_0}{q}t-\frac{\beta_aH}{v}$	q = adsorption capacity β_a = rate constant	C_0 = initial solute concentration C = effluent solute concentration H = bed depth v = linear flow velocity t = service time
Yoon–Nelson	$\ln\left(\frac{C}{C_0-C}\right)=K_{YN}t-K_{YN}\tau$	K_{YN} = rate constant τ = time when $C/C_0 = 0.5$	C_0 = initial solute concentration C = effluent solute concentration

TABLE 7.3
Some Examples of Data Reported on the Selective Recovery of Phenolic Compounds Extracted from Natural Sources

Source of Extracts or Liquid Stream	Adsorbent	Eluting Agent	References
Apple juice (caffeic acid, (+)-catechin, chlorogenic acid, *p*-coumaric acid, phloretin, phloridzin, quercetin, and quercetin glycosides)	Alimentech P-495	60% E 60%–80% M	Kammerer et al. (2007)
Apple pomace (phlorzidin, chlorogenic acid, quercetin and quercetin glycosides, and flavanol derivatives)	Amberlite XAD-16 HP	M	Schieber et al. (2003)
Black chokeberry (phenolics)	Amberlite XAD-7HP	70% E	D'Alessandro et al. (2013)
Peanut (resveratrol)	ADS-5	60% E	Xiong et al. (2014)
Blueberries (chlorogenic acid and anthocyanins: peonidin glycosides, cyanidin glycoside, petunidin glycosides, malvidin glycosides, myricetin glycosides, and quercetin glycosides)	FPX66	95% E	Buran et al. (2014)
Bayberry kernels (amygdalin)	D-101	40% E	Wang et al. (2015)
Blood oranges (anthocyanins)	NKA-9	50% E + CA (pH 2.5)	Shao-qian et al. (2010)
Berries waste (phenolics)	Amberlite XAD-7HP	70% E	Vauchel et al. (2015)
Broccoli (anthocyanins)	Amberlite XAD-7	70% E	Coutinho et al. (2004)
	Sephadex LH-20	70% E	
Chinese safflower seed (*N*-feruloylserotonin)	XAD	80% E	Jin et al. (2008)
Citrus aqueous (naringenin, hesperidin, narirutin, naringin, neohesperidin, tangeretin, nobiletin)	Amberlite XAD-16	96% E; EA	Garti et al. (2003)
Citrus peel (narirutin)	Amberlite XAD-7	80% M	Kim et al. (2007)
Citrus peel juice and molasses (phenolics)	Amberlite XAD-4, 7 and 16	dil. NaOH	Grohmann et al. (1999)
Brown algae (phlorotannins)	HP-20	40% E	Kim et al. (2014)
Marsh horsetail (flavonoids)	HPD-825	95% E	Qi et al. (2015)
Himalayan fleabane (scutellarein and apigenin glucuronides)	D-141	55% E	Zhang et al. (2009)

(Continued)

TABLE 7.3 (*Continued*)
Some Examples of Data Reported on the Selective Recovery of Phenolic Compounds Extracted from Natural Sources

Source of Extracts or Liquid Stream	Adsorbent	Eluting Agent	References
Burning bush (rutin, quercetin)	AB-8	60% E	Zhao et al. (2011)
Grape pomace aqueous extracts (phenolics)	Sepabeads SP700	E:W (82:18)	Díaz-Reinoso et al. (2010)
Chinese hawthorn (procyanidins)	LSA-10	50% E E:W (30–90:70–10)	Liu et al. (2010a)
Grape pomace (anthocyanins)	Nonpolar SDVB resin	M; E; 2-P	Kammerer et al. (2005)
Green tea leaves (polyphenols, caffeine)	XDA	85% E	Li et al. (2005)
	PA	E:EA:W 3:1:1	
Jerusalem artichoke (3-CQA)	ADS-21	60%–70% E	Sun et al. (2015)
Shimbillo (flavonoids)	Amberlite XAD-7	70% E	Silva et al. (2013)
Sweet potato (caffeoylquinic acids)	AB-8	70% E	Xi et al. (2015)
Jamun (anthocyanins)	Amberlite XAD-7HP	40% E	Jampani et al. (2014)
Kudzu (puerarin)	OCDPS resins	M:AA:W (5.0:6.6:88.4)	Yang and Tan (2008)
Kudzu (isoflavone)	Amberlite XAD-1	65% E	Fu et al. (2005)
Mango peel (mangiferin, isomangiferin, mangiferin gallate, isomangiferin gallate, quercetin and quercetin glycosides, rhamnetin glycoside)	Amberlite XAD 16 HP	M	Berardini et al. (2005)
Olive by-products, liquid from steam explosion (hydroxytyrosol)	Ion exchange + nonionic resin	M:W; 30% E	Fernández-Bolaños et al. (2002)
Orange juice (cyanidin-3-glucoside, hesperidin)	Relite EXA 118 ()	M, 75%–100%	Scordino et al. (2004)
Orange juice (anthocyanins, hesperidin, hydroxycinnamates, narirutin, limonin)	Relite EXA 118 ()	E, 75%–100%	Scordino et al. (2005)
Orange juice–processing wastewater (hesperidin, anthocyanins, hydroxycinnamic acids)	Kastell S-112	NaOH/10% E	Di Mauro et al. (2000)
Olive leaves (hydroxytyrosol, tyrosol, catechin, caffeic acid, vanillic acid, vanillin, rutin, luteolin-7-glucoside, verbascoside, apigenin-7-glucoside, diosmetin-7-glucoside, oleuropein, luteolin)	Silk fibroin	40% E	Altiok et al. (2008)

(*Continued*)

TABLE 7.3 (*Continued*)
Some Examples of Data Reported on the Selective Recovery of Phenolic Compounds Extracted from Natural Sources

Source of Extracts or Liquid Stream	Adsorbent	Eluting Agent	References
Olive mill wastewater (hydroxytyrosol, tyrosol)	Amberlite XAD-16 and 7HP	E:iP (1:1)	Agalias et al. (2007)
Olive mill wastewater (polyphenols)	Amberlite XAD-7, XAD-16	E	Bertin et al. (2011)
	Amberlite IRA-96	E+ 37% HCl	
Olive mill wastewater (phenols)	XAD-4	El	Zagklis et al. (2015)
Orange peel alkaline extract (hesperidin)	Kastell S-112	NaOH/10% E	Di Mauro et al. (1999)
Orange peel molasses (hydroxycinnamates, polymethoxylated flavone, hesperidin, and narirutin)	Sepabeads SP-70	W; A:W	Manthey (2004)
	DE52	KOH; AA	
	BioGel P2	15% E	
Pigeon pea roots (genistein, apigenin)	ADS-5 resin	50%–60% E	Liu et al. (2010b)
Preserved table olives ultrafiltered waters (phenolic compounds)	Amberlite XAD	E	Brenes and Castro (2003)
Rabdosia (phenolics and rosmarinic acid)	HP-20	70% E	Lin et al. (2012)
Red cabbage (anthocyanins)	Amberlite XAD-7HP	Abs E	Chandrasekhar et al. (2012)
Red wine (polyphenols)	Nonpolar SDVB resin	70% E	Chautard and Mayrand (2003)
Sea bilberry leaves (flavones)	NKA-2	50% E	Chen and Zhang (2014)
Spinach aqueous extract (spinacetin, patuletin)	EAC; PoAC; Amberlite XAD-7HP and 16HP	75% E; E	Aehle et al. (2004)
Sunflower (phenolic compounds)	Amberlite XAD-16HP	70% 2-P	Weisz et al. (2013)
Waste peanut shells (polyphenols)	NKA-9	90% E	Zhang et al. (2013)

Abbreviations: A, acetone; AA, acetic acid; E, ethanol; M, methanol; W, water; 2-P, 2-propanol; CA, citric acid; OCDPS, oligo-β-cyclodextrin-coupled polystyrene-based macroporous resins.

at the pilot scale (Kammerer et al., 2005, 2010; Weisz et al., 2010), with potential for industrial applications. Literature information can be found for the recovery of phenolic compounds from fruit juices and pomace as well as extracts from underutilized or waste materials.

7.3 MEMBRANE TECHNOLOGY

7.3.1 Fundamentals

A membrane can be defined as a semipermeable barrier separating two phases, and selectively limiting or facilitating the transport of one or more components between those phases. The stream passing through the membrane is named the permeate and the stream containing the rejected components is the retentate; membrane operation can result in the concentration of components in both of them (Coutinho et al., 2009).

The driving force responsible for the mass transport across the membrane can be a concentration, pressure, temperature, or electric potential gradient. Mass-transfer processes based on the application of a pressure gradient (microfiltration, ultrafiltration, nanofiltration, and reverse osmosis) have been used for the concentration and purification of phenolic components from natural sources. Figure 7.3 shows a classification of membrane technologies based on molecular size. Microfiltration (MF) membranes usually operate at pressures below 2 bar, allowing for the separation of suspended particles with sizes between 0.025 and 10 μm. Ultrafiltration (UF) membranes retain macromolecules and colloids with molecular masses between 1 and 300 kDa, whereas water and small molecules can permeate. The working pressure is in the range of 2–10 bar, in some cases reaching up to 30 bar (Nawaz et al., 2006). Nanofiltration (NF) membranes can separate compounds with molecular masses between 350 and 1000 Da at pressures in the range of 10–40 bar. Reverse osmosis (RO) membranes retain the solutes in solution, including small ions such as sodium (Cheryan, 1998; Coutinho et al., 2009; Murakami et al., 2011).

According to their morphology, membranes can be classified into symmetric (or isotropic) and asymmetric (or anisotropic) (Koops, 2002). Symmetric membranes have uniform structure and composition, and show excellent integrity, but low permeability (Baker, 2004; Mulder, 1997). Asymmetric membranes are formed by two layers of different porosity; the porous layer (50–150 μm) confers mechanical resistance

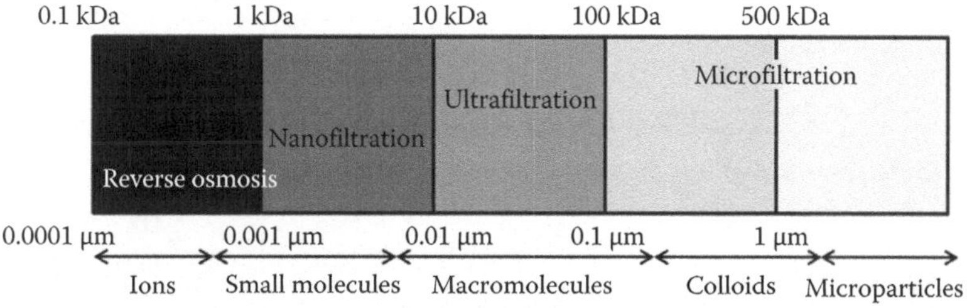

FIGURE 7.3 Membrane separation according to particle size and molecular weight.

without affecting permeability or retention, whereas the fine layer (0.1–0.5 μm) defines the selectivity and permeability (Mulder, 1997). Microfiltration and ultrafiltration membranes can be classified according to a number of characteristics, including the type of material (polymers, ceramics, metals, or glasses), module geometry (hollow fiber, capillary, tubular, flat sheets in plate-and-frame modules), or chemical nature (hydrophilic or hydrophobic).

Most commercial membranes are made of polymers (cellulose acetate, polyamide, polyethersulfone, polysulfone, polyacrylonitrile, or polyvinylidene fluoride), followed by ceramics, which are more expensive but possess excellent chemical, temperature, and mechanical stability (Baker, 2004; Mulder, 1997). Hydrophilic membrane surfaces are less susceptible to adsorption of organic species, and consequently less sensitive to fouling. Different techniques are available for introducing some degree of hydrophilicity, including chemical modification, plasma treatment, coating, grafting, or preparation of copolymers of hydrophobic and hydrophilic blocks.

7.3.2 Membrane Operation

The efficiency of a membrane is determined by two key parameters: flux and selectivity, which are closely related to the transport mechanism, morphology, and chemical nature of the membrane (Mulder, 1997). Flux (J) can be defined as the flow through the membrane *per* unit area and is usually measured in L/m^{2}h (LMH). Typical flux values range between 10 and 1000 LMH for MF and UF membranes, 30–150 LMH for NF, and 25–50 LMH for RO membranes (Baker, 2004; Scott and Hughes, 1996).

When pressure is the driving force for a membrane operation, the flux is an indicator of the hydraulic resistance of the membrane to the passage of a certain component. Flux depends on the characteristics of the feed stream and is proportional to ΔP between feed and effluent:

$$J_i = k \cdot \Delta P \tag{7.8}$$

where k is a constant representing membrane permeability. Permeability of pure water can be used to compare different membranes; low values mean that lower pressures are required to reach certain flux and the membrane is energetically more efficient. In most MF and UF processes, ΔP is essentially equal to the transmembrane pressure (TMP), defined as

$$TMP = \frac{P_1 + P_2}{2} - P_3 \tag{7.9}$$

where

P_1 is the inlet pressure
P_2 the outlet pressure
P_3 the permeate pressure (usually negligible)

However, if the osmotic pressure gradient between feed and permeate is significant ($\Delta\Pi$), the concentration gradients cause a significant reduction in the driving force:

$$\Delta P = TMP - \sigma\Delta\Pi \quad (7.10)$$

where σ is the Staverman coefficient measuring the membrane permeoselectivity (Mulder, 1997). For UF and MF processes, allowing the pass of salts, the osmotic effect is negligible and σ is usually zero. Operating with RO membranes, which retain salt almost quantitatively, σ is close to one, and overcoming the osmotic pressure is an important technical problem. The pressures employed in UF and MF are lower than those used for NF and RO operation, and the fluxes are higher.

The membrane selectivity for a given component i depends on the fraction rejected by the membrane, which is measured in terms of the rejection coefficient (R_i) defined as

$$R_i = \left(1 - \frac{C_{i,permeate}}{C_{i,feed}}\right) \quad (7.11)$$

where $C_{i,\,permeate}$ and $C_{i,\,feed}$ are the concentrations of component i in permeate and feed, respectively.

Depending on the transport mechanism, membranes can be classified into porous and dense. A membrane is usually considered porous if pore diameter is higher than 20 Å and dense if pore diameter is in the range 10–20 Å. In porous membranes (as the ones employed in MF, UF, and NF), the transport depends on the size, distribution, and structure of the pores. When the pore size is large, the exclusion mechanism is dominated by the pore morphology; whereas for progressively smaller pores, the chemical interactions between species in the mixture and the membrane become more influential on the transport mechanism. In dense membranes, the transport mechanism depends on the solubility and diffusivity of the molecules transported through the membrane, mainly dependent on the membrane material, the nature of the fluids, and the possible interactions between the permeated species and the membrane (Baker, 2004; Mulder, 1997).

7.3.2.1 Membrane Fouling

One of the major problems associated with the use of membranes is the progressive permeate flux decline caused by concentration polarization and membrane fouling, leading to increased operation costs due to the need of incorporation of cleaning routines in the process (Choi et al., 2005) and the reduction of the membrane useful life (Li and Chase, 2010).

The rapid reduction of flux during the early stages of the process can be ascribed to the rejected molecules reaching the surface by convective transport. These particles form a concentrated layer at the membrane surface, leading to a concentration gradient denoted as concentration polarization (Cheryan, 1998). The effect of the concentration polarization can be determined by comparing the flux for pure water (J_A) and the initial permeate flow (J_0) at the same transmembrane pressure. After the initial drop in flux, due to concentration polarization effects, the flux continue dropping due to membrane fouling, caused mainly by the deposition of macromolecules inside the pores of the membrane, which are blocked by adsorption of solutes on the membrane walls. Finally, flux can stabilize or reduce slowly due to either additional deposition on the membrane or consolidation of fouling. Membrane fouling depends on different factors, including the characteristics of the membrane, properties of feed stream, the molecular size of solutes and their interaction with the membrane surface, and other operational conditions (Turano et al., 2002). The flux reduction caused by concentration polarization can be reverted by modification of the operational parameters, such as the flow rate, feed concentration, and transmembrane pressure (TMP). Reversible fouling can be alleviated by backwashing, but irreversible fouling requires cleaning by chemical means (caustic or acid solutions) (Li and Chase, 2010).

7.3.2.2 Separation Efficiency

The separation efficiency depends on operational variables (such as cross-flow velocity or TMP) as well as on the properties of the feed (e.g., concentration or temperature). An increase in the cross-flow velocity causes an increase in the permeate flow, since the thickness of the polarization layer can be reduced by increasing turbulence that disperse the solute molecules from the membrane surface (Afonso et al., 2004; Coutinho et al., 2009). Values higher than 2 m/s are usually employed.

The flux increases with pressure, eventually leading to the formation of a gel layer on the membrane surface. At this point, further increases of pressure do not increase the flux and they could even reduce it due to compaction (Afonso et al., 2004). Increased temperatures result in increased permeate flux due to decreased viscosity and increased diffusivity. As a general trend, the permeate flow increases by 3%–4% *per* degree (Afonso et al., 2004). However, the negative effect of high temperatures on the stability of both membrane (especially on the organic ones) and solutes should be considered. Increased solute concentrations result in increased polarization effects, resulting in reduced permeate flux.

The efficiency of a membrane separation process is also affected by the configuration of the module, which should be designed to optimize productivity. A key aspect is the control of the fluid flow at both the feed and the permeate sides in order to generate a turbulent flow in the surface for reducing the mass transport resistance and the concentration polarization effects. The incorporation of baffles to create turbulence and the use of membrane modules allowing high flow rates (Choi et al., 2005) are among the most common strategies. Ultrasound, air, and injection of bubbles (Choi et al., 2005) or suspended solids (Noordman et al., 2003) have been proposed to prevent solute depositions on the membrane surface and to limit concentration polarization.

7.3.3 Equipment and Membrane Configurations

A simple setup for membrane separation consists of a pump causing circulation of feed stream over the surface of a membrane with the appropriate nature and

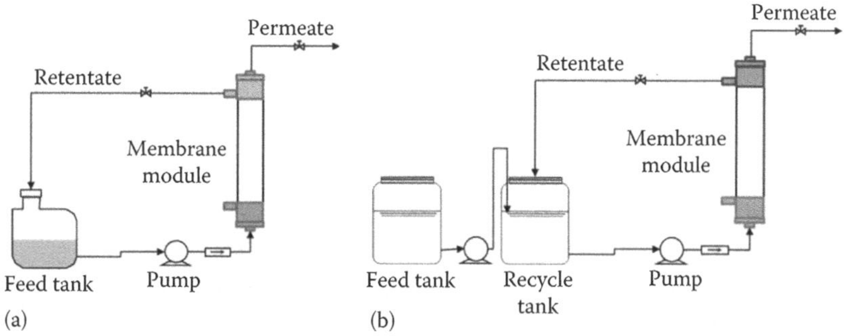

FIGURE 7.4 Schemes of the membrane operation in (a) batch and (b) fed-batch.

configuration (Figure 7.4). The permeate is enriched in solutes as the retentate volume is smaller than the one of the feed (Cheryan, 1998; Koros et al., 1996).

7.3.4 Examples

Membranes have been proposed for the concentration and selective fractionation of valuable compounds from aqueous and/or alcoholic streams (Abels et al., 2013; Li and Chase, 2010). Studies have been reported on the membrane concentration of raw extracts obtained with conventional solvents from different vegetal materials. Representative studies in this field are summarized in Table 7.4.

Membrane technology has also been proposed for the purification of extracts from emerging biomass fractionation technologies. González-López et al. (2012) considered the UF of autohydrolysis liquors from chestnut husks. Sarmento et al. (2008) used RO and NF for the concentration of phenolic extracts from cocoa beans obtained by conventional and supercritical CO_2 extraction. Membranes allow for the simple and rapid fractionation of phenolic compounds according to their molecular weight, as reported for crude extracts from almond skins (Prodanov et al., 2008), root cortices of mulberry (Yu et al., 2007), kaki (Gu et al., 2008), and procyanidins (Santamaría et al., 2002). Membrane technology succeeded in separating different components from the same material (Conidi et al., 2012; Li et al., 2005).

Two objectives are usually pursued when working with waste streams: reduction in chemical oxygen demand and recovery of valuable components (fats, sugars, polyphenols, etc.); as reported for wastewater from olive oil mills (Turano et al., 2002) or the cork industry (Bernardo et al., 2011). Combination of different techniques has succeeded in separating low-molecular-weight compounds from these effluents (Russo, 2007). For example, combinations of stages using MF, UF, NF, and RO (Cassano et al., 2011a, 2013; Conidi et al., 2015; Galanakis et al., 2010; García-Castello et al., 2010; Zagklis and Paraskeva, 2014) and adsorption/desorption resins have been proposed for the recovery of low-molecular-weight phenolics present in the retentate of RO (Zagklis et al., 2015).

TABLE 7.4
Examples of Application of Membrane Technology for the Separation and Concentration of Phenolic Compounds from Natural Sources

Source of Extracts or Liquid Stream	Type of Separation	Configuration	Material	Area (m^2)	Cut-Off (kDa) or Pore Size	Flow Velocity Feed Flow	TMP (Bar)	T (°C)	VRF	References
Almond skin	Centrifugal UF	—	—	—	10, 30, 50	—	—	—	—	Prodanov et al. (2008)
Artichoke wastewater	UF	Tubular	TiO_2	0.1	15	4 m^3/h	4.3	—		Conidi et al. (2015)
	NF	Spiral	PA	2.6	0.2–0.3	300 L/h	8	5		
Bark mate tree	NF	Spiral	—	0.9	0.15–0.30	—	60	22	1.5–6	Prudêncio et al. (2012)
Bergamot juice	UF	Hollow fiber	PS	0.16	100	114 L/h	0.7	24	7.8	Conidi et al. (2011)
	UF	Flat sheet	Fluoropolymer	0.0038	1	—	1–10	24	8	
	NF	Monotubular	TiO_2	0.005	0.75	97.6 L/h	7.5	24	—	
	NF	Monotubular	TiO_2	0.005	0.45	100 L/h	33	24	—	
Bilberry	AMC	Tubular central solid core	SRC + sulfonic acid groups	0.55	—	—	1	rT	—	Juadjur and Winterhalter (2012)
Brewery wastewater	UF	Flat sheet	PES	—	50	>250 L/(m^2·h)	1.5	22	—	Götz et al. (2014)
	NF	Flat sheet	PS	—	0.2	—	30	<70		
Pequi	NF	Flat sheet	PA	0.00159	0.2–0.3	—	0.8	25	2.5	Machado et al. (2013)

(Continued)

TABLE 7.4 (*Continued*)
Examples of Application of Membrane Technology for the Separation and Concentration of Phenolic Compounds from Natural Sources

Source of Extracts or Liquid Stream	Type of Separation	Configuration	Material	Area (m^2)	Cut-Off (kDa) or Pore Size	Flow Velocity Feed Flow	TMP (Bar)	T (°C)	VRF	References
Chestnut husk	NF	—	—	—	—	—	4	20	3	González-López et al. (2012)
	UF				—	20	3			
Chestnut leaves	UF	Flat sheet	PES	0.077	5, 10	—	2	20	3	Díaz-Reinoso et al. (2011)
Clarified beer dilute malt	MF	Tubular	Ceramic	0.0314	0.45 μm	0.85–0.9 m/s	0.5–1.1	3.5–7.1	—	Kazemi et al. (2013)
Clarified cane sugar juices	UF	Tubular	Ceramic	3.44	300	200–350 L/(m^2·h)	<4	98	2–10	Doherty et al. (2010)
	MF	Tubular	Ceramic	3.44	0.1 μm			80–98	2–10	
	MF	Tubular	Ceramic	3.44	0.45 μm			97–98	2–10	
	MF	Tubular	Ceramic	2.6	0.1 μm			96–98	2–10	
Cork-processing wastewater	UF	Flat sheet	Cellulose acetate	0.147	0.125–91	—	—	—	—	Bernardo et al. (2011)
	NF	Spiral	PS/PA	2.09	0.125					
Cranberry, blueberry, and apple juices	RO	Thin film	PA	2.6	—	133 L/h	25–35	2–40	—	Gunathilake et al. (2014)

(*Continued*)

TABLE 7.4 (*Continued*)
Examples of Application of Membrane Technology for the Separation and Concentration of Phenolic Compounds from Natural Sources

Source of Extracts or Liquid Stream	Type of Separation	Configuration	Material	Area (m^2)	Cut-Off (kDa) or Pore Size	Flow Velocity Feed Flow	TMP (Bar)	*T* (°C)	VRF	References
Defatted milled grape seeds	Diafiltration	Plate and frame	PS		DS-5 (180)	20 L/(m^2·h)	14			Beltrán et al. (2008)
	Diafiltration		RG		YM100 (100)	140 L/(m^2·h)	0.7			
	Diafiltration		DC	—	YC05 (0.5)	25.2 L/(m^2·h)	7	—	—	
	Diafiltration		—		MRC-1 (1)	3.5 L/(m^2·h)	3.5			
	Concentration		PS		DS-5 (18)	60.55 L/(m^2·h)	20			
Distilled grape pomace liquors	NF	Spiral	PA/PS		0.250, 0.35		8	20	3	Díaz-Reinoso et al. (2010)
	UF	Tubular	Ceramic		1		6		3	
Distilled grape pomace	NF	Spiral	PA/PS, TF		0.25–0.35, 0.15–0.30	2–3 m/s	—	20	5.5–6.5	Díaz-Reinoso et al. (2009)
	UF	Spiral, tubular	Titania, TF		1	2–3 m/s		20	5.5–6.5	
Eucalypt bark	UF	Flat sheet	PA	0.0014	1	4.5 L/min	3, 5, 8	35	—	Baptista et al. (2015)
			PES		5					
			PVDF		30					
			PS		60					
Flaxseed hulls	UF	Flat sheet	PES	3.1×10^{-4}	30	—	4	20	5	Loginov et al. (2013)
Blackberries	MF	Tubular	Ceramic	—	0.2 µm	—	—	—	—	Azofeifa et al. (2011)

(*Continued*)

TABLE 7.4 (*Continued*)
Examples of Application of Membrane Technology for the Separation and Concentration of Phenolic Compounds from Natural Sources

Source of Extracts or Liquid Stream	Type of Separation	Configuration	Material	Area (m^2)	Cut-Off (kDa) or Pore Size	Flow Velocity Feed Flow	TMP (Bar)	T (°C)	VRF	References
Grape seed	UF	—	PS, PVDF		8, 20, 200	—	0.2, 2, 5	—	—	Santamaría et al. (2002)
	NF	—	PA		60% $CaCl_2$ rejection		6			
Grape seed	UF	—	—	—	0.22, 0.45 µm		5	23	—	Nawaz et al. (2006)
Hart's tongue fern	UF	Flat sheet	PPEES	0.0028	1	(Water flux) 25–50/(E:W flux) 28–85 L/(m^2·h)	9	rT	—	Paun et al. (2015)
			PS			(Water flux)12–23/(E:W flux) 14–30 L/(m^2·h)				
Herb Robert	UF	—	RG PS-Pal	—	10	—	—	—	—	Neagu et al. (2010)
Herb Robert and sage	UF	Flat sheet	PS-PvP-MS	0.0028	10	—	—	—	2.5	Paun et al. (2011)
	NF				1	—	—			
	NF organic–inorganic				1	43.3 (Herb Robert)/47 (Sage) L/(m^2·h)	10			

(*Continued*)

TABLE 7.4 (*Continued*)
Examples of Application of Membrane Technology for the Separation and Concentration of Phenolic Compounds from Natural Sources

Source of Extracts or Liquid Stream	Type of Separation	Configuration	Material	Area (m^2)	Cut-Off (kDa) or Pore Size	Flow Velocity Feed Flow	TMP (Bar)	*T* (°C)	VRF	References
Mistletoe, thyme	MF	Flat sheet	PS-PvP	—	—	700 L/(m^2·h)	8		5	Simion et al. (2012)
	UF1				20	100 L/(m^2·h)				
	UF2				14	111 L/(m^2·h)				
	UF3				10	50 L/(m^2·h)				
Mulberry root cortices	MF		—		0.45 μm		5	25	4	Yu et al. (2007)
	UF		PS, PA		200, 50, 20, 10					
Olive mill wastewater	UF	Flat sheet	PS	—	17	—	0.5–3	30	—	Turano et al. (2002)
Olive mill wastewater	UF	Hollow fiber	PVDF	0.4	0.02 μm	433 L/h	4.3	18	3	Cassano et al. (2013)
	UF	Flat sheet	Composite fluoropolymer	0.0022	1	600 L/h	9	30	—	
	NF	Spiral	TFC	2.6	>97% salt rejection	—	5.9	22	—	
Olive mill wastewater	MF		Ceramic	0.35	0.1–1.4 μm	—	—	—	—	Pizzichini and Russo (2005)
	UF	Spiral	Polymeric		1–20					
	NF	Spiral	Polymeric		0.15–0.250					
	RO	Spiral	PA							

(*Continued*)

TABLE 7.4 (*Continued*)
Examples of Application of Membrane Technology for the Separation and Concentration of Phenolic Compounds from Natural Sources

Source of Extracts or Liquid Stream	Type of Separation	Configuration	Material	Area (m^2)	Cut-Off (kDa) or Pore Size	Flow Velocity Feed Flow	TMP (Bar)	*T* (°C)	VRF	References
Olive mill wastewater	Centrifugation	Rotating screen	Stainless steel	2.26	250 µm	1.5 m/h	—	—	—	Agalias et al. (2007)
		Rotating screen	Stainless steel	2.26	120 µm	—	—			
		Pall bagfilter	—	27 L capacity	25 µm	—	6.6			
Olive cake	NF	—	—	—	0.25	—	10	—	—	Nunes Da Ponte et al. (2007)
	RO				—		25			
Olive mill wastewater	MF	Tubular, spiral	Ceramic, PES	0.35, 3.8	0.8 µm, 0.35 µm, 500	4, 3.4 m^3/h	1.5, 2	20–25	3, 3.4	Russo (2007)
	UF	Spiral, tubular	PS, PES, Zirconium oxide	5, 8.36, 0.35	80, 20, 6, 1	3–5, 4 m^3/h	4.5, 1.5	20–25	3, 2.5	
	RO	Spiral	PA	7	99.5% salt rejection	1.2 m^3/h	2.5–10–15	20–25	2.5	

(*Continued*)

TABLE 7.4 (*Continued*)
Examples of Application of Membrane Technology for the Separation and Concentration of Phenolic Compounds from Natural Sources

Source of Extracts or Liquid Stream	Type of Separation	Configuration	Material	Area (m^2)	Cut-Off (kDa) or Pore Size	Flow Velocity Feed Flow	TMP (Bar)	T (°C)	VRF	References
Olive mill wastewater	MF	—	—	0.24	120	—	—	—	—	Villanova et al. (2008)
	UF			1.6	120–20, 20–1					
	NF			2.5	1–0.35					
	RO			2.5	0.35					
Olive mill wastewater	MF	Tubular	Al_2O_3	0.0048	200 nm	760 L/h	0.72	22		Garcia-Castello et al. (2010)
	NF	Spiral wound	PES	1.6	0.578	(Feed) 9.18 L/h	8	20	—	
	OD	Hollow fiber	Pt	1.4	30 nm	(Stripping) 26.76 L/h	4.08	1–40		
Olive mill wastewater	MEUF	Flat sheet	PVDF	0.00287	—	200–488 L/(m^2·h)	2–4	rT	—	El-Abbassi et al. (2011)
Olive mill wastewater	MF	Tubular	PP	0.036	0.2 μm	—	0.5	25	—	Cassano et al. (2011a)
	UF	Flat Sheet	PES-RC	0.00038465	4, 5, 10		5.2	25	3	
Olive mill wastewater	MF	Tubular	PP	8	0.1–0.3 μm	—	—	—	—	Servili et al. (2011)
	UF	Spiral	PA/PS	16	7					
	RO	Spiral	PS	9	0.10					
Olive mill wastewater	NF	Spiral wound	—	2.51	0.5 nm	600 L/h	<32	20	—	Stoller (2011)

(Continued)

TABLE 7.4 (*Continued*)
Examples of Application of Membrane Technology for the Separation and Concentration of Phenolic Compounds from Natural Sources

Source of Extracts or Liquid Stream	Type of Separation	Configuration	Material	Area (m^2)	Cut-Off (kDa) or Pore Size	Flow Velocity Feed Flow	TMP (Bar)	T (°C)	VRF	References
Olive mill wastewater	UF	Tubular	Ceramic	0.24	100 nm	—	1	—	—	Zagklis and Paraskeva (2014)
	NF	Spiral	Polymeric	2.4	95% $MgSO_4$ rejection	—	20	—	—	
	RO	Spiral	Polymeric	2.5	99% NaCl rejection	—	3.3	—	—	
Olive mill wastewater	UF	Hollow fiber	PVDF	0.4	0.02 µm	433 L/h	4.3	18	3	Cassano et al. (2013)
	UF	Flat sheet	Composite fluoropolymer	0.0022	1	600 L/h	9	30	—	
	NF	Spiral	TFC	2.6	>97% salt rejection	—	5.9	22	—	
Olive mill wastewater	UF	Filter sheets	PS, PES	—	100, 25, 10, 2	—	2, 3, 5, 8	—	—	Galanakis et al. (2010)
	NF	Filter sheets	PA	—	120	—	12			
Orange press liquor	NF	Spiral	PA/PS	2.1	0.18	—	20	20	3	Conidi et al. (2012)
			Polypiperazine amide	2.6	0.3		20	20	3	
			PES	1.6	0.4		6	20	3	
			PES	1.6	1		6	20	3	

(*Continued*)

TABLE 7.4 (*Continued*)
Examples of Application of Membrane Technology for the Separation and Concentration of Phenolic Compounds from Natural Sources

Source of Extracts or Liquid Stream	Type of Separation	Configuration	Material	Area (m^2)	Cut-Off (kDa) or Pore Size	Flow Velocity Feed Flow	TMP (Bar)	*T* (°C)	VRF	References
Persimmon pulp	UF	Hollow fiber	PS	—	10 L/h	10 L/h	0.5	—	—	Gu et al. (2008)
Pomegranate juice	UF-clarification	Hollow fiber	PEk+epoxy resin	0.46	68.8	66.96 L/h	0.96	25	—	Cassano et al. (2011b)
	OD	Tubular	PP	1.4	0.2 μm	(Feed) 30 L/h (Stripping) 20 L/h	0.4	25	—	
Pomegranate juice	MF	Flat sheet	CfDeP Ceramic	—	0.2 μm	—	—	—	—	Fischer et al. (2011)
Red grape must	NF	Flat sheet	PA	0.0066	0.3–0.4	300 L/h	35	20	—	Salgado et al. (2013)
Rough beer	MF	Tubular	Ceramic	0.00942	0.8 μm	6 m/s	3.74	10		Cimini et al. (2014)
Royal purple	OSN	Tubular elliptic	Polyimides	0.0051	0.3–0.5–0.9	8.5–14 L/(m^2·h)	20	30	—	Koleva and Simeonov (2014)
					0.3	1–12 L/(m^2·h)	30	30	25	
Rutin, chlorogenic acid, glucose solution	UF	Cassette	RC	0.1	1	90 L/h	4–5	30	—	Wei et al. (2010)

(*Continued*)

TABLE 7.4 (*Continued*)
Examples of Application of Membrane Technology for the Separation and Concentration of Phenolic Compounds from Natural Sources

Source of Extracts or Liquid Stream	Type of Separation	Configuration	Material	Area (m^2)	Cut-Off (kDa) or Pore Size	Flow Velocity Feed Flow	TMP (Bar)	T (°C)	VRF	References
Sage	MF	Flat sheet	PS	0.0028	0.45 μm	—	2–5	20	2	Paun et al. (2009)
	UF		PS-Pal							
St. John bush leaves	MF	Multichannel tubular	Ceramic	0.306	0.2 μm	5 m/s	0.6	30	—	Koffi et al. (2013)
	RO	Tubular	Polymeric	2.8	—	—	40	30	27.7	
Table olive–processing wastewater	UF	Flat sheet	PES	0.01	30	2.08 m/s	3	25	—	García-Ivars et al. (2015)
Thyme	NF	Flat sheet	Commercial (PS, PA)	0.00384	—	—	12	25	—	Achour et al. (2012)
	NF	Asymmetric	Synthetic (DC + FM + acetone)							
	UF	Asymmetric	Synthetic (DC + FM + acetone)							
Waste stream of a soymilk-making plant	RO	Plate and frame	—	—	98% salt rejection	3.2 L/min	14	20	—	Xu et al. (2004)

(*Continued*)

TABLE 7.4 (*Continued*)
Examples of Application of Membrane Technology for the Separation and Concentration of Phenolic Compounds from Natural Sources

Source of Extracts or Liquid Stream	Type of Separation	Configuration	Material	Area (m^2)	Cut-Off (kDa) or Pore Size	Flow Velocity Feed Flow	TMP (Bar)	*T* (°C)	VRF	References
White wine Kerner	MFa/MFb	Tubular	PP Accurel	13.4/17.3	0.2 μm	400–800 L/h	1.3	—	—	Ulbricht et al. (2009)
	MF	Capillary	PP CelgardX30	26.4	—	Module 19 m^2				
	MFa/MFb	Monolayer	PP Celgard24/25	37.3/48.2	0.043 μm/ 0.064 μm					
	MF	Capillary	PES	3.6	0.2 μm	—	—	—	—	
Vine shoot	UF	Flat sheet	PES	0.0025	30–100	—	1.5–4.5	22	7	Rajha et al. (2015)
Wine vinasses	MF	—	PES	0.46–2.3	0.1– 0.65 μm	—	—	<50	—	Moure et al. (2010)
	UF		Ceramic	—			8	30		

Abbreviations: AMC, adsorptive membrane chromatography; CfDeP, combination of cellulose fibers, diatomaceous earth, and perlite; DC, diacetate cellulose; E: Ethanol; FM, formamide; MEUF, Micellar-Enhanced Ultrafiltration; MS, Mesoporous silica; OD, osmotic distillation; OSN, organic solvent nanofiltration; PA, polyamide; Pal, Polyalanine; PEk, Poly (ether ketone) hollow fiber; PES, Polyether sulfone; PP, Polypropylene; PPEES, Poly phenylene ether ether sulfone; PS, Polysulfone; Pt, Polyethylene; PVDF, poly(vinyldene fluoride); PvP, Polyvinyl pyrrolidone; RG, regenerated cellulose; RO, reverse osmosis; rT, room temperature; SRC, Stabilized reinforced cellulose; TF, Thin Film; TFC, Thin-Film Composite; VRF, volume reduction factor; W, water.

7.4 CONCLUDING REMARKS

Membrane and adsorption technologies are environment friendly for recovery, concentration, and/or fractionation of valuable phenolic components found in process and waste streams. Both are particularly appropriate for separations of thermolabile compounds, which are difficult to obtain by conventional extraction or distillation.

ACKNOWLEDGMENTS

The authors acknowledge the Spanish Ministry of Science and Innovation (Research Project Reference CTM2012-38095, which had partial financial support from the FEDER funds of the European Union).

REFERENCES

Abels, C., Carstensen, F., and Wessling, M. 2013. Membrane processes in biorefinery applications. *Journal of Membrane Science* 444:285–317.

Achour, S., Khelifi, E., Attia, Y., Ferjani, E., and Noureddine Hellal, A. 2012. Concentration of antioxidant polyphenols from *Thymus capitatus* extracts by membrane process technology. *Journal of Food Science* 77(6):C703–C709.

Aehle, E., Raynaud-Le Grandic, S., Ralainirina, R., Baltora-Rosset, S., Mesnard, F., Prouillet, C., Mazière, J.C., and Fliniaux, M.A. 2004. Development and evaluation of an enriched natural antioxidant preparation obtained from aqueous spinach (*Spinacia oleracea*) extracts by an adsorption procedure. *Food Chemistry* 86:579–585.

Afonso, M.D., Ferrer, J., and Bórquez, R. 2004. An economic assessment of proteins recovery from fish meal effluents by ultrafiltration. *Trends in Food Science and Technology* 15:506–512.

Agalias, A., Magiatis, P., Skaltsounis, A.L., Mikros, E., Tsarbopoulos, A., Gikas, E., Spanos, I., and Manios, T. 2007. A new process for the management of olive oil mill waste water and recovery of natural antioxidants. *Journal of Agricultural and Food Chemistry* 55(7):2671–2676.

Ahmaruzzaman, M. 2008. Adsorption of phenolic compounds on low-cost adsorbents: A review. *Advances in Colloid and Interface Science* 143:48–67.

Aksu, Z. and Gönen, F. 2004. Biosorption of phenol by immobilized activated sludge in a continuous packed bed: Prediction of breakthrough curves. *Process Biochemistry* 39:599–613.

Alhamed, Y.A. 2009. Adsorption kinetics and performance of packed bed adsorber for phenol removal using activated carbon from date stones. *Journal of Hazardous Materials* 170:763–770.

Altiok, E., Baycin, D., Bayraktar, O., and Ulku, S. 2008. Isolation of polyphenols from the extracts of olive leaves (*Olea europaea* L.) by adsorption on silk fibroin. *Separation and Purification Technology* 62:342–348.

Arsuaga, J.M., Aguado, J., Arencibia, A., and López-Gutiérrez, M.S. 2014. Aqueous mercury adsorption in a fixed bed column of thiol functionalized mesoporous silica. *Adsorption* 20:311–319.

Azofeifa, G., Quesada, S., and Pérez, M. 2011. Effect of the microfiltration process on antioxidant activity and lipid peroxidation protection capacity of blackberry juice. *Brazilian Journal of Pharmacognosy* 21(5):829–834.

Baker, R.W. 2004. *Membrane Technology and Applications*. Menlo Park, CA: John Wiley & Sons.

Baptista, E.A., Pinto, P.C.R., Mota, I.F., Loureiro, J.M., and Rodrigues, A.E. 2015. Ultrafiltration of ethanol/water extract of *Eucalyptus globulus* bark: Resistance and cake build up analysis. *Separation and Purification Technology* 144:256–266.

Beltrán, M.T., Sanz, B., Santamaría, R., Murga, G., and Salazar, G. 2008. Recovery of anti-oxidants from grape products by using supercritical fluids and membrane technology. *Electronic Journal of Environmental, Agricultural and Food Chemistry* 7(8):3270–3278.

Berardini, N., Knödler, M., Schieber, A., and Carle, S. 2005. Utilization of mango peels as a source of pectin and polyphenolics. *Innovative Food Science and Emerging Technologies* 6:442–452.

Bernardo, M., Santos, A., Cantinho, P., and Minhalma, M. 2011. Cork industry wastewater partition by ultra/nanofiltration: A biodegradation and valorisation study. *Water Research* 45(2):904–912.

Bertin, L., Ferri, F., Scoma, A., Marchetti, L., and Fava, F. 2011. Recovery of high added value natural polyphenols from actual olive mill wastewater through solid phase extraction. *Chemical Engineering Journal* 171:1287–1293.

Brenes, M. and Castro, A. 2003. Procedure for obtaining phenolic extract with high concentration of anti-oxidants and involves ultra-filtration of solutions derived from preparation process of preserved table olives. Patent ES 2,186,467.

Buran, T.J., Amandeep, K., Sandhu, Z.L., Cheryl, R.R., Weihua, W.Y., and Liwei, G. 2014. Adsorption/desorption characteristics and separation of anthocyanins and polyphenols from blueberries using macroporous adsorbent resins. *Journal of Food Engineering* 128:167–173.

Cassano, A., Conidi, C., and Drioli, E. 2011a. Comparison of the performance of UF membranes in olive mill wastewaters treatment. *Water Research* 45(10):3197–3204.

Cassano, A., Conidi, C., and Drioli, E. 2011b. Clarification and concentration of pomegranate juice (*Punica granatum* L.) using membrane processes. *Journal of Food Engineering* 107(3–4):366–373.

Cassano, A., Conidi, C., Giorno, L., and Drioli, E. 2013. Fractionation of olive mill wastewaters by membrane separation techniques. *Journal of Hazardous Materials* 248–249(1):185–193.

Chandrasekhar, J., Madhusudhan, M.C., and Raghavarao, K.S.M.S. 2012. Extraction of anthocyanins from red cabbage and purification using adsorption. *Food and Bioproducts Processing* 90:615–623.

Chautard, A. and Mayrand, D. 2003. Vinegar manufacture with retention of tannins, colorings, and flavor. Patent WO 2,003,014,288.

Chen, Y. and Zhang, D. 2014. Adsorption kinetics, isotherm and thermodynamics studies of flavones from *Vaccinium bracteatum* Thunb leaves on NKA-2 resin. *Chemical Engineering Journal* 254(15):579–585.

Cheryan, M. 1998. *Ultrafiltration and Microfiltration Handbook*, 2nd ed. Boca Raton, FL: CRC Press.

Cimini, A., Marconi, O., Perretti, G., and Moresi, M. 2014. Novel procedure for lager beer clarification and stabilization using sequential enzymatic, centrifugal, regenerable PVPP and crossflow microfiltration processing. *Food and Bioprocess Technology* 7(11):3156–3165.

Conidi, C., Cassano, A., and Drioli, E. 2011. A membrane-based study for the recovery of polyphenols from bergamot juice. *Journal of Membrane Science* 375(1–2):182–190.

Conidi, C., Cassano, A., and Drioli, E. 2012. Recovery of phenolic compounds from orange press liquor by nanofiltration. *Food and Bioproducts Processing* 90(4):867–874.

Conidi, C., Rodríguez-López, A.D., García-Castello, A.M., and Cassano, A. 2015. Purification of artichoke polyphenols by using membrane filtration and polymeric resins. *Separation and Purification Technology* 144:153–161.

Corrêa, R.A., Calçada, L.A., and Peçanha, R.P. 2007. Development of a fluidized bed system for adsorption of phenol from aqueous solutions with commercial macroporous resins. *Brazilian Journal of Chemical Engineering* 24:15–28.

Coutinho, C.M., Chiu, M.C., Basso, R.C., Ribeiro, A.P.B., Gonçalves, L.A.G., and Viotto, L.A. 2009. State of art of the application of membrane technology to vegetable oils: A review. *Food Research International* 42:536–550.

Coutinho, M.R., Quadri, M.B., Moreira, R.F.P.M., and Quadri, M.G.N. 2004. Partial purification of anthocyanins from *Brassica oleracea* (red cabbage). *Separation Science and Technology* 39:3769–3782.

D'Alessandro, L.G., Vauchel, P., Przybylski, R., Chataigne, G., Nikov, I., and Dimitrov, K. 2013. Integrated process extraction–adsorption for selective recovery of antioxidant phenolics from *Aronia melanocarpa* berries. *Separation and Purification Technology* 120:92–101.

Del Rio, D., Rodriguez-Mateos, A., Spencer, J.P.E., Tognolini, M., Borges, G., and Crozier, A. 2013. Dietary (poly)phenolics in human health: Structures, bioavailability, and evidence of protective effects against chronic diseases. *Antioxidants and Redox Signaling* 18:1818–1892.

Di Mauro, A., Fallico, B., Passerini, A., and Maccarone, E. 2000. Waste water from citrus processing as a source of hesperidin by concentration on styrene–divinylbenzene resin. *Journal of Agricultural and Food Chemistry* 48:2291–2295.

Di Mauro, A., Fallico, B., Passerini, A., Rapisarda, P., and Maccarone, E. 1999. Recovery of hesperidin from orange peel by concentration of extracts on styrene–divinylbenzene resin. *Journal of Agricultural and Food Chemistry* 47:4391–4397.

Díaz-Reinoso, B., González-López, N., Moure, A., Domínguez, H., and Parajó, J.C. 2010. Recovery of antioxidants from industrial waste liquors using membranes and polymeric resins. *Journal of Food Engineering* 96:127–133.

Díaz-Reinoso, B., Moure, A., Domínguez, H., and Parajó, J.C. 2009. Ultra- and nanofiltration of aqueous extracts from distilled fermented grape pomace. *Journal of Food Engineering* 91:587–593.

Díaz-Reinoso, B., Moure, A., Domínguez, H., and Parajó, J.C. 2011. Membrane concentration of antioxidants from *Castanea sativa* leaves aqueous extracts. *Chemical Engineering Journal* 175:95–102.

Doherty, W.O.S., Rackemann, D.W., and Steindl, R.J. 2010. Fouling of tubular ceramic membranes during processing of cane sugar juice. *Desalination and Water Treatment* 16(1–3):45–56.

Duong, D. 1998. *Adsorption Analysis: Equilibria and Kinetics*. Singapore: Imperial College Press.

El-Abbassi, A., Khayet, M., Hafidi, A. 2011. Micellar-enhanced ultrafiltration process for the treatment of olive mill wastewater. *Water Research* 45(15):4522–4530.

Fernández-Bolaños, J., Heredia, A., Rodríguez, G., Rodríguez, R., Jiménez, A., and Guillén, R. 2002. Method for obtaining purified hydroxytyrosol from products and by-products derived from the olive tree. WO 2002064537.

Fischer, U.A., Dettmann, J.S., Carle, R., and Kammerer, D.R. 2011. Impact of processing and storage on the phenolic profiles and contents of pomegranate (*Punica granatum* L.) juices. *European Food Research and Technology* 233(5):797–816.

Fu, B.Q., Liu, J., Li, H., Li, L., Lee, F.S.C., and Wang, X.R. 2005. The application of macroporous resins in the separation of licorice flavonoids and glycyrrhizic acid. *Journal of Chromatography A* 1089:18–24.

Galanakis, C.M., Tornberg, E., and Gekas, V. 2010. Clarification of high-added value products from olive mill wastewater. *Journal of Food Engineering* 99:190–197.

Garcia-Castello, E., Cassano, A., Criscuoli, A., Conidi, C., and Drioli, E. 2010. Recovery and concentration of polyphenols from olive mill wastewaters by integrated membrane system. *Water Research* 44(13):3883–3892.

García-Ivars, J., Iborra-Clar, M.I., Alcaina-Miranda, M.I., Mendoza-Roca, J.A., and Pastor-Alcañiz, L. 2015. Treatment of table olive processing wastewaters using novel photo-modified ultrafiltration membranes as first step for recovering phenolic compounds. *Journal of Hazardous Materials* 290:51–59.

Garti, N., Agmon, G., and Pintus, E. 2003. Method for selectively obtaining antioxidant rich extracts from citrus fruits. US 6,528,099.

Geng, X., Ren, P., Pi, G., Shi, R., Yuan, Z., and Wang, C. 2009. High selective purification of flavonoids from natural plants based on polymeric adsorbent with hydrogen-bonding interaction. *Journal of Chromatography A* 1216:8331–8338.

Ghasemi, K., Ghasemi, Y., Ehteshamnia, A., Nabavi, S.M., Nabavi, S.F., Ebrahimzadeh, M.A., and Pourmorad, F. 2011. Influence of environmental factors on antioxidant activity, phenol and flavonoids contents of walnut (*Juglans regia* L.) green husks. *Journal of Medicinal Plants Research* 5(7):1128–1133.

González-López, N., Moure, A., Domínguez, H., and Parajó, J.C. 2012. Valorization of chestnut husks by non-isothermal hydrolysis. *Industrial Crops and Products* 36:172–176.

Götz, G., Geißen, S.U., Ahrens, A., and Reimann, S. 2014. Adjustment of the wastewater matrix for optimization of membrane systems applied for water reuse in breweries. *Journal of Membrane Science* 465:68–77.

Grohmann, K., Manthey, J.A., Cameron, R.G., and Buslig, B.S. 1999. Purification of citrus peel juice and molasses. *Journal of Agricultural and Food Chemistry* 47:4859–4867.

Gu, H.F., Li, C.M., Xu, Y.J., Hu, W.F., Chen, M.H., and Wan, Q.H. 2008. Structural features and antioxidant activity of tannin from persimmon pulp. *Food Research International* 41:208–217.

Gunathilake, K.D.P.P., Yu, L.J., and Rupasinghe, H.P.V. 2014. Reverse osmosis as a potential technique to improve antioxidant properties of fruit juices used for functional beverages. *Food Chemistry* 148:335–341.

Hall, K.R., Eagleton, L.C., Acrivos, A., and Vermeulen, T. 1966. Pore and solid diffusion kinetics in fixed bed adsorption under constant pattern conditions. *Industrial and Engineering Chemistry Fundamentals* 5:212–223.

Han, R., Ding, D., Xu, Y., Zou, W., Wang, Y., Li, Y., and Zou, L. 2008. Use of rice husk for the adsorption of congo red from aqueous solution in column mode. *Bioresource Technology* 99:2938–2946.

Ho, Y.S., Ng, J.C.Y., and McKay, G. 2000. Kinetics of pollutant sorption by biosorbents: Review. *Separation and Purification Methods* 29:189–232.

Jampani, C., Naik, A., and Raghavarao, K.S.M.S. 2014. Purification of anthocyanins from jamun (*Syzygium cumini* L.) employing adsorption. *Separation and Purification Technology* 125:170–178.

Jin, Q., Yue, J., Shan, L., Tao, G., Wang, X., and Qiu, A. 2008. Process research of macroporous resin chromatography for separation of N-(p-coumaroyl)serotonin and N-feruloylserotonin from *Chinese safflower* seed extracts. *Separation and Purification Technology* 62:370–375.

Juadjur, A., and Winterhalter, P. 2012. Development of a novel adsorptive membrane chromatographic method for the fractionation of polyphenols from bilberry. *Journal of Agricultural and Food Chemistry* 60(10):2427–2433.

Kammerer, D., Claus, A., Schieber, A., Carle, R. 2005. A novel process for the recovery of polyphenols from grape (*Vitis vinifera* L.) pomace. *Journal of Food Science* 70: C157–C163.

Kammerer, D.R., Saleh, Z.S., Carle, R., and Stanley, R.A. 2007. Adsorptive recovery of phenolic compounds from apple juice. *European Food Research and Technology* 224:605–613.

Kammerer, J., Kammerer, D.R., Jensen, U., and Carle, R. 2010. Interaction of apple polyphenols in a multi-compound system upon adsorption onto a food-grade resin. *Journal of Food Engineering* 96:544–554.

Kazemi, M.A., Soltanieh, M., Yazdanshenas, M., and Fillaudeau, L. 2013. Influence of crossflow microfiltration on ceramic membrane fouling and beer quality. *Desalination and Water Treatment* 51(22–24):4302–4312.

Kim, J., Yoon, M., Yang, H., Jo, J., Han, D., Jeon, Y.J., and Cho, S. 2014. Enrichment and purification of marine polyphenol phlorotannins using macroporous adsorption resins. *Food Chemistry* 162:135–142.

Kim, M.R., Kim, W.C., Lee, D.Y., and Kim, C.W. 2007. Recovery of narirutin by adsorption on a non-ionic polar resin from a water-extract of *Citrus unshiu* peels. *Journal of Food Engineering* 78:27–32.

Knaebel, K.1999. The basics of adsorbers design. *Chemical Engineering* 4:92–101.

Koffi, E.N., Le Guernevé, C., Lozano, P.R., Meudec, E., Adjé, F.A., Bekro, Y.A., and Lozano, Y.F. 2013. Polyphenol extraction and characterization of *Justicia secunda* Vahl leaves for traditional medicinal uses. *Industrial Crops and Products* 49:682–689.

Koleva, V. and Simeonov, E. 2014. Solid liquid extraction of phenolic and flavonoid compounds from *Cotinus coggygria* and concentration by nanofiltration. *Chemical and Biochemical Engineering Quarterly* 28(4):545–551.

Koops, G.H. 2002. Preparation and characterization of micro- and ultrafiltration membranes: Encyclopedia of desalination and water resources. *Membrane Processes* II 9–28.

Koros, W.J., Ma, Y.H., and Shimidzu, T. 1996. Terminology for membranes and membrane processes. *Pure and Applied Chemistry* 68(7):1479–1489.

Li, J. and Chase, H.A. 2010. Applications of membrane techniques for purification of natural products. *Biotechnology Letters* 32(5):601–608.

Li, P., Wang, Y., Ma, R., and Zhang, X. 2005. Separation of tea polyphenol from Green Tea Leaves by a combined CATUFM-adsorption resin process. *Journal of Food Engineering* 67:253–260.

Lin, L., Zhao, H., Dong, Y., Yang, B., and Zhao, M. 2012. Macroporous resin purification behavior of phenolics and rosmarinic acid from *Rabdosia serra* (MAXIM.) HARA leaf. *Food Chemistry* 130:417–424.

Liu, T., Cao, Y., and Zhao, M. 2010a. Extraction optimization, purification and antioxidant activity of procyanidins from hawthorn (*C. pinnatifida* Bge. var. major) fruits. *Food Chemistry* 119(4):1656–1662.

Liu, W., Zhang, S., Zu, Y.-G., Fu, Y.-J., Ma, W., Zhang, D.-Y., Kong, Y., and Li, X.-J. 2010b. Preliminary enrichment and separation of genistein and apigenin from extracts of pigeon pea roots by macroporous resins. *Bioresource Technology* 101(12):4667–4675.

Loginov, M., Boussetta, N., Lebovka, N., and Vorobiev, E. 2013. Separation of polyphenols and proteins from flaxseed hull extracts by coagulation and ultrafiltration. *Journal of Membrane Science* 442:177–186.

Machado, M.T.C., Mello, B.C.B.S., and Hubinger, M.D. 2013. Study of alcoholic and aqueous extraction of pequi (*Caryocar brasiliense* Camb.) natural antioxidants and extracts concentration by nanofiltration. *Journal of Food Engineering* 117(4):450–457.

Manthey, J.A. 2004. Fractionation of orange peel phenols in ultrafiltered molasses and mass balance studies of their antioxidant levels. *Journal of Agricultural and Food Chemistry* 52 (25):7586–7592.

Mijangos, F., Navarro, A., and Jodra, Y. 2001. Kinetic analysis of phenol adsorption from aqueous systems. *Canadian Journal of Chemical Engineering* 79:737–743.

Moure, A., Domínguez H., Parajó, J.C., González-Muñoz, M.J., Díaz-Reinoso, B., Conde, E., Conde, M.J., González-López, N., Levoso, A., Castro, M., Vidal, E., and González, J. 2010. Procedimientos de recuperación de compuestos antioxidantes presentes en efluentes de destilería de vino. Patent ES 2,345,748 A2.

Mulder, M. 1997. *Basic Principles of Membrane Technology*. Dordrecht, the Netherlands: Kluwer Academic Publishers.

Murakami, A.N.N., Amboni, R.D.M.C., Prudêncio, E.S., Amante, E.R., Zanotta, L.M., Maraschin, M., Petrus, J.C.C., and Teófilo, R.F. 2011. Concentration of phenolic compounds in aqueous mate (*Ilex paraguariensis* St. Hil) extract through nanofiltration. *LWT: Food Science and Technology* 44(10):2211–2216.

Nawaz, H., Shi, J., Mittal, G.S., and Kakuda, Y. 2006. Extraction of polyphenols from grape seeds and concentration by ultrafiltration. *Separation and Purification Technology* 48(2):176–181.

Neagu, E., Roman, G.P., Radu, G.L., and Nechifor, G. 2010. Concentration of the bioactive principles in *Geranium robertianum* extracts through membranare procedures (ultrafiltration). *Romanian Biotechnology Letters* 15(1):5042–5048.

Negrea, A., Lupa, L., Ciopec, M., and Negrea, P. 2011. Experimental and modelling studies on As(III) removal from aqueous medium on fixed bed column. *Chemical Bulletin of "Politehnica" University of Timisoara, Romania, Series of Chemistry and Environmental Engineering* 56(70):89–93.

Noordman, T.R., Kooiker, K., Bel, W., Dekker, M., and Wesselingh, J.A. 2003. Concentration of aqueous extracts of defatted soy flour by ultrafiltration—Effect of suspended particles on the filtration flux. *Journal of Food Engineering* 58(2):135–141.

Nunes Da Ponte, M.L.M., Dos Santos, J.L.C., Matias, A.A.F., Nunes, A.V.M.M., Duarte, C.M.M., Crespo, J.P.S.G. 2007. Method of obtaining a natural hydroxytyrosol-rich concentrate from olive tree residues and subproducts using clean technologies. U.S. Patent WO/2007/013032.

Otero, M., Zabkova, M., and Rodrigues, A.E. 2005. Comparative study of the adsorption of phenol and salicylic acid from aqueous solution onto nonionic polymeric resins. *Journal of Separation and Purification Technology* 45:86–95.

Paraskeva, P., Kalderis, D., Diamadopoulos, E. 2008. Production of activated carbon from agricultural by-products. *Journal of Chemical Technology and Biotechnology* 83:581–592.

Paun, G., Neagu, E., Albu, C., and Radu, G.L. 2015. Application of the polyphenylene ether-ether-sulfone ultrafiltration membrane for concentration of antioxidants from the *Phyllitis scolopendrium* L. extract. *New Journal of Chemistry* 39:1154–1160.

Paun, G., Neagu, E., and Radu, G.L. 2009. Antiradical activities of *Salvia officinalis* and *Viscum album* L. extracts concentrated by ultrafiltration process. *Acta Scientiarum Polonorum Technologia Alimentaria* 8(3):47–58.

Paun, G., Neagu, E., Tache, A., Radu, G.L., and Parvulescu, V. 2011. Application of the nanofiltration process for concentration of polyphenolic compounds from *Geranium robertianum* and *Salvia officinalis* extracts. *Chemical and Biochemical Engineering Quarterly* 25(4):453–460.

Pizzichini, M. and Russo, C. 2005. Process for recovering the components of olive mill wastewater with membrane technologies. Patent WO 2005/123603.

Plazinski, W., Rudzinski, W., and Plazinska, A. 2009. Theoretical models of sorption kinetics including a surface reaction mechanism: A review. *Advances in Colloid and Interface Science* 152:2–13.

Prodanov, M., Garrido, I., Vacas, V., Lebrón-Aguilar, R., Dueñas, M., Gómez-Cordovés, C., and Bartolome, B. 2008. Ultrafiltration as alternative purification procedure for the characterization of low and high molecular-mass phenolics from almond skins. *Analytica Chimica Acta* 609:241–251.

Prudêncio, A.P.A., Prudêncio, E.S., Amboni, R.D.M.C., Murakami, A.N.N., Maraschin, M., Petrus, J.C.C., Ogliari, P.J., and Leite, R.S. 2012. Phenolic composition and antioxidant activity of the aqueous extract of bark from residues from mate tree (*Ilex paraguariensis* St. Hil.) bark harvesting concentrated by nanofiltration. *Food and Bioproducts Processing* 90:399–405.

Qi, X.L., Peng, X., Huang, Y.Y., Li, L., Wei, Z.F., Zu, Y.G., and Fu, Y.J. 2015. Green and efficient extraction of bioactive flavonoids from *Equisetum palustre* L. by deep eutectic solvents-based negative pressure cavitation method combined with macroporous resin enrichment. *Industrial Crops and Products* 70:142–148.

Rajha, H.N., Boussetta, N., Louka, N., Maroun, R.G., and Vorobiev, E. 2015. Effect of alternative physical pretreatments (pulsed electric field, high voltage electrical discharges and ultrasound) on the dead-end ultrafiltration of vine-shoot extracts. *Separation and Purification Technology* 146:243–251.

Rudzinski, W. and Plazinski, W. 2007. Theoretical description of the kinetics of solute adsorption at heterogeneous solid/solution interfaces: On the possibility of distinguishing between the diffusional and the surface reaction kinetics models. *Applied Surface Science* 253:5827–5840.

Russo, C. 2007. A new membrane process for the selective fractionation and total recovery of polyphenols, water and organic substances from vegetation waters (VW). *Journal of Membrane Science* 288(1–2):239–246.

Ruthven, D.M. 2005. *Adsorption in Van Nostrand's Encyclopedia of Chemistry*. John Wiley & Sons. Published online July 15, 2005. DOI: 10.1002/0471740039.vec0079.

Salgado, C., Palacio, L., Carmona, F.J., Hernández, A., and Prádanos, P. 2013. Influence of low and high molecular weight compounds on the permeate flux decline in nanofiltration of red grape must. *Desalination Journal* 315:124–134.

Santamaría, B., Salazar, G., Beltrán, S., and Cabezas, J.L. 2002. Membrane sequences for fractionation of polyphenolic extracts from defatted milled grape seeds. *Desalination Journal* 148(1–3):103–109.

Sarmento, L.A.V., Machado, R.A.F., Petrus, J.C.C., Tamanini, T.R., and Bolzan, A. 2008. Extraction of polyphenols from cocoa seeds and concentration through polymeric membranes. *Journal of Supercritical Fluids* 45:64–69.

Schieber, A., Hilt, P., Endreß, H.U., Rentschler, C., and Carle, R. 2003. A new process for the combined recovery of pectin and phenolic compounds from apple pomace. *Innovative Food Science and Emerging Technologies* 4:99–107.

Scordino, M., Di Mauro, A., Passerini, A., and Maccarone, E. 2004. Adsorption of flavonoids on resins: Cyanidin 3-glucoside. *Journal of Agricultural and Food Chemistry* 52:1965–1972.

Scordino, M., Di Mauro, A., Passerini, A., and Maccarone, E. 2005. Selective recovery of anthocyanins and hydroxycinnamates from a by-product of citrus processing. *Journal of Agricultural and Food Chemistry* 53:651–658.

Scott, K. and Hughes, R. 1996. *Industrial Membrane Separation Technology*. London, UK: Blackie Academic & Professional.

Servili, M., Esposto, S., Veneziani, G., Urbani, S., Taticchi, A., Di Maio, I., Selvaggini, R., Sordini, B., and Montedoro, G. 2011. Improvement of bioactive phenol content in virgin olive oil with an olive-vegetation water concentrate produced by membrane treatment. *Food Chemistry* 124(49):1308–1315.

Shafeeyan, M.S., Wan Daud, W.M.A., and Shamiri, A. 2014. A review of mathematical modeling of fixed-bed columns for carbon dioxide adsorption. *Chemical Engineering Research and Design* 92:961–988.

Shao-qian, C., Si-Yi, P., Xiao-Lin, Y., and Hong-Fei, F. 2010. Isolation and purification of anthocyanins from blood oranges by column chromatography. *Agricultural Sciences in China* 9(2):207–215.

Silva, E.M., Rogez, H., Da Silva, I.Q., and Larondelle, Y. 2013. Improving the desorption of *Inga edulis* flavonoids from macroporous resin: Towards a new model to concentrate bioactive compounds. *Food and Bioproducts Processing* 91:558–564.

Simion, D., Gaidau, C., Paun, G., Koleva, M., and Cupara, S. 2012. Modern techniques for concentration of active principles from plant extracts, based on membranary technologies. *Leather Footwear Journal* 12(3):223–230.

Soto, M.L., Moure, A., Dominguez, H., and Parajó, J.C. 2011. Recovery, concentration and purification of phenolic compounds by adsorption. *Journal of Food Engineering* 105:1–27.

Soto, M.L., Moure, A., Domínguez, H., and Parajó, J.C. 2008. Charcoal adsorption of phenolic compounds present in distilled grape pomace. *Journal of Food Engineering* 84:156–163.

Stoller, M. 2011. Effective fouling inhibition by critical flux based optimization methods on a NF membrane module for olive mill wastewater treatment. *Chemical Engineering Journal* 168(3):1140–1148.

Sun, P.C., Liu, Y., Yi, Y.T., Li, H.Y., Fan, P., and Xia, C.H. 2015. Preliminary enrichment and separation of chlorogenic acid from *Helianthus tuberosus* L. leaves extract by macroporous resins. *Food Chemistry* 68:55–62.

Thomas, W.J. and Crittenden, B.D. 1998. *Adsorption Technology and Design*. Oxford, UK: Butterworth & Heinemann.

Toth, J. (ed.). 2002. *Adsorption: Theory, Modeling, and Analysis*. Surfactant Science Series, Vol. 107. New York: Marcel Dekker.

Turano, E., Curcio, S., Paola, M.G., Calabrò, V., and Lorio, G. 2002. An integrated centrifugation–ultrafiltration system in the treatment of olive mill wastewater. *Journal of Membrane Science* 209:519–531.

Ulbricht, M., Ansorge, W., Danielzik, I., König, M., and Schuster, O. 2009. Fouling in microfiltration of wine: The influence of the membrane polymer on adsorption of polyphenols and polysaccharides. *Separation and Purification Technology* 68(3):335–342.

Vauchel, P., D'Alessandro, L.G., Dhulster, P., Nikov, I., and Dimitrov, K. 2015. Pilot scale demonstration of integrated extraction–adsorption eco-process for selective recovery of antioxidants from berries wastes. *Journal of Food Engineering* 158:1–7.

Villanova, L., Villanova, L., Fasiello, G., and Merendino, A. 2008. Process for the recovery of tyrosol and hydroxytyrosol from oil mill wastewaters and catalytic oxidation method in order to convert tyrosol in hydroxytyrosol. U.S. Patent 7,427,358.

Wang, T., Lu, S., Xia, Q., Fang, Z., and Johnson, S. 2015. Separation and purification of amygdalin from thinned bayberry kernels by macroporous adsorption resins. *Journal of Chromatography* 975:52–58.

Weber, W.J. and Morris, J.C. 1962. Advances in water pollution research: Removal of biologically resistant pollutants from waste waters by adsorption. *Proceedings of the International Conference on Water Pollution Symposium*, Vol. 2, pp. 231–266, Pergamon, Oxford, UK.

Wei, S., Hossain, Md.M., and Saleh, Z.S. 2010. Concentration of rutin model solutions from their mixtures with glucose using ultrafiltration. *International Journal of Molecular Sciences* 11(2):672–690.

Weisz, G.M., Carle, R., and Kammerer, R. 2013. Sustainable sunflower processing. Recovery of phenolic compounds as a by-product of sunflower protein extraction *Innovative Food Science and Emerging Technologies* 17:169–179.

Weisz, G.M., Schneider, L., Schweiggert, U., Kammerer, D.R., and Carle, R. 2010. Sustainable sunflower processing: Development of a process for the adsorptive decolorization of sunflower (*Helianthus annuus* L.) protein extracts. *Innovative Food Science and Emerging Technologies* 11(4):733–741.

Wu, F.C., Tseng, R.L., and Juang, R.S. 2009. Initial behavior of intraparticle diffusion model used in the description of adsorption kinetics. *Chemical Engineering Journal* 153:1–8.

Xi, l., Mu, T., and Sun, H. 2015. Preparative purification of polyphenols from sweet potato (*Ipomoea batatas* L.) leaves by AB-8 macroporous resins. *Food Chemistry* 172:166–174.

Xiong, Q., Zhang, Q., Zhang, D., Shi, Y., Jiang, C., and Shi, X. 2014. Preliminary separation and purification of resveratrol from extract of peanut (*Arachis hypogaea*) sprouts by macroporous adsorption resins. *Food Chemistry* 145:1–7.

Xu, L., Lamb, K., Layton, L., and Kumar, A. 2004. A membrane-based process for recovering isoflavones from a waste stream of soy processing. *Food Research International* 37:867–874.

Yang, L. and Tan, T. 2008. Enhancement of the isolation selectivity of isoflavonoid puerarin using oligo-b-cyclodextrin coupled polystyrene-based media. *Biochemical Engineering Journal* 40:189–198.

Yang, R. 2003. *Adsorbents: Fundamentals and Applications.* Hoboken, NJ: John Wiley & Sons.

Yu, Z.R., Hung, C.C., Weng, Y.M., Su, C.L., and Wang, B.J. 2007. Physiochemical, antioxidant and whitening properties of extract from root cortices of mulberry as affected by membrane process. *Food Science and Technology* 40:900–907.

Zagklis, D.O., Vavourakia, A.I., Kornarosa, M.E., and Paraskevaa, C.A. 2015. Purification of olive mill wastewater phenols through membrane filtration and resin adsorption/desorption. *Journal of Hazardous Materials* 285:69–76.

Zagklis, D.P. and Paraskeva, C.A. 2014. Membrane filtration of agro-industrial wastewaters and isolation of organic compounds with high added values. *Water Science and Technology* 69: 202–207.

Zhang, G., Hu, M., He, L., Fu, P., Wang, L., Zhou, J. 2013. Optimization of microwave-assisted enzymatic extraction of polyphenols from waste peanut shells and evaluation of its antioxidant and antibacterial activities *in vitro*. *Food and Bioproducts Processing* 91:158–168.

Zhang, Z.F., Liu, Y., Luo, P., and Zhang, H. 2009. Separation and purification of two flavone glucuronides from *Erigeron multiradiatus* (Lindl.) Benth with macroporous resins. *Journal of Biomedicine and Biotechnology* 2009:875629.

Zhao, Z., Dong, L., Wu, Y., and Lin, F. 2011. Preliminary separation and purification of rutin and quercetin from *Euonymus alatus* (Thunb.) Siebold extracts by macroporous resins. *Food and Bioproducts Processing* 89:266–272.

8 Fractionation of Polyphenols

Peter Winterhalter and Ulrich H. Engelhardt

CONTENTS

8.1 INTRODUCTION

One of the few liquid chromatographic techniques that can be predictably scaled up from an analytical to production scale is countercurrent chromatography (CCC) (Wood et al., 2007). CCC is a descriptive term for continuous liquid–liquid partition processes that do not employ any solid supporting matrix. Liquid–liquid partitioning

TABLE 8.1
Countercurrent Chromatography Techniques

Technique	Authors	Year
Craig Systems	Craig et al.	1951
Helix CCC	Ito and Bowman	1970
Rotation Locular CCC	Ito and Bowman	1970
Droplet CCC	Tanimura et al.	1970
Flowthrough CPC	Ito and Bowman	1971
Toroidal Coil Centrifuge	Ito	1980
Centrifugal DCCC (CPC)	Murayama et al.	1982
Coil Planet Centrifuge (MLCCC, HSCCC)	Ito et al.	1982
Low-Speed Rotary CCC	Du et al.	2000
Spiral-Coil LSRCCC	Köhler et al.	2004
Industrial Scale-Up	Sutherland	2007

methods have been available for many years (Craig et al., 1951). However, due to long separation times, inconvenience, and enormous solvent consumption, their application has been very limited. In recent years, the introduction of innovative CCC techniques (cf. Table 8.1) has led to a renaissance in their use for the separation and purification of natural products (Conway, 1990; Ito and Conway, 1996).

In many applications, these methods offer additional or alternative procedures to the more extensively employed chromatographic separations on solid adsorbents. Compared to high-performance liquid chromatography (HPLC), CCC offers several advantages. The gentle operating conditions, especially the lack of active surfaces, ensure an isolation of labile compounds and guarantee a total recovery of the sample. The sample load is significantly higher (>10 g up to kg scale), and the separation requires only solvents instead of expensive solid-phase columns. Since an almost unlimited number of solvent systems are available for CCC applications, this technique is extremely versatile. It should also be emphasized that CCC gives a completely different pattern of selectivity compared to HPLC due to its solely partition-based mechanism of separation (Conway and Petroski, 1995; Ito, 1981, 1986; Ito and Conway, 1996).

For efficient fractionation of a complex natural mixture of phenolics by CCC, enrichment of the phenolics on organic resins and prefractionation steps are recommended. In many cases, the crude mixtures include the following three groups of phenolics: anthocyanins, colorless copigments (e.g., flavonols, flavanols, cinnamates, and stilbenes), and polymers. Strategies for the straightforward purification of these different groups of phenolics will be presented.

8.2 ENRICHMENT AND PREFRACTIONATION OF CRUDE PHENOLIC EXTRACTS PRIOR TO CCC

Solvent extraction is frequently used for the recovery of polyphenols from different natural resources, such as fruits, vegetables, and other plant materials, as well as by-products and side streams of the food industry (e.g., pomace). For extraction, solvents such as water and organic solvents (e.g., methanol, ethanol) are widely applied often in combination with acids (e.g., citric acid) (Brazinha et al., 2014). Alternatively, high temperature and pressure (subcritical water extractions) are employed (Ju and Howard, 2005). In order to enhance extraction yield, enzyme treatment (pectinolytic and cellulolytic enzymes) as well as pulsed electric field treatment or high hydrostatic pressure have been applied (Corrales et al., 2008; Kammerer et al., 2005). After the extraction process, crude phenolic mixtures are obtained, which can be concentrated and prefractionated by using different strategies. Examples are the use of polyamide and polyvinylpolypyrrolidone (PVPP), as well as organic resins and ion-exchange processes (resins and membranes).

8.2.1 The Use of Polyamide and Polyvinylpolypyrrolidone

8.2.1.1 Basic Principles

Polyamide column chromatography has been applied since the 1970s for the cleanup of samples prior to spectrophotometry (Herrmann, 1973). PVPP and polyamide have also been used for a long time to isolate certain phenolic compounds for analytical and other purposes. Both polyamide and PVPP interact with phenolic hydroxyl groups by hydrogen bonds. Such interactions are mainly present in aqueous media, whereas in organic solvents the bonds are cleaved. Depending on the strength of hydrogen bonds, different solvent mixtures are applied enabling the fractionated elution of phenolic compounds. In the case of polyamide, there are different types available. Polyamide 6 (polycaprolactam) is most widely used; however, there are also other materials, such as polyamide 11 or polyamide 6.6 (Figure 8.1).

Both polyamide and PVPP can be used as batch material or in a chromatographic column for solid-phase extraction (SPE). Usually, polyamide is pretreated and stored in an organic solvent–water mixture or swollen in water for at least 2 h prior to use. After filling it into a suitable column, the organic solvent has to be removed before the application of an aqueous solution, for example, a plant extract. In some cases, for example, for the elution of catechins, it might be advantageous to use elevated temperatures (e.g., 40°C) instead of ambient temperatures. In some cases, phenolic compounds cannot be eluted simply by methanol, ethanol, or acetone but by using organic solvents with added ammonia or formic acid (Risch and Herrmann, 1988; Wildanger and Herrmann, 1973).

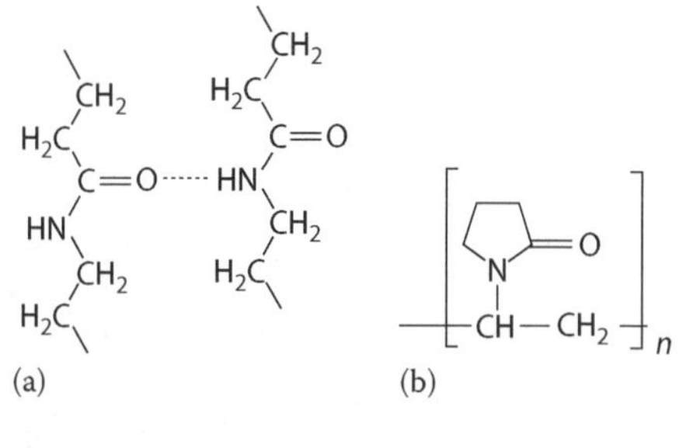

FIGURE 8.1 Polyamide (a) and polyvinylpolypyrrolidone (b).

Polyamide and PVPP are both solid materials and consequently irreversible adsorption of some of the analytes is possible, contrary to the support-free liquid–liquid techniques of CCC described also in this chapter. Sometimes, this becomes obvious when the material retains a dark color, even after elution with strong solvents. The eluotropic order for polyamide is water < methanol < ethanol < isopropanol < acetone < formamide < dimethylformamide.

8.2.1.2 Mechanistic Studies

This section reviews selected studies with a focus on the mechanism rather than the application. The importance of hydrogen bonds for the separation on polyamide was already stated (Gao et al., 2011). It was also shown that within the group of phenolic acids, gallic acid was less retarded compared to benzoic acid, 4-hydroxy-, and 3,5-dihydroxybenzoic acid. The unusual behavior of gallic acid has been explained by the assumption that gallic acid forms intramolecular hydrogen bonds, which reduce the binding strength to polyamide. The elution behavior of the other compounds tested was determined by the number of free OH groups.

The adsorption of catechins (flavanols) and caffeine on PVPP was studied by Dong et al. (2011). Caffeine seemed to be bound to catechins rather than to the PVPP itself and was nearly completely recovered by elution with water, while the catechins were eluted with DMSO/ethanol. Magalhães et al. (2010) used PVPP to adsorb phenolic compounds from a hop extract. Mechanistic studies were also included, which confirmed that usually the number of OH groups determines the strength of binding, whereas derivatization, such as methylation of the OH group, weakens the interaction. Moreover, it is discussed here that not only hydrogen bonds but also the number of aromatic rings (via an interaction with the PVPP ring) and steric hindrance will affect the elution behavior. Another mechanistic study was centered on flavonol glycosides (FOGs) and concluded that the interaction of the free aglycones is stronger compared to the glycosides (Laborde et al., 2006).

8.2.1.3 Selected Applications

A well-known application of polyamide/PVPP is their use in brewing technology to remove the phenolics in order to avoid a sediment formation during storage. Polyamide/PVPP can also be used to isolate polyphenols from side streams of the brewing industry. Barbosa-Pereira et al. (2014) used PVPP to develop a suitable extraction procedure including an ethyl acetate extraction to get partially purified antioxidant mixtures. Li et al. (2014) also used polyamide pretreatment to isolate flavonol glycosides on a semipreparative scale from the peel of *Trichosanthes kirilowii*. A recent paper describes the isolation of flavonol glycosides from green tea using this approach. The phenolic compounds of green tea are bound to polyamide filled in a chromatographic column and separated by a stepwise elution with aqueous methanol and methanol (Jiang et al., 2015). This enabled the separation of the vast majority of the catechins that are much more abundant than the FOG in green tea.

Another more recent application was the isolation of phenolic compounds, among others, from an ethyl acetate extract obtained from thermally treated two-phase olive pomace ("alperujo"). Minor compounds could be detected after polyamide column using fractionated elution with H_2O, 25% MeOH (v/v), 50% MeOH (v/v), 75% MeOH (v/v), and 100% MeOH (Rubio-Senent et al., 2015). Peng et al. (2014) made use of polyamide to isolate apigenin, luteolin, and rosmarinic acid from aerial parts of *Elsholtzia splendens*. Here, polyamide separation was part of the cleanup and fractionated elution using dichloromethane/methanol/formic acid mixtures was employed.

Often, polyamide column chromatography is used as part of a more complex isolation scheme for phenolics. One example is the combination of polyamide column chromatography and size-exclusion chromatography on Sephadex for flavonol glycosides purification from an extract of *Adenanthera pavonina* L. leaves (Mohammed et al., 2014). In a study with cocoa flavonoids, which led to the detection of an unusual catechin, polyamide purification was essential. In this case, a polyamide cartridge was conditioned with DMSO/formic acid and washed with water. The elution step was again DMSO/formic acid (Kofink et al., 2007).

The aforementioned examples may illustrate the performance of polyamide/PVPP for the isolation and the cleanup prior to chromatographic separations. Both can be easily applied prior to CCC techniques to reduce the complexity of the matrix.

8.2.2 Organic Resins

Nonionic polymeric resins are widely applied for the recovery of flavonoids from plant extracts. The resins, which can be packed in columns of large diameters, have the advantage to enrich the phenolics in a very gentle and efficient way, whereas sugars, acids, and other polar constituents pass the column unretained. After rinsing the column with water, elution of the retained phenolics (glycosides and aglycones) is then performed with water–alcohol mixtures. For food-grade status, use of ethanol is essential. A comparison of the recovery of flavonoids from aqueous plant extracts by use of different Amberlites available on the market suggested that the polystyrene resins (XAD-2, XAD-4, and XAD-16) were more suitable for flavonoids than the polyacrylic resins XAD-7 and XAD-8 (Tomás-Barberán et al., 1992). Especially, XAD-2 has the advantage of permitting the stepwise elution of flavonoids according to their polarity. Nevertheless, in the case of anthocyanins, one of the most widely applied adsorbents is the polyacrylic resin Amberlite XAD-7 (Schwarz et al., 2003), which consists of ethylene glycol dimethyl acrylate (EGDMA). Most recently, preparation and performance of modified macroporous "XAD-7 materials" have been reported (Yao et al., 2015). These new resins, which are not commercially available yet, consist of divinylbenzene and EGDMA. In the case of the copolymerizate with 20% EGDMA, optimal adsorption capacity and selectivity with regard to anthocyanins have been reported (Yao et al., 2015).

8.2.3 Ion-Exchange Processes (Resins and Membranes)

Ion-exchange processes are mainly used for the enrichment of anthocyanins because under acidic conditions the anthocyanins are present as flavylium cations, thus enabling the use of cation exchangers. He and Giusti (2011) applied a SPE technique employing an Oasis® MCX SPE cartridge (Waters Corp., Milford, Massachusetts). The MCX sorbent features combined mechanisms of strong cation-exchange and reversed-phase adsorption. Kammerer et al. (2011) on the other hand used anion-exchange resins for enrichment and fractionation of major apple flavonoids, phenolic acids, and dihydrochalcones. In addition, membrane technology has been used for the purification of anthocyanins from various fruits including side streams of the food industry (Juadjur, 2012; Juadjur and Winterhalter, 2012). The membrane adsorber Sartobind® S 75 (Sartorius Stedim Biotech, Göttingen, Germany) consists of a stabilized cellulose membrane with negatively charged sulfonic acid groups. Under acidic conditions, only the positively charged flavylium cations are retarded (*retentate fraction*) on the membrane surface and the other phenolic compounds pass the adsorber unretained (*permeate fraction*). After rinsing of the adsorber with ethanol/acetic acid (19:1, v/v), elution of anthocyanins is performed with a mixture of 1 M aqueous NaCl solution/methanol (50:50, v/v). In this way, a pure anthocyanin fraction free of any other phenolics can be obtained. This method has first been developed for bilberry anthocyanins and later *inter alia* adapted to anthocyanins from grape pomace. Depending on the size of the membrane adsorber cartridge, separation in the 10 g to 1 kg scale can be carried out.

From the permeate fraction that contains the copigments and polymeric phenolics, the low-molecular-weight phenolic constituents are easily liquid–liquid extracted with ethyl acetate. In this way, three clearly separated fractions, that is, anthocyanins, copigments, and polymers, are obtained. Although anthocyanins and copigments can be directly used for nutraceutical or pharmaceutical purposes, the polymer fraction may require a further treatment, for example, acid-catalyzed depolymerization (see Section 8.7.2).

8.3 FRACTIONATION BY COUNTERCURRENT CHROMATOGRAPHY

CCC is an automated version of liquid–liquid extraction, comparable to the repeated partitioning of an analyte between two immiscible phases by vigorous mixing in a separator funnel. The real breakthrough of this technique came with the invention of the "Coil Planet Centrifuge" that had been developed in the laboratory of Dr. Yoichiro Ito at the National Institutes of Health (Bethesda, Maryland) in the 1980s (Ito, 1981). Today, this technique is known as high-speed CCC (HSCCC). HSCCC uses a seal-free planetary motion of the separating column, which is usually a long PTFE tubing (a so-called coil). Due to the two-axis gyration mechanism, the centrifugal field rapidly changes in intensity and direction. This results in alternate mixing and settling of the biphasic solvent system on the column, which finally leads to separation of the analytes. In addition to "hydrodynamic" techniques, "hydrostatic" techniques are today increasingly used in CCC. The *hydrostatic* technique, which normally refers to "centrifugal partition chromatography" (CPC) (Foucault, 1995; Murayama et al., 1982), operates under a constant gravitational field that is produced by a single-axis rotation mechanism. Details of hydrostatic and hydrodynamic CCC techniques will be presented later. First, the common principle of CCC will be explained.

8.3.1 Theoretical Aspects of CCC

CCC is complementary to HPLC; in the latter, the separation occurs between a chemically or physically bonded stationary phase, which occupies less than 10% of the column volume, and the respective mobile phase. In both hydrostatic and hydrodynamic CCCs, the stationary phase is a liquid, which can amount up to 80%–90% of the total column volume. This is the secret behind the observation that although the number of theoretical plates (N) in CCC is much lower compared to HPLC (order of hundreds compared to 10,000 in HPLC), resolution (R_s) of the peaks is quite high.

This is apparent from the following equation:

$$R_s = \frac{V_{R_2} - V_{R_1}}{\left(\frac{W_2 + W_1}{2}\right)} = S_f \cdot \frac{\sqrt{N}}{4} \cdot \frac{K_{D_2} - K_{D_1}}{1 - S_f \cdot \left[1 - \left(\frac{K_{D_2} + K_{D_1}}{2}\right)\right]}$$

where

S_f is the stationary-phase retention ratio
N is the peak efficiency expressed in plate number
V_R is the peak retention volume
W is the peak with at base
K_D is the solute distribution constant or partition coefficient
1 and 2 refer to the first and second eluted peaks

Hence, the unique feature of CCC (good resolution R_s with a low plate number N) is a result of the fact that the chromatographic resolution factor R_s is dependent on S_f, the amount of liquid stationary phase retained by the column (Conway, 1990). CCC is furthermore the only chromatographic separation technique in which the stationary-phase volume contained in the column is an adjustable parameter depending on operational factors, such as flow rate of mobile phase and rotational speed (Du et al., 1999; Wood et al., 2003). In a recent paper, Berthod and Faure (2015) have studied the influence of different tubing parameters on peak resolution R_s. They found that for producing a maximum possible resolution factor R_s for a given solute polarity, the optimum tubing bore was within the 0.8–3 mm bore range, which is the range of the most intermediate volume CCC apparatuses commercially available (Berthod and Faure, 2015).

8.3.2 Solvent System Selection for CCC Separations

The crucial point for a successful separation by CCC is the choice of a suitable solvent system. The basic requirement for such a system is that it consists of two immiscible phases and that the target compounds have different partition coefficients in this system. Over the last few years, numerous solvent mixtures have been successfully employed to separate a wide variety of phenolics (Berthod et al., 2009; Ito, 2005; Oka et al., 1991). It is therefore recommended to start with these approved systems and to optimize one of them according to the specific requirements of the actual separation. In order to do this, a so-called partition study must be carried out.

This involves dissolution of a small amount of the phenolic extract in the biphasic solvent system (e.g., few milligrams of the sample in 2 mL of each of the two selected phases in a test tube), shaking them, and allowing the system to equilibrate (recommended time <30 s). The concentration of the analytes in each of the layers is then determined by spectrophotometry, HPLC, gas chromatography, or thin-layer chromatography. Alternatively, the solvents can be evaporated and the masses of the residues determined by gravimetry. The partition coefficient K_D is calculated by dividing the concentration of the target compounds in the upper phase by the concentration in the lower phase. The optimum partition coefficients are in the range of 0.5–1.0. In the case of lower K_D values, compounds elute too quickly and are not well separated, whereas higher values will give long retention times and considerable peak broadening. Depending on the outcome of the "partition study," the polarity of the two-phase mixture has to be modified. Useful information about the selection and optimization of CCC solvent systems can be found in the literature (Friesen and Pauli, 2005, 2007; Friesen et al., 2015; Ito, 2005; Oka et al., 1991). Oka et al. (1991) have proposed 24 biphasic solvent systems composed of *n-hexane*, *ethyl acetate*, *n-butanol*, *methanol*, *water*, *acetic acid*, and *chloroform* in varying proportions. The suggested solvent systems have suitable settling times and cover a broad polarity range. Examples for typical biphasic solvent mixtures that are widely applied are *butanol/water*, *ethyl acetate/butanol/water*, *hexane/ethyl acetate/methanol/water* (HEMWat system), *heptane/ethyl acetate/methanol/water* (Arizona system), or *chloroform/methanol/water*.

Many of the CCC fractionations can be run with food-grade solvents except for the chlorinated solvents, which must be avoided in case of food applications (Faure et al., 2014a). Attempts have been made to replace organic solvents by "greener" solvents, such as limonene, which can be obtained in large amounts from citrus essential oil (Faure et al., 2014b,c). Limitations arise due to the high boiling point of limonene hampering its removal from the purified CCC fractions.

8.3.3 Modes of Operation

In a successful CCC separation, the elution mode plays an important role when the instrument and the biphasic solvent system are defined. The following possibilities for elution modes have to be considered (Conway, 1990):

1. *Reversed phase*: The aqueous phase is pumped through as the mobile phase.
2. *Normal phase*: The nonaqueous phase is used as the mobile phase.
3. *Dual mode*: Mobile and stationary phases are reversed partway through the run.
4. *Gradient mode*: The concentration of one or more components in the mobile phase is varied throughout the run to achieve optimal resolution across a wider range of polarities. For example, a methanol/water gradient may be employed using pure heptane as the stationary phase.
5. *Elution extrusion mode*: This mode consists of two steps—The first step is a classical elution. After a certain time, the mobile phase is changed to the phase originally used as the stationary phase. In this way, all the retained solutes are eluted within one column volume.

8.4 INSTRUMENTATION

8.4.1 Hydrodynamic Techniques

8.4.1.1 High-Speed Countercurrent Chromatography and High-Performance Countercurrent Chromatography

Separation in HSCCC takes place in a so-called multilayer coil that is made by wrapping an inert Teflon or stainless steel tubing around a holder in multiple layers. Consequently, this technique is also known as multilayer coil CCC (MLCCC). The tubing usually has an inner diameter between 1.6 and 2.6 mm and a length of up to 160 m (Figure 8.2). Multiple coils can be connected in series to increase the total volume of the instrument and the sample capacity.

During separation, the coil is rotated in a planetary fashion. It rotates at 800–1000 rpm around its own "planetary" axis and simultaneously around the "solar" axis. This planetary motion has two effects: (1) *retention of the stationary phase* and (2) *partitioning of solutes.*

Retention of the stationary phase. During rotation of a coil filled with immiscible liquids, the two phases move toward opposite ends of the coil, known as head and tail. Generally, the less dense phase displaces the heavier phase toward the tail, but the orientation is influenced by viscosity and interfacial tension. This phenomenon, known as "hydrodynamic equilibrium," requires one to choose the elution mode carefully, and it gives the analyst the choice to select either the lighter or heavier layer as the mobile or stationary phase. In any case, one of the two immiscible phases is retained in the instrument, while the second phase is pumped through the column as the mobile phase.

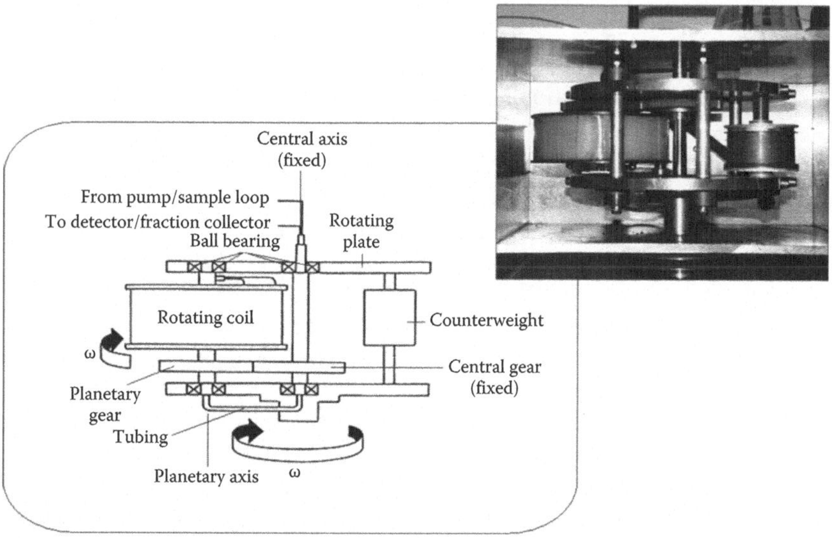

FIGURE 8.2 Principal setup of a single-coil high-speed countercurrent chromatograph. (Photo courtesy of P. Winterhalter.)

Partitioning of solutes. The planetary rotation creates a fluctuating acceleration field, which enables vigorous mixing of the two phases followed by settling within the coil. In areas of the coil, which are close to the center of rotation, the force field is weak. Consequently, the phases are mixed. At a further point of their orbit, when they are far away from the center of rotation, the force field becomes stronger and the two phases are separated. Alternate mixing and settling is repeated with each rotation, resulting up to 60,000 partitioning steps per hour.

In new instruments, the rotational speed can be increased up to 1600 rpm. These so-called high-performance CCC (HPCCC) apparatuses generate an approximately a fourfold stronger centrifugal force field (240 × *g*) than conventional HSCCC. This allows for the use of higher mobile phase flow rates, significantly reducing separation time without compromising chromatographic resolution (DeAmicis et al., 2011).

8.4.1.2 Operation of a Countercurrent Chromatograph

With the optimized solvent system in hand and the proper elution mode decided, the separation can be started. The operation steps for a CCC separation consist of (1) complete loading of the column with the stationary phase using a HPLC pump, (2) start of rotation and concomitant pumping of the mobile phase, (3) equilibration of the system (characterized by termination of stationary-phase displacement), (4) injection of the sample in an equal amount of the biphasic solvent system via a sample loop, and (5) elution of compounds according to their coefficient values. The separation is monitored by a UV/Vis detector and is usually accomplished within a few hours (HSCCC) or 1–2 h (HPCCC). More recently, coupling of CCC with mass spectrometry was described (das Neves Costa et al., 2015; Jerz et al., 2012).

8.4.2 Hydrostatic Techniques

8.4.2.1 Centrifugal Partition Chromatography

Conceptually, this technique is a further development of droplet CCC (DCCC) (Tanimura et al., 1970). However, the long separation times of the solely gravitationally operated DCCC apparatus were reduced by applying a centrifugal force field. Consequently, the first prototype of this type of CCC device has been named centrifugal DCCC (CDCCC) (Murayama et al., 1982). Today, the technique is generally known as CPC (Foucault, 1995). In a CPC device, the column is mounted in a centrifuge rotor. The column consists of a series of partition compartments connected in series by ducts, engraved on disks. Depending on the number of disks used, analytical and preparative CPC apparatuses are available. Due to the single-axis rotation, CPC is considered as a hydrostatic technique. For the separation, the column is filled with one phase of the biphasic solvent system and, when rotated, the other phase can be pumped through. Two modes of operation can be chosen: either the heavy (*descending mode*) or lighter (*ascending mode*) phase can be used as the mobile phase. Solvent system selection is the same as in HSCCC.

8.5 SCALE-UP OF COUNTERCURRENT CHROMATOGRAPHY

Strategies for the scale-up of CCC have been presented by Wood et al. (2007) and Sutherland (2007). Established protocols recommend transposing the results obtained on a small column to a large volume column working with the same relative amount of stationary phase (S_f) retained by the column. A geometrical scale-up factor is defined as the ratio of the large to small CCC column volume. The strategy is to increase the mobile phase flow rate and solute mass load by the scale-up factor. In a certain range (up to kilograms), this is possible today, and examples for successful industrial separations were compiled by Sutherland (2007). In both hydrostatic and hydrodynamic systems, preparative columns are available that allow separations in the 100 g to kg scale. The biggest column in HSCCC available today has a total volume of 18 L (Maxi-DE, Dynamic Extractions Ltd., Berkshire, UK, cf. Figure 8.3), whereas the biggest CPC centrifuge consists of a 25 L rotor (Partus Technologies, Reims, France). Information about manufacturers of commercial preparative CCC devices has been gathered by Sutherland (2007).

A different approach for industrial scale-up of CCC has been presented by Du et al. (2005). This group has presented a prototype of a so-called low-speed rotary countercurrent chromatograph (LSRCCC). In LSRCCC, a cylindrical column rotates slowly around a single axis. The use of a special convoluted tubing enabled

FIGURE 8.3 Preparative high-speed countercurrent chromatograph with a total coil volume of 18 L from Dynamic Extractions Ltd., Berkshire, United Kingdom. (Photo courtesy of Prof. I. Sutherland.)

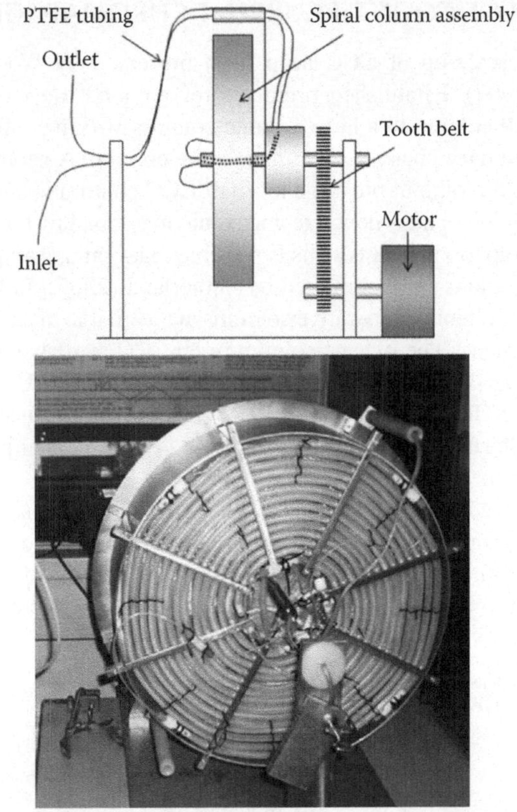

FIGURE 8.4 Prototype of a spiral-coil low-speed rotary countercurrent chromatograph. (Reproduced from Köhler, N. et al., *J. Liq. Chromatogr Relat. Technol.*, 27, 2547, 2004. With permission. Photo courtesy of P. Winterhalter.)

sufficient retention of stationary phase at a rotational speed of only 50–100 rpm (Du et al., 2000, 2003). In experiments with an instrument equipped with a 10 L column, sample load was in the range of 500 g per separation and 63 g of salicin (purity 95.3%) has been obtained (Du et al., 2005). Another development in this field was the "spiral-coil LSRCCC " (Köhler et al., 2004). In *spiral-coil LSRCCC*, the coil is spirally mounted on a holder, and due to the additional spiral force, stationary-phase retention is considerable increased. This allows higher flow rates and hence reduces the separation times (cf. Figure 8.4).

8.6 APPLICATIONS OF CCC FOR THE SEPARATION OF POLYPHENOLS

Strategies of solvent system selection for the isolation of flavonoids by CCC have recently been published by das Neves Costa and Leitão (2010). This review compiles numerous solvent systems that have been successfully applied to the separation of free and glycosylated flavonoids. HSCCC was by far the most used technique, being reported in nearly 90% of all papers dealing with the separation of flavonoids. Each of the hydrostatic techniques of CPC and DCCC covered another 5% of the applications. The most frequently cited solvent system for the isolation of free flavonoids was hexane/ethyl acetate/MeOH/water. For the isolation of glycosylated flavonoids (flavones, flavonols, isoflavones, and flavanones), EtOAc/BuOH/water as well as $CHCl_3$/MeOH/water were the most widely applied solvent systems (das Neves Costa and Leitão, 2010). Separation of anthocyanins by CCC is also well established (Schwarz et al., 2003). The anthocyanins are first isolated by adsorption on Amberlite XAD-7 and, after elution and concentration, are separated by HSCCC on a preparative scale. The first fractionation of the XAD-7 isolate can be achieved with the solvent system MTBE/*n*-butanol/acetonitrile/water (2:2:1:5, v/v/v/v; acidified with 0.1% TFA, a less dense layer acting as the stationary phase). This system is optimized for the isolation/separation of monoglucosylated anthocyanins (e.g., malvidin 3-glucoside). More polar pigments such as the diglucosides and the polymeric fraction are separated in a second step by application of either ethyl acetate/*n*-butanol/water (2:3:5, v/v/v, acidified with 0.1% TFA) or a modified MTBE/*n*-butanol/acetonitrile/water system (1:3:1:5, v/v/v/v; acidified with 0.1% TFA). For the less polar acylated anthocyanins, successful separations have been obtained by using ethyl acetate/water (1:1, v/v; acidified with 0.1% TFA) (Degenhardt et al., 2000; Schwarz et al., 2003). Other examples include *inter alia* isoflavones (Stürtz et al., 2006), proanthocyanidins (PAs) (Köhler et al., 2008a), and stilbenes (Macke et al., 2012).

8.7 FRACTIONATION AND DEPOLYMERIZATION OF POLYMERIC PHENOLICS

8.7.1 Centrifugal Precipitation Chromatography

Centrifugal precipitation chromatography was applied to the separation of polymeric phenolics in wine (Degenhardt et al., 2001). This continuous precipitation technique generates solvent gradients through a long separation channel under a centrifugal force field. The wine polymers are first precipitated with MTBE and subsequently exposed to a gradually increasing ethanol concentration. This causes repetitive precipitation and dissolution of the polymers along the separation channel. Eventually, the polymers are eluted in the order of their solubility in the organic solvent. The monomers and oligomers elute first followed by different classes of fractionated polymers. Further structural data can then be obtained by mass spectrometry or phloroglucinolysis (Degenhardt et al., 2001).

8.7.2 Depolymerization of Polymeric Proanthocyanidins

Bioavailability of proanthocyanidins is restricted to oligomers with up to three flavan-3-ol units. In order to convert the polymeric PAs into bioavailable ones, an acid-catalyzed depolymerization process has to be applied. There are two possibilities. The first one uses a random cleavage of the flavan-3-ol chain under acidic conditions and gives rise to short-chain cleavage products. The second approach uses a more controlled degradation, the so-called semisynthesis (cf. Figure 8.5).

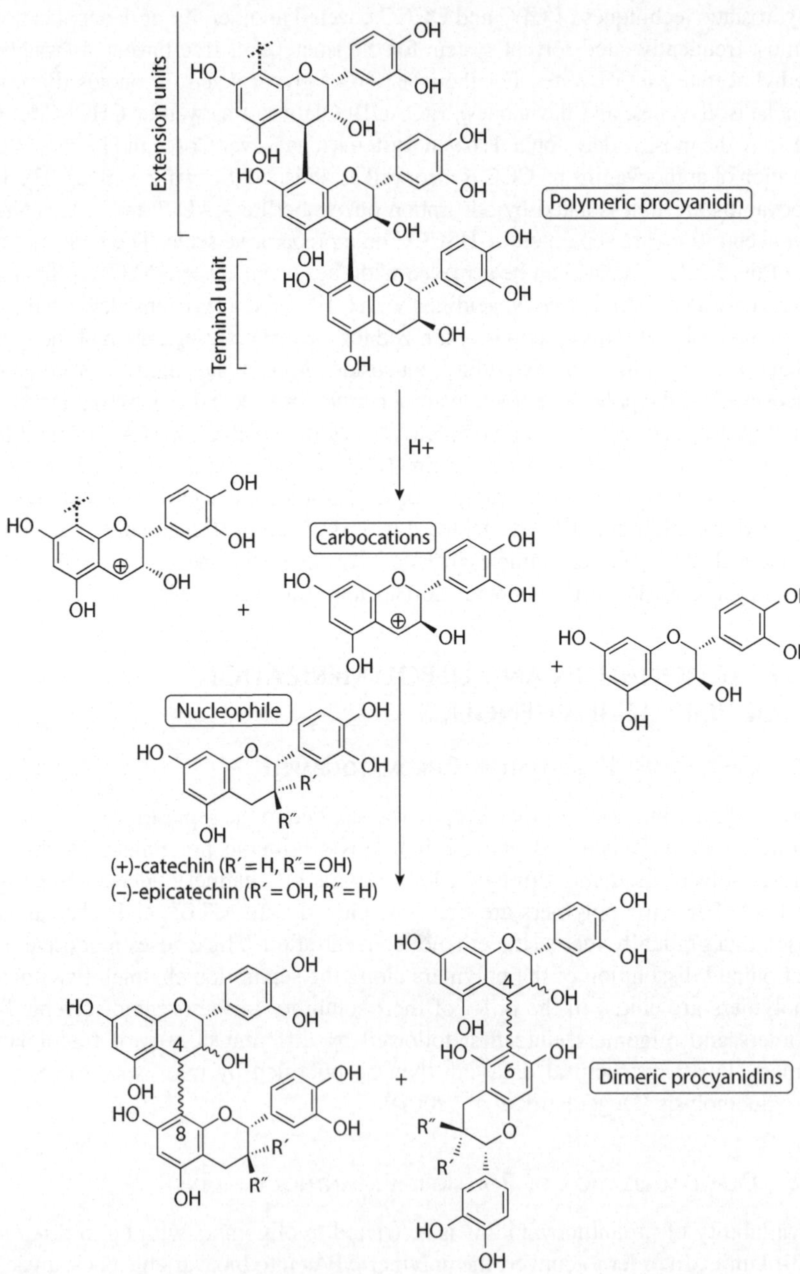

FIGURE 8.5 Mechanism of the directed depolymerization reaction of polymeric PA.

Polymeric PAs are treated with 1 N methanolic HCl. During this process, the interflavanoid linkage of polymeric PA is cleaved by releasing the flavan-3-ol moieties of the extension unit as carbocation together with an uncharged terminal unit. The liberated carbocation can immediately react with nucleophiles added in excess, such as (+)-catechin or (−)-epicatechin, giving rise to the formation of dimeric procyanidins. During this process, the polymeric procyanidins are depolymerized and bioavailable dimeric procyanidins are formed. Depending on the nucleophiles chosen, the reaction products can be tailored in a simple way. Isolation of the dimeric reaction products is then performed by preparative CCC (Esatbeyoglu et al., 2014; Köhler et al., 2008b).

8.8 CONCLUDING REMARKS

Since its invention in the 1970s (Tanimura et al., 1970), CCC has slowly but steadily grown to be a mature preparative technique. Nevertheless, this technique is still less widely known and used than conventional HPLC. The modern commercially manufactured CCC devices use high gravitational force fields, which allow high flow rates and reduce the separation times to 1–2 h. Preparative sample loads are feasible, and the technique is especially suited to the separation of polar substances such as polyphenols. This is demonstrated by an increasing number of cases in which very successful preparative separations of polyphenols have been reported.

REFERENCES

Barbosa-Pereira, L., Bilbao, A., Vilches, P. et al. 2014. Brewery waste as a potential source of phenolic compounds: Optimisation of the extraction process and evaluation of antioxidant and antimicrobial activities. *Food Chemistry* 145:191–197.

Berthod, A. and Faure, K. 2015. Revisiting resolution in hydrodynamic countercurrent chromatography: Tubing bore effect. *Journal of Chromatography A* 1390:71–77.

Berthod, A., Ruiz-Angel, M.J., and Carda-Broch, S. 2009. Countercurrent chromatography: People and applications. *Journal of Chromatography A* 1216:4206–4217.

Brazinha, C., Cadima, M., and Crespo, J.G. 2014. Optimization of extraction of bioactive compounds from different types of grape pomace produced at wineries and distilleries. *Journal of Food Science* 79:E1142–E1149.

Conway, W.D. 1990. *Countercurrent Chromatography: Apparatus, Theory, and Applications.* New York: VCH Publishers Inc.

Conway, W.D. and Petroski, R.J. (eds.). 1995. *Modern Countercurrent Chromatography.* ACS Symposium Series 593. Washington, DC: American Chemical Society.

Corrales, M., Toepfl, S., Butz, P., Knorr, D., and Tauscher, B. 2008. Extraction of anthocyanins from grape by-products assisted by ultrasonics, high hydrostatic pressure or pulsed electric fields: A comparison. *Innovative Food Science and Emerging Technologies* 9:85–91.

Craig, L.C., Hausmann, W., Ahrens, E.H. Jr., and Harfenist, E.J. 1951. Automatic countercurrent distribution equipment. *Analytical Chemistry* 23:1236–1244.

das Neves Costa, F. and Leitão, G.G. 2010. Strategies of solvent system selection for the isolation of flavonoids by countercurrent chromatography. *Journal of Separation Science* 33:336–347.

das Neves Costa, F., Jerz, G., de Souza Figueiredo, F., Winterhalter, P., and Guimarães Leitão, G. 2015. Solvent system selectivities in countercurrent chromatography using *Salicornia gaudichaudiana* metabolites as practical example with off-line ESI-MS/MS injection profiling. *Journal of Chromatography A* 1385:20–27.

DeAmicis, C., Edwards, N., Giles, M.B. et al. 2011. Comparison of preparative reversed phase liquid chromatography and countercurrent chromatography for the kilogram scale purification of crude spinetoram insecticide. *Journal of Chromatography A* 1218:6122–6127.

Degenhardt, A., Engelhardt, U.H., Winterhalter, P., and Ito, Y. 2001. Centrifugal precipitation chromatography—A novel chromatographic system for fractionation of polymeric pigments from black tea and red wine. *Journal of Agricultural and Food Chemistry* 49:1730–1736.

Degenhardt, A., Hofmann, S., Knapp, H., and Winterhalter, P. 2000. Preparative isolation of anthocyanins by high-speed countercurrent chromatography and application of the color activity concept to red wine. *Journal of Agricultural and Food Chemistry* 48:5812–5818.

Dong, Z.-B., Liang, Y.-R., Fan, F.-Y., Ye, J.-H., Zheng, X.-Q., and Lu, J.-L. 2011. Adsorption behavior of the catechins and caffeine onto polyvinylpolypyrrolidone. *Journal of Agricultural and Food Chemistry* 59:4238–4247.

Du, Q., Jerz, G., He, Y. et al. 2005. Semi-industrial isolation of salicin and amygdalin from plant extracts using slow rotary counter-current chromatography. *Journal of Chromatography A* 1074:43–46.

Du, Q., Winterhalter, P., and Ito, Y. 2003. Large convoluted tubing for scale-up of slow rotary countercurrent chromatograph. *Journal of Liquid Chromatography and Related Technologies* 26:1991–2002.

Du, Q., Wu, C., Qian, G., Wu, P., and Ito, Y. 1999. Relationship between the flow-rate of the mobile phase and retention of the stationary phase in counter-current chromatography. *Journal of Chromatography A* 835:231–235.

Du, Q., Wu, P., and Ito, Y. 2000. Low-speed rotary countercurrent chromatography using a convoluted multilayer helical tube for industrial separation. *Analytical Chemistry* 72:3363–3365.

Esatbeyoglu, T., Juadjur, A., Wray, V., and Winterhalter, P. 2014. Semisynthetic preparation and isolation of dimeric procyanidins B1–B8 from roasted hazelnut skins (*Corylus avellana* L.) on a large scale using countercurrent chromatography. *Journal of Agricultural and Food Chemistry* 62:7101–7110.

Faure, K., Bouju, E., Doby, J., and Berthod, A. 2014a. Limonene in Arizona liquid systems used in countercurrent chromatography. II. Polarity and stationary-phase retention. *Analytical and Bioanalytical Chemistry* 406:5919–5926.

Faure, K., Bouju, E., Suchet, P., and Berthod, A. 2014b. Limonene in Arizona liquid systems used in countercurrent chromatography. I. Physicochemical properties. *Analytical and Bioanalytical Chemistry* 406:5909–5917.

Faure, K., Mekaoui, N., and Berthod, A. 2014c. Saving solvents in chromatographic purifications: The counter-current chromatography technique. In *Green Chromatographic Techniques: Separation and Purification of Organic and Inorganic Analytes*, eds. Inamuddin, Dr. and A. Mohammad, pp. 1–18. Dordecht, the Netherlands: Springer.

Foucault, A.P. (ed.). 1995. *Centrifugal Partition Chromatography*. Chromatographic Science 68. New York: Dekker.

Friesen, J.B., McAlpine, J.B., Chen, S.-N., and Pauli, G.F. 2015. Countercurrent separation of natural products: An update. *Journal of Natural Products* 78:1765–1796.

Friesen, J.B. and Pauli, G.F. 2005. G.U.E.S.S.—A generally useful estimate of solvent systems for CCC. *Journal of Liquid Chromatography and Related Technologies* 28:2777–2806.

Friesen, J.B. and Pauli, G.F. 2007. Rational development of solvent system families in counter-current chromatography. *Journal of Chromatography A* 1151:51–59.

Gao, M., Wang, X., Gu, M., Su, Z., Wang, Y., and Janson, J.-C. 2011. Separation of polyphenols using porous polyamide resin and assessment of mechanism of retention. *Journal of Separation Science* 34:1853–1858.

He, J. and Giusti, M.M. 2011. High-purity isolation of anthocyanins mixtures from fruits and vegetables—A novel solid-phase extraction method using mixed mode cation-exchange chromatography. *Journal of Chromatography A* 1218:7914–7922.

Herrmann, K. 1973. Quantitative Bestimmung der Catechine und Hydroxyzimtsäuren in pflanzlichen Lebensmitteln. *Fresenius' Zeitschrift für Analytische Chemie* 267:361–365.

Ito, Y. 1980. The toroidal coil planet centrifuge without rotating seals applied to countercurrent chromatography. *Analytical Biochemistry* 102:150–152.

Ito, Y. 1981. Efficient preparative counter-current chromatography with a coil planet centrifuge. *Journal of Chromatography A* 214:122–125.

Ito, Y. 1986. High-speed countercurrent chromatography. *CRC Critical Reviews in Analytical Chemistry* 17:65–143.

Ito, Y. 2005. Golden rules and pitfalls in selecting optimum conditions for high-speed counter-current chromatography. *Journal of Chromatography A* 1065:145–168.

Ito, Y. and Bowman, R.L. 1970. Countercurrent chromatography: Liquid-liquid partition chromatography without solid support. *Science* 167(3916):281–283.

Ito, Y. and Bowman, R.L. 1971. Countercurrent chromatography with flow-through coil planet centrifuge. *Science* 173(3995):420–422.

Ito, Y. and Conway, W.D. (eds.). 1996. *High-speed Countercurrent Chromatography.* Chemical Analysis 132. New York: Wiley.

Ito, Y., Sandlin, J., and Bowers, W.G. 1982. High-speed preparative counter-current chromatography with a coil planet centrifuge. *Journal of Chromatography A* 244:247–258.

Jerz, G., Murillo-Velásquez, J.A., Skrjabin, I., Gök, R., and Winterhalter, P. 2012. Anacardic acid profiling in cashew nuts by direct coupling of preparative high-speed counter-current chromatography and mass spectrometry (*prep*HSCCC-ESI-/APCI-MS/MS). In *Recent Advances in the Analysis of Food and Flavors*, Vol. 1098, eds. S. Toth, and C. Mussinan, pp. 145–165. ACS Symposium Series 1098. Washington, DC: American Chemical Society.

Jiang, H., Engelhardt, U.H., Thräne, C., Maiwald, B., and Stark, J. 2015. Determination of flavonol glycosides in green tea, oolong tea and black tea by UHPLC compared to HPLC. *Food Chemistry* 183:30–35.

Ju, Z. and Howard, L.R. 2005. Subcritical water and sulfured water extraction of anthocyanins and other phenolics from dried red grape skin. *Journal of Food Science* 70:S270–S276.

Juadjur, A. 2012. Entwicklung einer membranchromatographischen Methode zur Isolierung von Anthocyanen aus Heidelbeeren und weiteren Früchten (in German). PhD dissertation, Technische Universität Braunschweig, Braunschweig, Germany.

Juadjur, A. and Winterhalter, P. 2012. Development of a novel adsorptive membrane chromatographic method for the fractionation of polyphenols from bilberry. *Journal of Agricultural and Food Chemistry* 60:2427–2433.

Kammerer, D., Claus, A., Schieber, A., and Carle, R. 2005. A novel process for the recovery of polyphenols from grape (*Vitis vinifera* L.) pomace. *Journal of Food Science* 70:C157–C163.

Kammerer, J., Boschet, J., Kammerer, D.R., and Carle, R. 2011. Enrichment and fractionation of major apple flavonoids, phenolic acids and dihydrochalcones using anion exchange resins. *LWT: Food Science and Technology* 44:1079–1087.

Kofink, M., Papagiannopoulos, M., and Galensa, R. 2007. (–)-Catechin in cocoa and chocolate: Occurrence and analysis of an atypical flavan-3-ol enantiomer. *Molecules* 12:1274–1288.

Köhler, N., Chou, E., Ito, Y., and Winterhalter, P. 2004. Development of a new preparative spiral-coil low-speed rotary countercurrent chromatographic (spiral-coil LSRCCC) method. *Journal of Liquid Chromatography and Related Technologies* 27:2547–2560.

Köhler, N., Wray, V., and Winterhalter, P. 2008a. Preparative isolation of procyanidins from grape seed extracts by high-speed counter-current chromatography. *Journal of Chromatography A* 1177:114–1125.

Köhler, N., Wray, V., and Winterhalter, P. 2008b. New approach for the synthesis and isolation of dimeric procyanidins. *Journal of Agricultural and Food Chemistry* 56:5374–5385.

Laborde, B., Moine-Ledoux, V., Richard, T., Saucier, C., Dubourdieu, D., and Monti, J.-P. 2006. PVPP-polyphenol complexes: A molecular approach. *Journal of Agricultural and Food Chemistry* 54:4383–4389.

Li, A., Sun, A., Liu, R., Zhang, Y., and Cui, J. 2014. An efficient preparative procedure for main flavonoids from the peel of *Trichosanthes kirilowii* Maxim. using polyamide resin followed by semi-preparative high performance liquid chromatography. *Journal of Chromatography B: Analytical Technologies in the Biomedical and Life Sciences* 965:150–157.

Macke, S., Jerz, G., Empl, M.T., Steinberg, P., and Winterhalter, P. 2012. Activity-guided isolation of resveratrol oligomers from a grapevine-shoot extract using countercurrent chromatography. *Journal of Agricultural and Food Chemistry* 60:11919–11927.

Magalhães, P.J., Vieira, J.S., Gonçalves, L.M., Pacheco, J.G., Guido, L.F., and Barros, A.A. 2010. Isolation of phenolic compounds from hop extracts using polyvinylpolypyrrolidone: Characterization by high-performance liquid chromatography-diode array detection-electrospray tandem mass spectrometry. *Journal of Chromatography A* 1217:3258–3268.

Mohammed, R.S., Abou Zeid, A.H., El-Kashoury, E.A., Sleem, A.A., and Waly, D.A. 2014. A new flavonol glycoside and biological activities of *Adenanthera pavonina* L. leaves. *Natural Product Research* 28:282–289.

Murayama, W., Kobayashi, T., Kosuge, Y., Yano, H., Nunogaki, Y., and Nunogaki, K. 1982. A new centrifugal countercurrent chromatograph and its application. *Journal of Chromatography A* 239:643–649.

Oka, F., Oka, H., and Ito, Y. 1991. Systematic search for suitable two-phase solvent systems for high-speed counter-current chromatography. *Journal of Chromatography A* 538:99–108.

Peng, H., Xing, Y., Gao, L., Zhang, L., and Zhang, G. 2014. Simultaneous separation of apigenin, luteolin and rosmarinic acid from the aerial parts of the copper-tolerant plant *Elsholtzia splendens*. *Environmental Science and Pollution Research* 21:8124–8132.

Risch, B. and Herrmann, K. 1988. Hydroxyzimtsäure-Verbindungen in Citrus-Früchten. *Zeitschrift für Lebensmittel-Untersuchung und Forschung* 187:530–534.

Rubio-Senent, F., Martos, S., Lama-Muñoz, A., Fernández-Bolaños, J.G., Rodríguez-Gutiérrez, G., and Fernández-Bolaños, J. 2015. Isolation and identification of minor secoiridoids and phenolic components from thermally treated olive oil by-products. *Food Chemistry* 187:166–173.

Schwarz, M., Hillebrand, S., Habben, S., Degenhardt, A., and Winterhalter, P. 2003. Application of high-speed countercurrent chromatography to the large-scale isolation of anthocyanins. *Biochemical Engineering Journal* 14:179–189.

Stürtz, M., Lander, V., Schmid, W., and Winterhalter, P. 2006. Preparative isolation of isoflavones from soy and red clover. *Molecular Nutrition and Food Research* 50:356–361.

Sutherland, I.A. 2007. Recent progress on the industrial scale-up of counter-current chromatography. *Journal of Chromatography A* 1151:6–13.

Tanimura, T., Pisano, J.J., Ito, Y., and Bowman, R.L. 1970. Droplet countercurrent chromatography. *Science* 169 (3940):54–56.

Tomás-Barberán, F.A., Blázquez, M.A., Garcia-Viguera, C., Ferreres, F., and Tomás-Lorente, F. 1992. A comparative study of different amberlite XAD resins in flavonoid analysis. *Phytochemical Analysis* 3:178–181.

Wildanger, W. and Herrmann, K. 1973. Qualitativer Nachweis und quantitative Bestimmung von Flavonolen und Flavonen. *Journal of Chromatography* 76:433–440.

Wood, P., Ignatova, S., Janaway, L. et al. 2007. Counter-current chromatography separation scaled up from an analytical column to a production column. *Journal of Chromatography A* 1151:25–30.

Wood, P.L., Hawes, D., Janaway, L., and Sutherland, I.A. 2003. Stationary phase retention in CCC: Modelling the J-type centrifuge as a constant pressure drop pump. *Journal of Liquid Chromatography and Related Technologies* 26:1373–1396.

Yao, L., Zhang, N., Wang, C., and Wang, C. 2015. Highly selective separation and purification of anthocyanins from bilberry based on a macroporous polymeric adsorbent. *Journal of Agricultural and Food Chemistry* 63:3543–3550.

9 Drying and Preservation of Polyphenols

Paz Robert, Paula García, and Carolina Fredes

CONTENTS

9.1 INTRODUCTION

Polyphenol-rich species have long been thought to have considerable health benefits mainly due to their total antioxidant activity against cellular oxidation reactions (Bouaziz et al., 2009; Janicsak et al., 2010; Xia et al., 2010). Moreover, cardioprotective (Cale et al., 2010; Xia et al., 2010), antimicrobial (Bouaziz et al., 2009; Xia et al., 2010), anticancer, and anti-inflammatory (Xia et al., 2010) properties have also been reported in both animal and *in vitro* models. These benefits have stimulated regular research regarding the phenolic status and antioxidant activity of various plant species and new ecotypes/varieties in different countries.

The most common method to determine total phenolic content is the Folin–Ciocalteu method (Singleton and Rossi, 1965) by spectrometry. Other methods to determine the total polyphenol content by class include (1) total catechin content obtained by treatment with a vanillin reagent (Swain and Hillis, 1959), (2) total proanthocyanidin content obtained by acid hydrolysis (Porter et al., 1986), and (3) total anthocyanin content obtained by a pH differential method (Lee et al., 2005). Nevertheless, these methods cannot provide any information regarding the compositional profile of the polyphenols or the relative amounts of these compounds.

Several thousand molecules with a polyphenol structure (i.e., several hydroxyl groups on aromatic rings) have been identified in higher plants. Based on the number of phenol rings that they contain and the structural elements that bind these rings together, polyphenols have been classified into phenolic acids, flavonoids, stilbenes, and lignans (Manach et al., 2004). Flavonoids are the most common polyphenols in higher plants, sharing a common structure consisting of two aromatic rings (A and B) that are bound by three carbon atoms that form an oxygenated heterocycle (ring C). Flavonoids may be grouped into six subclasses as a function of the type of heterocycle involved: flavonols, flavones, isoflavones, flavanones, anthocyanidins, and flavanols (catechins) (Manach et al., 2004). In addition, anthocyanins correspond to the glycoside or acyl-glycoside of the anthocyanidins. The basic structure of an anthocyanidin is a flavylium ion with different hydroxyl or methoxyl substitutions; the six anthocyanidins (pelargonidin, cyanidin, delphinidin, peonidin, petunidin, and malvidin) are the most common in vascular plants (Clifford, 2000). Another group of flavonoids, not included in the previous classification, is proanthocyanidins, also called procyanidins, condensed tannins, or oligomeric procyanidins. Proanthocyanidins are polymeric compounds made of polyhydroxyflavan-3-ol units linked together by carbon–carbon bonds (Gabetta et al., 2000).

To identify the best plant sources of polyphenols, several studies comparing the total phenolic content in consumed plant species have been published, where measures of antioxidant capacity obtained by different methods are commonly included (Araya et al., 2006; Fredes et al., 2014; Halvorsen et al., 2002; Speisky et al., 2008, 2012). In this context, several species from diverse botanical families have been described as polyphenol-rich sources and have been used in drying studies.

In addition, by-products, especially those from plant-based materials, present several advantages, such as low cost, reduced storage and elimination costs, environmental friendliness, and revalorization of both the process and its by-products (Devesa-Rey et al., 2011). The main limitation of using polyphenol-rich species as raw materials is that polyphenols are susceptible to degradation reactions when they are extracted from the cellular vacuole. Polyphenols show limited stability against environmental conditions (temperature, oxygen, pH, and light) and limited solubility. Some polyphenols also involve potential sensory problems, such as astringency (catechins, isoflavones, and proanthocyanidins), bitterness (isoflavones and proanthocyanidins), and unpleasant flavors (lignans) (Fang and Bhandari, 2010). Encapsulation technology may be a strategy to improve polyphenol stability and extend their shelf life (Champagne and Fustier, 2007; Desai and Park, 2005; Fang and Bhandari, 2010; Gibbs et al., 1999; Gouin, 2004; Nazzaro et al., 2012; Robert and Fredes, 2015).

The first studies on the microencapsulation of polyphenols began in the 2000s, and the search for different designs of polyphenol microparticles for potential uses in the food and nutraceutical industries has been an important issue. In this context, the design of natural colors has been focused on anthocyanin microparticles with both high anthocyanin retention and storage stability (Kandansamy and Somasundaram, 2012). However, the design of health/functional ingredients has been focused on polyphenol microparticles not only with a high polyphenol retention and high antioxidant activity but also with controlled release in the gastrointestinal tract. The objective of this chapter is to compile the scientific research on the encapsulation of polyphenols in order to discuss the main variables in the formulation and process that affect the characteristics of the polyphenol microparticles obtained by spray drying. Stability during microparticle storage and spray-drying technology for polyphenols without encapsulation is also analyzed.

9.2 ENCAPSULATION TECHNOLOGY

In recent years, encapsulation technology has increased in importance in the food industry, particularly in the development of functional and healthy foods. Encapsulation is a technique where an active compound (liquid or gas) is introduced into or entrapped in a matrix (homogeneous system) or polymeric wall (heterogeneous system) to protect it from environmental conditions. In addition, encapsulation prevents interactions with other food components, limits gastrointestinal tract exposure, and controls (time and/or place) active compounds release (Desai and Park, 2005).

Scientific publications regarding the encapsulation of polyphenols from different plant sources have shown a growing trend over the last decade. Since 2002, 131 published articles have been indexed in the ISI Web of Science™ database. In all of these studies, polyphenol encapsulation has been primarily focused on providing protection from environmental conditions (light, oxygen, and temperature), avoiding oxidation, and increasing shelf life.

Currently, some polyphenol encapsulation studies are focusing on their release within a simulated gastrointestinal tract (Betz et al., 2012; Flores et al., 2014; Frank et al., 2012; Oidtmann et al., 2012; Wang et al., 2013; Zheng et al., 2011) in order to define their bioaccessibility/bioavailability. Different encapsulation methods for polyphenols have been described, such as oil-dispersed phase, water/oil/water (w/o/w), extrusion, emulsification/heat gelation, microgel synthesis, freeze drying, supercritical CO_2, and adsorption (Table 9.1), with spray drying being the most common method used to encapsulate polyphenols (Table 9.2).

9.3 MICROENCAPSULATION BY SPRAY DRYING

Spray drying is useful for processing heat-sensitive materials because of its short drying times (5–30 s) (Desai and Park, 2005). This method produces powder microparticles with low water activity (a_w), simplifying transport, handling, and storage, as well as ensuring microbiological quality (Gouin, 2004). However, encapsulation by spray drying is considered an immobilization technology rather than a true encapsulation technology because some active compounds remain exposed on the microparticle surface (Voos et al., 2010).

Spray drying is a process for turning a feed solution into a dry powder in a single, continuous step. First, the feed (solution, dispersion, or emulsion) is atomized into

TABLE 9.1
Different Encapsulation Methods of Polyphenols from Different Plant Sources

Raw Material	Encapsulation Method	Encapsulating Agent	References
Bayberry fruit extract	Oil dispersed phase—spray drying	Ethyl cellulose	Zheng et al. (2011)
Bilberry fruit extract	Emulsion	Whey protein isolate	Betz et al. (2012)
Bilberry fruit extract	Double emulsion (w/o/w)	Pectin (calcium chloride), polyglycerin-polyricinoleat (PGPR)	Frank et al. (2012)
Bilberry pomace extract	(a) Extrusion (b) Emulsification/heat gelation	(a) Amidated pectin (b) Whey protein isolate	Baum et al. (2014)
Bilberry pomace extract	(a) Emulsification/heat gelation (b) Extrusion	(a) Whey protein isolate (b) Amidated pectin	Kropat et al. (2013)
Bilberry pomace extract	(a) Extrusion (b) Emulsification/heat gelation	(a) Amidated pectin (b) Whey protein isolate	Oidtmann et al. (2012)
Blueberry pomace extract	Microgel synthesis	Oxidized potato starch + sodium trimetaphosphate	Wang et al. (2013)
Grapefruit juice	Freeze drying	Maltodextrin + gum arabic	Gurak et al. (2013)
Green tea extract	Freeze drying	Maltodextrin, β-cyclodextrin, or maltodextrin + β-cyclodextrin	Pasrija et al. (2015)
Jaboticaba peel extract	(a) Rapid extraction of supercritical solution (b) Ionic gelification	(a) Polyethyleneglycol (b) Ca-alginate	Santos et al. (2013)
Olive leaf extract	Freeze drying	β-Cyclodextrin	Mourtzinos et al. (2007)
Olive leaf infusion	Electrostatic extrusion	Alginate–chitosan	Belščak-Cvitanović et al. (2011)
Yerba maté leaf powder	Adsorption	High hydrostatic pressure (HHP)-treated cornstarch	Deladino et al. (2015)
Yerba maté leaf powder	Adsorption	HHP-treated cornstarch	Texeira et al. (2015)

fine drops in a hot-drying medium (air or nitrogen). When the droplets are exposed to the drying medium, rapid evaporation of the solvent (mainly water) occurs and a powder is instantaneously produced (Masters, 1972). Spray drying includes four stages: (1) atomization of the feed solution into a spray, (2) spray–air contact, (3) drying of the spray (moisture evaporation), and (4) separation of the dried product from the air (Masters, 1972). Encapsulation by spray drying involves another essential stage: the preparation and homogenization of a feed solution.

TABLE 9.2
Encapsulation of Polyphenols by Spray Drying from Different Plant Sources

Raw Material	Encapsulating Agent	References
Andes berry fruit extract	Maltodextrin, gum arabic, cornstarch, yucca starch, Capsul® TA (Ingredion Inc., Westchester, IL), Hi-CA™100 (Ingredion Inc., Westchester, IL)	Villacrez et al. (2014)
Apple skin extract, grape seed extract, and olive leaf extract	Caseinate-lecithin	Kosaraju et al. (2008)
Artichoke extract	Lactose or hypromellose (HPMC)	Gavini et al. (2005)
Bayberry fruit juice	Maltodextrin	Fang and Bhandari (2011)
Bilberry pomace extract	Maltodextrin + pectin	Baum et al. (2014)
Bilberry pomace extract	Maltodextrin + pectin	Oidtmann et al. (2012)
Black carrot tuber extract	Maltodextrin	Ersus and Yurdagel (2007)
Blackcurrant pomace extract	Maltodextrin or inulin	Bakowska-Barczak and Kolodziejczyk (2011)
Black glutinous rice bran extract	Maltodextrin	Laokuldilok and Kanha (2015)
Blueberry fruit/pomace extract	Whey protein isolate or gum arabic	Flores et al. (2014)
Blueberry fruit extract	Mesquite gum	Jimenez-Aguilar et al. (2011)
Corozo fruit extract	Maltodextrin	Osorio et al. (2010)
Grape seed extract	Gum arabic and maltodextrin	Zhang et al. (2007)
Green tea extract	Maltodextrin, β-cyclodextrin, or maltodextrin + β-cyclodextrin	Pasrija et al. (2015)
Jaboticaba peel extract	Maltodextrin, gum arabic + maltodextrin or Capsul™ + maltodextrin	Silva et al. (2013)
Kokum fruit extract	Maltodextrin, gum acacia, or tricalcium phosphate	Nayak and Rastogi (2010a)
Maté tea leaf decoction	Colloidal silicon dioxide	Yatsu et al. (2011)
Mengkudu unripe pulp extract	Carrageenan	Krishnaiah (2009a)
Noni fruit extract	k-Carrageenan	Krishnaiah (2009b)
Noni fruit extract	k-Carrageenan and maltodextrin	Krishnaiah et al. (2012)
Olive leaf extract	Chitosan	Kosaraju et al. (2006)
Pomegranate fruit juice or fruit extract	Maltodextrin or soybean protein isolate	Robert et al. (2010)
Pomegranate peel extract	Maltodextrin	Çam et al. (2014)
Roselle flower extract	Citrus fiber	Chiou and Langrish (2007)
Roselle flower extract	Maltodextrin, gum arabic, or soluble starch	Idham et al. (2012)
White oak leaf infusion	Lactose and sodium caseinate	Rocha-Guzmán et al. (2010)

Different atomization mechanisms can be applied in spray drying, where pressure nozzles (pressure energy) and rotary atomizers (centrifugal energy) are commonly used (Masters, 1972). When atomization of the feed solution occurs at higher atomization energies, smaller droplets are produced, increasing the heat and mass transfer between the atomized droplets and the drying medium due to the large contact surface (Gharsallaoui et al., 2007). The droplets are dehydrated by the hot drying medium when the solvent (mainly water) rapidly evaporates in a heated chamber. Due to the quick moisture evaporation, a dry crust is formed over the droplet's surface and the moisture diffuses out from the inside of the particles. The dehydrated powdered particles are then transported to a cyclone separator for recovery. Drying is finished when the particle temperature equals the air temperature. This process takes a few seconds, and the liquid feed is converted into powdered particles without any appreciable loss (Gibbs et al., 1999; Gouin, 2004; Mahdavi et al., 2014).

There are many spray dryer designs available for various production levels, from the laboratory and pilot scale to the industrial scale. In addition, there are three mechanisms for the movement of spray air in the drying chamber (cocurrent, countercurrent, or mixed flow) depending on the direction of both the hot air and the feed atomization streams through the drying chamber (Masters, 1972).

The most common parameter employed in the characterization of the microparticles obtained by spray drying is the determination of their polyphenol (i.e., catechin, proanthocyanidin, and/or anthocyanin) content. Nevertheless, this parameter does not produce information about the microencapsulation process. To assess the effectiveness of this process, several parameters are used, such as yield (powder recovery), polyphenol retention, and encapsulation efficiency (EE). The calculation of these parameters is detailed in Equations 9.1 through 9.3:

$$\text{Yield \%} = \frac{\text{Gram of powder after spray drying}}{\text{Gram of solids in the feed solution}} \times 100 \tag{9.1}$$

$$\text{Retention \%} = \frac{\text{Polyphenols in the powder (mg g}^{-1}\text{)}}{\text{Polyphenols in the feed solution (mg g}^{-1}\text{)}} \times 100 \tag{9.2}$$

$$\text{EE \%} = \frac{\text{Polyphenols in the powder} - \text{Surface polyphenols in the powder}}{\text{Polyphenols in the powder}} \times 100 \tag{9.3}$$

The yield is the ratio of the solids in the feed solution before drying and the solids obtained after drying. Polyphenol retention corresponds to the loss of polyphenols during the drying process, and it is determined as the relation between the polyphenols in the feed solution and the polyphenols in the microparticles. EE represents polyphenol–polymer interaction due to electrostatic interactions, hydrogen bonding, hydrophobic interactions, or Van der Waals forces (Robert et al., 2010). The polyphenols located on the surface of the microparticles are usually determined using a solvent or a mixture of solvents, where the polymer(s) used as the encapsulating agent (EA) is insoluble. The determination of total polyphenols from microparticles can vary, depending on the polymer used as the EA. For example, when both the polyphenols and the EA are soluble in the same solvent(s), the polyphenols are released and only one step is required. In some cases, it is necessary to destroy the microparticles by acid or base hydrolysis before extracting the polyphenols with solvents.

The determination of polyphenols is primarily achieved through either colorimetry (Folin–Ciocalteu) or high-performance liquid chromatography (HPLC) by quantifying total phenolic content. Most of the times only total anthocyanins are determined either by spectrometry (Fang and Bhandari, 2011; Jimenez-Aguilar et al., 2011; Nayak and Rastogi, 2010a; Osorio et al., 2010; Robert et al., 2010; Silva et al., 2013) or HPLC-Diode-Array Detection (DAD) (Oidtmann et al., 2012). Therefore, information regarding specific anthocyanin levels is difficult to find. This is a serious limitation when searching for the primary bioactivity or specific health effect of some plant species.

To obtain polyphenol microparticles with desirable characteristics through spray drying, it is necessary to optimize processing conditions and some parameters (Silva et al., 2013). Generally, this optimization is achieved by trial and error. The formulation of the feed solution (nature of the EA, extract type, polyphenol type, active/EA ratio) and the process variables (feed temperature, inlet air temperature, outlet air temperature, and inlet/outlet air temperature ratio) has been investigated as independent variables (Gharsallaoui et al., 2007). Optimization studies have included dependent or response variables such as total phenolic content, surface polyphenols, EE, polyphenol recovery, powder yield, antioxidant activity, moisture content, water activity, hygroscopicity, color, particle size, powder morphology, bulk density, solubility, and glass transition temperature (T_g), (Bakowska-Barczak and Kolodziejczyk, 2011; Ersus and Yurdagel, 2007; Jimenez-Aguilar et al., 2011; Nayak and Rastogi, 2010a; Robert et al., 2010; Villacrez et al., 2014). Statistical methodologies allow handling multiple responses simultaneously in order to obtain the best balance among the response variables. The desirability approach transforms multiple responses into a single response, facilitating the interpretation of the results. Several experimental designs have been performed to assess the effect of independent variables (formulation and process) on response variables to optimize the encapsulation of polyphenols from different plant sources (Robert et al., 2010; Silva et al., 2013; Villacrez et al., 2014). For example, a 2^2 central composite rotable design has been applied to evaluate the effects of EA content and inlet air temperature on the EE of polyphenols and anthocyanins from pomegranate juice and hydroalcoholic extract microparticles (Robert et al., 2010). Silva et al. (2013) studied different EAs and inlet air temperatures on anthocyanin retention, moisture content, total solids, hygroscopicity, and antioxidant activity. In both studies, the response surface methodology was used to define the optimal conditions by applying a multiple response optimization using the desirability function to determine the optimal conditions. The experimental data were fitted to a second-order regression model considering linear, quadratic, and cross-product forms of the independent variables (Robert et al., 2010; Silva et al., 2013). Villacrez et al. (2014) applied a factorial screening design (9×2) to estimate the effects of nine EAs and nozzles with two different internal diameters on water activity, moisture and anthocyanin contents.

9.4 FORMULATION VARIABLES IN THE MICROENCAPSULATION OF POLYPHENOLS BY SPRAY DRYING

9.4.1 Types of Polyphenols and Preparation of Feed Solutions

Several plant species, such as plant herbs, fruits, flowers, tubers, and plant by-products, have been reported as sources of polyphenols in encapsulation studies (Tables 9.1 and 9.2). The polyphenol extraction methods for each plant species will be described. The identification and quantification of polyphenols in plant pulp, juices, and/or extracts have usually been performed by liquid chromatography (LC) and spectrometry (the Folin–Ciocalteu method).

The feed solution includes the mixture of polyphenol extract and the EA; following a homogenization step, it produces a solution, dispersion, or emulsion depending on the solubility of the polyphenol and the EA.

9.4.1.1 Herbal Infusions as Sources of Polyphenols

Plant infusions are deeply rooted in American indigenous cultures, such as the Mapuche (Montes and Wilkomirsky, 1987). Additionally, the drinking of herbal infusion beverages is a very common custom in many countries in the Americas. Yatsu et al. (2011) studied the encapsulation of polyphenols from maté tea (*Ilex paraguariensis* St. Hil., Aquifoliaceae). The use of LC to identify polyphenols in dried leaves revealed the presence of chlorogenic acid, cryptochlorogenic acid, neochlorogenic acid, P4, P5, and P6 (isomeric dicaffeoyl esters of quinic acid), and rutin. The treatment of maté as a raw material included the decoction of dried leaves in water (1.5:10 at 96°C for 15 min), cooling the solution down to 40°C and filtration. The feed solution was obtained by adding the EA to the resulting maté tea infusion. Similarly, Rocha-Guzmán et al. (2010) studied the encapsulation of polyphenols from white oak (*Quercus resinosa* Liemb, Fagaceae) leaf infusions. The treatment of white oak leaves includes a final freeze-drying process after the preparation of the infusion. The feed solution was formed by the addition of the freeze-dried infusion to the EA solution.

Pasrija et al. (2015) studied the encapsulation of green tea extract (*Camellia sinensis* var. pubilimba). The treatment of green tea dried leaves involved extraction using hot water (80°C) in a water/green tea ratio of 20:1 (mL/g) for 30 min (Vuong et al., 2011). Then, the green tea extract was homogenized with the corresponding EAs.

9.4.1.2 Berry-Type Fruits, Tubers, Flowers, and Other Sources of Anthocyanins

Anthocyanins are responsible for the color of many fruits, flowers, and tubers and are recognized in anthocyanin-rich berry-type fruits such as blackberries (*Rubus* sp., Rosaceae), blueberries (*Vaccinium corymbosum* L., Ericaceae), bilberries (*Vaccinium myrtillus* L., Ericaceae), blackcurrants (*Ribes nigrum* L., Grossulariaceae), and pomegranate (*Punica granatum* Linn., Punicaceae), which have been reported in microencapsulation studies. In addition, novel species such as Andes berry (*Rubus glaucus* Benth, Rosaceae), bayberry (*Myrica gole* L., Myricaceae), black mulberry (*Morus nigra*, Moraceae), corozo (*Bactris guineensis* [L.] H.E. Moore, Arecaceae), jaboticaba (*Myrciaria jaboticaba* [Vell.] O. Berg, Myrtaceae), and kokum (*Garcinia*

TABLE 9.3
Anthocyanidins, Colors, and Plant Sources

Anthocyanidin	Color	Plant Sources	References
Delphinidin	Blue, red	Bilberry, blackcurrant, black mulberry, blueberry, jaboticaba, pomegranate, red grape, roselle	Casinesse et al. (2007), Manhita et al. (2006), Verberic et al. (2015), Lohachoompol et al. (2008), Leite-Legatti et al. (2012), Radunic et al. (2015), Huang et al. (2009), Zhang et al. (2014)
Cyanidin	Orange, red	Andes berry, bayberry, bilberry, blackberry, black carrot, blackcurrant, blueberry, corozo, kokum, jaboticaba, pomegranate, red grape, roselle	Osorio et al. (2012), Bao et al. (2005), Casinesse et al. (2007), Koca and Karadeniz (2009), Cuevas Montilla et al. (2011), Manhita et al. (2006), Lohachoompol et al. (2008), Osorio et al. (2010), Nayak et al. (2010b), Leite-Legatti et al. (2012), Radunic et al. (2015), Huang et al. (2009), Zhang et al. (2014)
Petunidin	Blue, red	Bilberry, blackcurrant, blueberry, red grape	Casinesse et al. (2007), Manhita et al. (2006), Lohachoompol et al. (2008), Huang et al. (2009)
Peonidin	Orange, red	Bilberry, blackberry, black carrot, blackcurrant, blueberry, corozo, red grape, roselle	Casinesse et al. (2007), Koca and Karadeniz (2009), Cuevas Montilla et al. (2011), Manhita et al. (2006), Lohachoompol et al. (2008), Osorio et al. (2010), Huang et al. (2009), Zhang et al. (2014)
Malvidin	Blue, red	Bilberry, blueberry, red grape	Casinesse et al. (2007), Lohachoompol et al. (2008), Huang et al. (2009)

indica Choisy, Guttiferae) have been used as raw materials. The main anthocyanidins described in these species are summarized in Table 9.3.

Several methods of treating berry fruits as raw materials for anthocyanin encapsulation have been described. The simplest method is to employ pulp or fruit juice without any additional solvent (Fang and Bhandari, 2011; Nayak and Rastogi, 2010a; Robert et al., 2010); a filtration process is followed to eliminate solids before the preparation of a feed solution for encapsulation. The feed solution is obtained by homogenizing the pulp or juice with one or more EAs. Other studies on anthocyanin encapsulation have described different methods of anthocyanin extraction from fruits, before the preparation of the feed solution, that add steps to the encapsulation process. For example, Robert et al. (2010) and Jimenez-Aguilar et al. (2011) described the use of ethanol/water (1:1) for pomegranate arils and ethanol (96%) for blueberry fruits, respectively. An Amberlite XAD16N column has been used to perform a solid–liquid fractionation of anthocyanins from whole blueberries and blueberry pomace using three different solvent systems (acetonic, ethanolic, and methanolic) (Flores et al., 2014) and from corozo fruits (without seeds) using methanol/acetic acid (Osorio et al., 2010).

Anthocyanins accumulate primarily in the fruit epicarp (peel) (Wang et al., 2012a). Thus, pomace extract from bilberry (Baum et al., 2014; Oidtmann et al., 2012), blackcurrant (Bakowska-Barczak and Kolodziejczyk, 2011), and peel extract from jaboticaba (Silva et al., 2013) have also been used in the microencapsulation of anthocyanins by spray drying. As expected, more steps are included when either fruit pomace (PO) or peel (PE) are used as raw materials in encapsulation studies. Bilberry PO (Baum et al., 2014; Oidtmann et al., 2012) commonly comes from pomace extraction with methanol following filtration, evaporation, and freeze-drying. Bilberry PO also contains other polyphenols, tannins, carbohydrates, and roughage (Baum et al., 2014; Oidtmann et al., 2012). Jaboticaba PE (Silva et al., 2013) comes from extraction with ethanol (70%) that is acidified (pH 2.0) with hydrochloric acid following filtration and evaporation processes.

Other plant sources studied in the encapsulation of anthocyanins include black carrot (*Daucus carota* ssp. *sativus* var. *atrorubens* Alef., Umbelliferae) (Ersus and Yurdagel, 2007) and roselle flowers (*Hibiscus sabdariffa* L.; Malvaceae) (Chiou and Langrish, 2007; Idham et al., 2012); these anthocyanidins are summarized in Table 9.3. The black carrot tubers' pretreatment included anthocyanin extraction with ethanol/HCl (85:15 v/v), followed by rotary evaporation and filtration. The dry powdered roselle calyces' pretreatment included extraction with water/acidified ethanol (2% citric acid at pH 2) (1:1), followed by adsorption/desorption on Amberlite XAD-16 (Idham et al., 2012).

Laokuldilok and Kanha (2015) studied the encapsulation of anthocyanins from black glutinous rice (*Oryza sativa* L. var. glutinosa). For this purpose, the bran of the rice grain was separated and then extracted with 70% ethanol followed by concentration in a rotary evaporator.

9.4.1.3 Plant By-Products

9.4.1.3.1 Olive Leaves as a Source of Oleuropein and Other Polyphenols

Extra virgin olive oil is extracted from fresh olive fruits (*Olea europaea* L.), and its production is associated with the generation of large quantities of waste (Meksi et al., 2012; Nasini et al., 2012), including the loss of polyphenols in olive oil by-products (Frankel et al., 2013). Olive leaves constitute 5% of the by-product of the *almazaras*. The main bioactive compounds described in olive leaves are polyphenols (El Sedef and Karakaya, 2009; Mylonaki et al., 2008), where oleuropein is one of the most important phenolic compounds (19% w/w). Other polyphenols described in olive leaves include verbascoside, oleuroside, ligstroside, and glycosylated flavonoids (mainly luteolin-7-glycoside, apigenin-7-glycoside, rutin [quercetin-3-rutinoside]), and, less frequently, apigenin-7-rutinoside, luteolin-7-rutinoside and luteolin-4-glycoside, flavonoid aglycones (apigenin, quercetin, kaempferol, hesperidin, luteolin) and ferulic, caffeic, *p*-coumaric, chlorogenic, and vanillin acids (Jimenez et al., 2011). Kosaraju et al. (2006, 2008) studied the encapsulation of polyphenols from commercially available olive leaf extracts (OLEs), where the feed solution includes the OLEs and the corresponding EA solutions.

9.4.1.3.2 Grape Seeds as a Source of Proanthocyanidins

In the production of red wine, the solids remaining after the fermentation of red grapes are usually known as wine pomace, which mainly includes skin, seeds, remaining pulp, and small pieces of stalk. Wine pomace also contains residual yeasts and bacteria, which were the main agents of alcoholic and malolactic fermentation (García-Lomillo et al., 2014). Proanthocyanidins are the major constituent of grape seed extracts (GSEs) (Park et al., 2014). Zhang et al. (2007) studied the use of grape seeds from a wine operation as raw material for the encapsulation of proanthocyanidins using a Supercritical Fluid Extractor (SCFE) -CO_2 followed by extraction with acetic acid and a purification process with macroporous adsorbent resins in order to isolate proanthocyanidins. The feed solution was obtained by the addition of the extracted proanthocyanidins (core material) to the solution of selected EAs. In contrast, Kosaraju et al. (2008) studied the encapsulation of polyphenols from GSE, where a commercial freeze-dried extract was suspended in water and distributed in an aqueous ethanol solution EA.

9.4.1.4 Other Plant Species as Sources of Polyphenols

Gavini et al. (2005) reported the phenolic composition of artichoke (*Cynara scolymus* L. var. Spinoso sardo) flower extract, where chlorogenic acid, cynarin, caffeic acid, isochlorogenic acids, cynaroside (luteolin-7-*O*-glucoside), and scolymoside (luteolin-7-*O*-rutinoside) were identified. The treatment of artichokes as raw material includes extraction of polyphenols with ethanol (96%) in a Soxhlet apparatus followed by filtration, concentration, and purification by column chromatography (Celite).

Krishnaiah (2009a,b) and Krishnaiah et al. (2012) studied the encapsulation of polyphenols from noni (*Morinda citrifolia* L., Rubiaceae) fruit. The authors used both unripe fruit dried pulp and dried fruit as raw materials. In both cases, ethylacetate extraction was followed by filtration.

9.5 ENCAPSULATING AGENTS

There are several biopolymers available for use as EAs. The nature and properties of these biopolymers (film capacity, oxygen permeability, and hygroscopicity, among others) play important roles in the stability of the active compounds (polyphenols). A variety of EAs have been studied in polyphenol encapsulation by spray drying (Table 9.2). Several natural gums have been used, such as gum arabic (Flores et al., 2014; Idham et al., 2012; Villacrez et al., 2014; Zhang et al., 2007), mesquite gum (Jimenez-Aguilar et al., 2011), and acacia gum (Nayak and Rastogi, 2010a). In addition, whey proteins (Flores et al., 2014) and soy proteins (Robert et al., 2010) are effective EAs. Carbohydrates are also useful, like starches (Villacrez et al., 2014), maltodextrins (MD) of different dextrose equivalents (DE) (Bakowska-Barczak and Kolodziejczyk, 2011; Ersus and Yurdagel, 2007; Fang and Bhandari, 2011; Idham et al., 2012; Nayak and Rastogi, 2010a; Osorio et al., 2010; Robert et al., 2010; Silva et al., 2013; Villacrez et al., 2014; Zhang et al., 2007) and inulin (Bakowska-Barczak and Kolodziejczyk, 2011). Mixtures of EAs have also been described, such as AG + MD, Capsul™ + MD (Silva et al., 2013), and MD + pectin (Baum et al., 2014; Oidtmann et al., 2012).

The choice of EA is very important for proper EE, the stability of the active compounds in the microparticles during storage and release properties in foods and the gastrointestinal tract. Biopolymers that have high water absorption capacities can increase the water availability in the microparticles for chemical reactions, which is more important for polyphenols that are degraded by hydrolysis reactions. However, low-molecular-weight biopolymers or biopolymers that contain low polymerization degree saccharides (i.e., some MDs) have been reported to cause oxygen permeability, which is detrimental when the polyphenols are degraded by oxygen. An ideal EA for water-soluble polyphenols should have film-forming properties, low viscosity at high solid contents, low hygroscopicity, and be low cost. Moreover, in the case of water-insoluble polyphenols or more hydrophobic polyphenols, the EA should have emulsifying properties.

The EA nature, EA content, and/or polyphenols/EA ratio have been considered as variables that affect the properties of polyphenol microparticles, such as polyphenol retention, moisture content, hygroscopicity, color, antioxidant activity, and morphology (Bakowska-Barczak and Kolodziejczyk, 2011; Robert et al., 2010; Silva et al., 2013; Villacrez et al., 2014).

MD is obtained by the hydrolysis of different starch sources (e.g., corn, potato), and it is the most common biopolymer used as an EA to encapsulate polyphenols by spray drying (Table 9.2), mainly because MD is highly soluble in water. Many studies have focused on the effects of the DE of MD on the properties of polyphenol microparticles (Bakowska-Barczak and Kolodziejczyk, 2011; Ersus and Yurdagel, 2007; Nayak and Rastogi, 2010a). In encapsulation of black carrot with MDs of varying DE (10, 20–23, and 28–31), the highest anthocyanin content (630 mg anthocyanin/100 g dry matter of powder) was observed using an MD with a DE of 20–23 (Ersus and Yurdagel, 2007). Similarly, in the encapsulation of anthocyanins from kokum fruit extract with MD of various DE (6, 19, 21, and 33), anthocyanin retention increased (65%–86.1%) with increasing DE (6–21). The enhancement of the barrier properties of the microparticles was attributed to a greater proportion of short-chain saccharides in MD with high DE (Nayak and Rastogi, 2010a). On the contrary, Bakowska-Barczak and Kolodziejczyk (2011) did not find a significant effect of DE on the antioxidant activity (ABTS and DPPH) nor on the content of anthocyanins and polyphenols in microencapsulated blackcurrant. However, the drying yield was lower with high DEs. The DE of MD also affects the moisture content and hygroscopicity of polyphenol microparticles (Ersus and Yurdagel, 2007; Nayak and Rastogi, 2010a). Hygroscopicity was found to increase with DE values because MD with higher DE contains short-chain saccharides and more hydroxyl groups and develops stickiness, yielding a slower drying rate and resulting in higher moisture content (Ersus and Yurdagel, 2007; Ferrari et al., 2012; Nayak and Rastogi, 2010a).

Gum arabic is a polymer consisting of D-glucuronic acid, L-rhamnose, D-galactose, and L-arabinose, with approximately 2% protein (Dickinson, 2003). The protein fraction is responsible for emulsification, whereas the filming properties are attributed to the carbohydrate fraction. It has high solubility and low viscosity in aqueous solutions. Gum arabic is usually used for encapsulation purposes; however, its drawbacks are that it is expensive and varies in quality (Gharsallaoui et al., 2007; Kenyon, 1995).

Inulin (In) is a biopolymer less commonly applied in the encapsulation of polyphenols by spray drying. It is a fructooligosaccharide composed of fructose units with β-(1,2) links to glucose at the end of the chain. Interest in inulin is due to its properties: it is a nontoxic, biocompatible, water-soluble, biodegradable, and cheap polymer (Palma et al., 2014; Robert et al., 2012). Inulin was less effective than MD in the encapsulation of anthocyanins and polyphenols from blackcurrant, leading to lower polyphenols content and antioxidant activity in the microparticles (Bakowska-Barczak and Kolodziejczyk, 2011).

Mixtures of different biopolymers are used to shape the properties of the EA. Caseinate-lecithin emulsions produced encapsulated polyphenol extracts with significant retention of antioxidant capacity (DPPH) in GSE, apple extract, and OLE (Korasaju et al., 2008). The addition of gum acacia and tricalcium phosphate to MD with a DE of 21 resulted in high anthocyanin retention (Nayak and Rastogi, 2010a). Villacrez et al. (2014) compared the encapsulation of Andes berry aqueous extracts using nine EAs: MD (DE 20), AG, cornstarch, yucca starch, Capsul® TA, Hi-Cap™100, MD DE(20)/AG (1:1 w/w), MD DE(20)/cornstarch (1:1 w/w), and MD DE20/yucca starch (1:1 w/w). MD, Hi-Cap 100 and MD DE20/yucca starch (1:1 w/w) had the highest significant anthocyanin content (0.54–0.59 mg cy-3-glu equivalents/g of powder). The a_w and moisture (%) ranged between 0.199–0.760 and 3.43–12.56, respectively, and they were below the accepted maximum of 0.6. Silva et al. (2013) studied the encapsulation of jaboticaba (*Myrciaria jaboticaba*) peel extracts with MD (30%), MD (5%) + gum arabic (25%), and MD (5%) + Capsul (25%) at inlet air temperatures of 140°C, 160°C, and 180°C on anthocyanin retention, moisture content, total solids, hygroscopicity, color, and antioxidant activity. The results showed that anthocyanin retention was high for all these systems (over 80%), with higher retention for the MD and GA/MD at 160°C. There was no difference in antioxidant activity among systems. Microparticles with C/MD showed powders with the lowest hygroscopicity, which was explained by a lipophilic component (octenyl succinate) in its structure.

The external morphology of the microparticles can be affected by EA composition, as was reported by Villacrez et al. (2014). Andes berry microparticles with nine EAs showed different levels of dented surfaces because the EAs have varying compositions of low-molecular-weight sugars acting as plasticizers and prevent shrinkage. Krishnaiah (2009b) studied the morphology of mengkudu microparticles with carrageenan at different mengkudu/carrageenan ratios, resulting in abundant spherical and smooth particles. In another study, the polyphenols/EA ratio has been used as an independent variable to examine different powder properties. Krishnaiah (2009a) studied unripe mengkudu pulp extract/carrageenan ratios ranging from 1:2 to 1:8, and found that a ratio of 1:2 at an inlet air temperature of 150°C increases polyphenol and flavonoid content as well as antioxidant capacity, as measured by DPPH.

The efficiency of different EA to microencapsulate a pure polyphenol (vanillin) was studied by Sun-Waterhouse et al. (2013); the following EE order was observed: inulin > methyl β-cyclodextrin (Mβ-CD) > hydroxypropylmethylcellulose (HPMC) > sodium alginate. On the other hand, encapsulated quercetin had similar EE values to those of inulin, Mβ-CD, and HPMC (Sun-Waterhouse et al., 2013). These results were attributed to the lower thermal stability of quercetin compared to vanillin. Lauro et al. (2007) showed that the EE of naringin (rhamnoglucoside) and naringenin was higher with cellulose acetate phthalate (CAP) than with HPMC. In this same study, the EE was higher for naringin (glycosylated form) (78%–80%) than for naringenin (60%–70%) when the EA was CAP; the opposite effect on the EE (naringin [80%] and naringenin [92%]) was reported when the EA was HPMC for a polyphenol/HPMC ratio of 1:3 (Lauro et al., 2007). Both studies showed that the type of polyphenol, especially its structural features, influences the EE when comparing the same EA. However, the magnitude of the EE depends on the biopolymer used as an EA (Lauro et al., 2007; Sun-Waterhouse et al., 2013).

Flores et al. (2014) studied the effect of the extraction solvent (acetonic, ethanolic, and methanolic) on the physicochemical properties of blueberry extract microencapsulated with gum arabic and whey protein isolate. They concluded that anthocyanin retention was affected by both the extraction solvent and the EA (whey protein isolate or gum arabic), whereas polyphenol retention was affected only by the solvent type. In addition, they observed that acetone preferentially extracts malvidin, whereas alcohol extraction concentrates petunidin, delphinidin, cyanidin, and peonidin; however, the retention of each anthocyanin has not been determined. Moreover, alcohol-based solvents showed greater levels of antioxidant activity, as measured by FRAP, for both EAs (Flores et al., 2014).

9.6 PROCESS VARIABLES STUDIED IN THE MICROENCAPSULATION OF POLYPHENOLS BY SPRAY DRYING

9.6.1 Inlet Air Temperature

It has been demonstrated that operating temperatures (inlet, outlet, and ratio of inlet/outlet air temperatures) are important factors in the spray drying of heat-sensitive materials. Fang and Bhandari (2011) reported a high retention rate of polyphenols (96%) and anthocyanins (94%) at an inlet and outlet air temperatures of 15°C and 80°C, respectively, in the encapsulation of bayberry juice. Higher inlet (>160°C – 180°C) and outlet (107°C – 131°C) air temperatures caused large anthocyanin losses in black carrot powder microencapsulated with MD (Ersus and Yurdagel, 2007). Thus, when the outlet air temperature decreases, keeping the inlet air temperature constant, the amounts of active compounds in the powder increase (Ersus and Yurdagel, 2007). For blueberry extract microencapsulated with mesquite gum, Jimenez-Aguilar et al. (2011) observed high losses of phenolic compounds (26%) and anthocyanins (24%) with an inlet air temperature of 160°C, while at 140°C the losses were moderate (4% and 12%, respectively).

High inlet air temperatures affect some properties of the powder, such as moisture content, hygroscopicity, and T_g. In blueberry (Jiménez-Aguilar et al., 2011) and blackcurrant (Bakowska-Barczak and Kolodziejczyk, 2011) extracts, the higher the inlet air temperature, the lower the moisture content of the powder. The resulting large temperature differences between the surrounding air temperature and the droplet surface enhance the heat transfer rate, favoring the evaporation of water. Moisture acts as a plasticizer in materials and affects the T_g (Fang and Bhandari, 2011). Decreasing MD DE (higher molecular weight) leads to higher T_gs of the powder. Moreover, higher inlet air temperatures improve the drying yield of unripe mengkudu pulp extract (Krishnaiah, 2009a). In this study, the highest antioxidant capacity (DPPH) was observed at 130°C (Krishnaiah, 2009a) for inlet air temperatures between 110°C and 150°C.

At low and high inlet air temperatures, the shrinkage of the particles during the drying process can yield powders with indented surfaces, as revealed by scanning electron microscopy (SEM). At low inlet air temperatures, water diffusion is slow; hence, the particles have more time to shrink. At high inlet air temperatures, rapid evaporation and high pressure inside the particles also produce shrinkage (Alamilla-Beltran et al., 2005). SEM microphotography has shown that the degree of shrinkage depends on the type of EA, although it is independent of the type of encapsulated polyphenol (Robert et al., 2012; Sun-Waterhouse et al., 2013).

9.7 STABILITY OF MICROPARTICLES DURING STORAGE

It is known that the encapsulation of polyphenols from plant extracts greatly improves their stability. The main storage conditions that affect the stability of encapsulated polyphenols are light, temperature, oxygen, and relative humidity (RH). This latter condition is considered a critical factor for dehydrated products. Polyphenols' degradation in microparticles usually follows first-order degradation kinetics. The degradation rate constant is obtained in a graph from the slope of the ln(C/Co) versus time (Jiménez-Aguilar et al., 2011; Nayak and Rastogi, 2010a; Osorio et al., 2010).

Fang and Bhandari (2011) studied the stability of individual phenolics (gallic acid, cyanidin 3-glucoside, quercetin-3-galactoside, quercetin-3-glucoside, and quercetin deoxyhexoside) using HPLC-DAD-ESIMS after 6 months of storage. Just after spray drying, the five phenolic compounds were found in the bayberry powders with high retention (93%–97%). Nevertheless, after 6 months, the phenolic contents declined at different rates, depending on the storage conditions. Cyanidin 3-glucoside was not detected when the powder was stored at 40°C for 6 months, suggesting that cyanidin 3-glucoside in bayberry powder is less stable than gallic acid and flavonols under the study conditions.

The properties of the EAs (i.e., film capacity, oxygen permeability, and hygroscopicity) and the plant extract components (i.e., sugar content and mucilage) play important roles in the stability of encapsulated polyphenols. Bakowska-Barczak and Kolodziejczyk (2011) studied the impact of EA on the stability of blackcurrant encapsulated with different MDs (DE 11, DE 18, and DE 21) and inulin during storage at 8°C and 25°C. Anthocyanins were significantly more stable in inulin, and no effect of the DE of the MD was found. MD showed the best anthocyanin protective effect in pomegranate juice encapsulated with MD (DE 12–20) and soybean protein isolate stored at 60°C (Robert et al., 2010).

A high water absorption capacity of the microparticles could increase the water available for hydrolysis reactions. It has been reported that the glass transition temperature of an amorphous powder determines the stability of polyphenols during storage. If the storage temperature is less than T_g, the powder exists in a glass state, molecular mobility decreases, and, therefore, degradation of active compounds decreases. This glass state produces stable powders during long periods of storage.

In a study by Fang and Bhandari (2011), bayberry juice encapsulated with MD was stored at different temperatures (5°C, 25°C, and 40°C) and a_w (0.11–0.44) for 6 months. They reported that, at the same storage temperature, higher a_w led to a decrease in the T_g and thus to higher losses of polyphenols and anthocyanins. The water present in the powder will act like a plasticizer, leading to a rubbery state wherein physicochemical changes, such as caking, collapse, agglomeration, browning, and oxidation, could occur. Moreover, low-molecular-weight biopolymers, such as saccharides, have been reported to cause permeability and allow oxygen diffusion into the microparticles, favoring polyphenol oxidation reactions. No effect of the DE of MD was reported for microencapsulated black carrot during the storage period, although the hygroscopicity of the powders increased with increasing DE values of MD (Ersus and Yourdagel, 2007).

Villacrez et al. (2014) studied the storage of Andes berry microparticles in MD (DE 20) and Hi-Cap at temperatures of 4°C and 18°C and RH of 75% and 95% over 5 days because the powders lost their fluidity after this time. Losses of anthocyanin reached 1.5% and 4.4% at RH of 75% and 95%, respectively, for both storage temperatures. The control sample was stored at 18°C and 60% RH for 3 months, showing a constant anthocyanin content.

Osorio et al. (2010) reported the effects of ethanolic/citric acid extraction (osmotic of dehydration [OC] and soxhlet extraction [SC]) on anthocyanin content using thermal analysis with a thermogravimetric analyzer (TGA) and differential scanning calorimeter as well as on the storage stability of corozo (*Bactris guineensis*) fruit extracts microencapsulated with MD (DE 19–20). The results suggest that the MSC was more thermally stable, based on the TGA and lower degradation rate constants during storage at different humidity (76% and 96%) and temperature (18°C, 50°C, and 70°C) conditions.

9.8 SPRAY DRYING OF POLYPHENOLS

Spray drying for several polyphenol-rich species has been used as a drying technology without the inclusion of an encapsulation process, for example, in drying açai (Tonon et al., 2010), blackberry (Ferrari et al., 2012), bayberry (Fang and Bhandari, 2012), and black mulberry (Fazaeli et al., 2012). The focus of these studies was to increase the shelf life of polyphenols in powder formulations. Biopolymers were used as carriers, and MD was the most commonly used (Fang and Bhandari, 2012; Fazaeli et al., 2012; Ferrari et al., 2012; Tonon et al., 2010). Other carriers used were whey protein isolates (Fang and Bhandari, 2012) and gum arabic (Fazaeli et al., 2012). The carrier concentration is a relevant factor in spray-drying polyphenols. High carrier concentrations reduce the moisture content and hygroscopicity of the powders (Fazaeli et al., 2012; Ferrari et al., 2012). In addition, since less water is evaporated, drying costs are lower.

9.9 FUTURE TRENDS

Existing studies on the encapsulation of polyphenols mainly cover research in basic science. More work is required to transfer this knowledge into practical rules to guide the design of effective and efficient commercial scale polyphenol encapsulation processes. The global market offers a good opportunity for polyphenol microparticles as food ingredients. The global market for food encapsulation is estimated to reach US$ 42 billion by 2017, growing at an annual rate of 6.4% between 2012 and 2017. The United States (44%) and Europe (34%) are the main actors in the world market for the encapsulation of food ingredients. In Latin America, Brazil is the only participant in this market, with annual growth rates of approximately 7% (Robert et al., 2013). Hence, encapsulation could become a valuable tool to add value and provide competitive advantages to food manufacturers. The increasing international regulation in the use of synthetic colorants and the growing awareness of informed consumers represent a good opportunity for natural food colorants based on polyphenol pigments. In addition, the demand for novel healthy ingredients will push the food industry to design and develop polyphenol microparticles with controlled release in the gastrointestinal tract. In these applications, microencapsulation is a key technology to obtain stable products.

ACKNOWLEDGMENT

The authors acknowledge FONDECYT Project Grant No. 3150342, CYTED 415RT0495 and ANILLO ACT 1105 (Conicyt–Chile).

REFERENCES

Araya, H., Clavijo, C., and Herrera, C. 2006. Capacidad antioxidante de frutas y verduras cultivadas en Chile. *Archivos Latinoamericanos de Nutrición* 56(4):361–365.

Alamilla-Beltrán, L., Chanona-Pérez, J.J., Jimenez-Aparicio, A.R., and Gutierrez-López, G.F. 2005. Description of morphological changes of particles along spray drying. *Journal of Food Engineering* 67:179–184.

Bakowska-Barczak, A.M. and Kolodziejczyk, P.P. 2011. Black currant polyphenols: Their storage stability and microencapsulation. *Industrial Crops and Products* 34:1301–1309.

Bao, J., Cai, Y., Sun, M., Wang, G., and Corke, H. 2005. Anthocyanins, flavonols, and free radical scavenging activity of Chinese bayberry (*Myrica rubra*) extracts and their color properties and stability. *Journal of Agricultural and Food Chemistry* 53:2327–2332.

Baum, M., Schantz, M., Leick, S. et al. 2014. Is the antioxidative effectiveness of a bilberry extract influenced by encapsulation? *Journal of the Science of Food and Agriculture* 94:2301–2307.

Belščak-Cvitanović, A., Stojanović, R., Manojlović, V., Komes, D., Juranović Cindrić, I., Nedović, V., and Bugarski, B. 2011. Encapsulation of polyphenolic antioxidants from medicinal plant extracts in alginate–chitosan system enhanced with ascorbic acid by electrostatic extrusion. *Food Research International* 44:1094–1101.

Betz, M., Steiner, B., Schantz, M. et al. 2012. Antioxidant capacity of bilberry extract microencapsulated in whey protein hydrogels. *Food Research International* 47:51–57.

Bouaziz, M., Dhouib, A., Loukil, S., Boukhris, M., and Sayadi, S. 2009. Polyphenols content, antioxidant and antimicrobial activities of extracts of some wild plants collected from the south of Tunisia. *African Journal of Biotechnology* 8:7017–7027.

Cale, J.M., Li, S.H., Warnock, M., Su, E.J., North, P.R., Sanders, K.L., Puscau, M.M., Emal, C.D., and Lawrence, D.A. 2010. Characterization of a novel class of polyphenolic inhibitors of plasminogen activator inhibitor-1. *Journal of Biological Chemistry* 285:7892–7902.

Çam, M., Cihat Içyer, N., and Erdogan, F. 2014. Pomegranate peel phenolics: Microencapsulation, storage stability and potential ingredient for functional food development. *LWT: Food Science and Technology* 55:117–123.

Cassinese, C., de Combarieu, E., Falzoni, M., Fuzzati, N., Pace, R., and Sardone, N. 2007. New liquid chromatography method with ultraviolet detection for analysis of anthocyanins and anthocyanidins in *Vaccinium myrtillus* fruit dry extracts and commercial preparations. *Journal of AOAC International* 90:911–920.

Champagne, C.P. and Fustier, P. 2007. Microencapsulation for the improved delivery of bioactive compounds into foods. *Current Opinion in Biotechnology* 18:184–190.

Chiou, D. and Langrish, T.A.G. 2007. Development and characterisation of novel nutraceuticals with spray drying technology. *Journal of Food Engineering* 82:84–91.

Clifford, M.N. 2000. Review anthocyanins—Nature, occurrence and dietary burden. *Journal of the Science of Food and Agriculture* 80:1063–72.

Cuevas Montilla, E., Rodriguez Arzaba, M., Hillebrand, S., and Winterhalter, P. 2011. Anthocyanin composition of black carrot (*Daucus carota* ssp. sativus var. atrorubens Alef.) cultivars Antonina, beta sweet, deep purple, and purple haze. *Journal of Agricultural and Food Chemistry* 59:3385–3390.

Deladino, L., Teixeira, A.S., Navarro, A.S., Alvarez, I., and Molina-García, A.D., and Martino, M. 2015. Corn starch systems as carriers for yerba mate (*Ilex paraguariensis*) antioxidants. *Food and Bioproducts Processing* 94:463–472.

Desai, K.G.H. and Jin Park, H. 2005. Recent developments in microencapsulation of food ingredients. *Dry Technology* 23:1361–1394.

Devesa-Rey, R., Vecino, X., Varela-Alende, J.L., Barral, M.T., Cruz, J.M., and Moldes, A.B. 2011. Valorization of winery waste vs. the costs of not recycling. *Waste Management* 31:2327–2335.

Dickinson, E. 2003. Hydrocolloids at interfaces and the influence on the properties of dispersed systems. Food Hydrocolloids 17:25–39.

El Sedef, N. and Karakaya, S. 2009. Olive tree (*Olea europaea*) leaves: Potential beneficial effects on human health. *Nutrition Reviews* 67:632–638.

Ersus, S. and Yurdagel, U. 2007. Microencapsulation of anthocyanin pigments of black carrot (*Daucus carota* L.) by spray drier. *Journal of Food Engineering* 80:805–812.

Fang, Z. and Bhandari, B. 2010. Encapsulation of polyphenols–A review. *Trends in Food Science and Technology* 21:510–523.

Fang, Z. and Bhandari, B. 2011. Effect of spray drying and storage on the stability of bayberry polyphenols. *Food Chemistry* 129:1139–1147.

Fang, Z. and Bhandari, B. 2012. Comparing the efficiency of protein and maltodextrin on spray drying of bayberry juice. *Food Research International* 48:478–483.

Fazaeli, M., Emam-Djomeh, Z., Ashtari, A.K., and Omid, M. 2012. Food and bioproducts processing effect of spray drying conditions and feed composition on the physical properties of black mulberry juice powder. *Food and Bioproducts Processing* 90:667–675.

Ferrari, C.C., Pimentel, S., Germer, M., and de Aguirre, J.M. 2012. Effects of spray-drying conditions on the physicochemical properties of blackberry powder effects of spray-drying conditions on the physicochemical properties of blackberry powder. *Dry Technology* 30:154–163.

Flores, F.P., Singh, R.K., Kerr, W.L., Pegg, R.B., and Kong, F. 2014. Total phenolics content and antioxidant capacities of microencapsulated blueberry anthocyanins during in vitro digestion. *Food Chemistry* 153:272–278.

Frank, K., Walz, E., Gräf, V., Greiner, R., Köhler, K., and Schuchmann, H.P. 2012. Stability of anthocyanin-rich w/o/w-emulsions designed for intestinal release in gastrointestinal environment. *Journal of Food Science* 77:N50–N57.

Frankel, E., Bakhouche, A., Lozano-Sánnchez, J., Segura-Carretero, A., and Fernández-Gutiérrrez, A. 2013. Literature review on production process to obtain extra virgin olive oil enriched in bioactive compounds. Potential use of by-products as alternative sources of polyphenols. *Journal of Agricultural and Food Chemistry* 61:5179–5188.

Fredes, C., Montenegro, G., Zoffoli, J., Santander, F., and Robert, P. 2014. Comparison of total phenolic, total anthocyanin and antioxidant activity of polyphenol-rich fruits grown in Chile. *Ciencia e investigación agraria* 41:49–60.

Gabetta, B., Fuzzati, N., Griffini, A., Lolla, E., Pace, R., Ruffilli, T., and Peterlongo, F. 2000. Characterization of proanthocyanidins from grape seeds. *Fitoterapia* 71:162–175.

García-Lomillo, J., González-San Jose, M.L., Del Pino-García, R., Rivero-Pérez, M.D., and Muñiz-Rodríguez, P. 2014. Antioxidant and antimicrobial properties of wine by-products and their potential uses in the food industry. *Journal of Agricultural and Food Chemistry* 62:12595–12602.

Gavini, E., Alamanni, M.C., Cossu, M., and Giunchedi, P. 2005. Tabletted microspheres containing *Cynara scolymus* (var. Spinoso sardo) extract for the preparation of controlled release nutraceutical matrices. *Journal of Microencapsulation* 22(5):487–499.

Gharsallaoui, A., Roudaut, G., Chanbin, O., Voilley, A., and Saurel, R. 2007. Applications of spray-drying in microencapsulation of food ingredients: An overview. *Food Research International* 40:1107–1121.

Gibbs, B., Kermasha, S., Alli, I., and Mulligan, C. 1999. Encapsulation in the food industry: A review. *International Journal of Food Science and Nutrition* 50:213.224.

Gouin, S. 2004. Microencapsulation: Industrial appraisal of existing technologies and trends. *Trends in Food Science and Technology* 15:330–347.

Gurak, P.D., Correa Cabral, L.M., and Rocha-Leão, M.H. 2013. Production of grape juice powder obtained by freeze-drying after concentration by reverse osmosis. *Brazilian Archives of Biology and Technology* 56:1011–1017.

Halvorsen, B.L., Holte, K., Myhrstad, M.C.W. et al. 2002. A Systematic screening of total antioxidants in dietary plants. *Journal of Nutrition* 132:461–471.

Huang, Z., Wang, B., Williams, P., and Pace, R.D. 2009. Identification of anthocyanins in muscadine grapes with HPLC-ESI-MS. *LWT: Food Science and Technology* 12, 819–824.

Idham, Z., Muhamad, I.I., and Sarmidi, M.R. 2012. Degradation kinetics and color stability of spray-dried encapsulated anthocyanins from *Hibiscus sabdariffa* L. *Journal of Food Process Engineering* 35(4):522–542.

Janicsak, G., Zupko, I., Mathe, I., and Hohmann, J. 2010. Comparative study of the antioxidant activities of eleven salvia species. *Natural Product Communications* 5:227–230.

Jiménez, P., Masson, L., Barriga, A., Chávez, J., and Robert, P. 2011. Oxidative stability of oils containing olive leaf extracts obtained by pressure, supercritical and solvent extraction. *European Journal of Lipid Science and Technology* 113:497–505.

Jiménez-Aguilar, D.M., Ortega-Regules, A.E., Lozada-Ramírez, J.D., Pérez-Pérez, M.C.I., Vernon-Carter, E.J., and Welti-Chanes, J. 2011. Color and chemical stability of spray-dried blueberry extract using mesquite gum as wall material. *Journal of Food Composition and Analysis* 24:889–894.

Kandansamy, K. and Somasundaram, P.D. 2012. Microencapsulation of colors by spray drying—A review. *International Journal of Food Engineering* 8:1–15.

Kenyon, M. Modified starch, maltodextrin and corn syrup solids as wall material for food encapsulation. 1995. In *Encapsulation and Controlled Release of Food Ingredients*, eds. S. Risch and G. Reineccius, pp. 42–50. Washington, DC: American Chemical Society.

Koca, I. and Karadeniz, B. 2009. Antioxidant properties of blackberry and blueberry fruits grown in the Black Sea Region of Turkey. *Scientia Horticulturae* 121:447–450.

Kosaraju, S.L., D'ath, L., and Lawrence, A. 2006. Preparation and characterisation of chitosan microspheres for antioxidant delivery. *Carbohydrate Polymers* 64:163–167.

Kosaraju, S.L., Labbett, D., Emin, M., Konczak, I., and Lundin, L. 2008. Delivering polyphenols for healthy ageing. *Nutrition and Dietetics* 65:S48–S52.

Krishnaiah, D. 2009a. Study on retention of bioactive components of *Morinda citrifolia* L. using spray-drying. *Journal of Applied Sciences* 9:3092–3097.

Krishnaiah, D. 2009b. Optimal operating conditions of spray dried noni fruit extract using k-Carrageenan as adjuvant. *Journal of Applied Sciences* 9:3062–3067.

Krishnaiah, D., Sarbatly, R., and Nithyanandam, R. 2012. Microencapsulation of *Morinda citrifolia* L. extract by spray-drying. *Chemical Engineering Research and Design* 90:622–632.

Kropat, C., Betz, M., Kulozik, U. et al. 2013. Effect of microformulation on the bioactivity of an anthocyanin-rich bilberry pomace extract (*Vaccinium myrtillus* L.) in vitro. *Journal of Agricultural and Food Chemistry* 61:4873–4881.

Laokuldilok, T. and Kanha, N. 2015. Effects of processing conditions on powder properties of black glutinous rice (*Oryza sativa* L.) bran anthocyanins produced by spray drying and freeze drying. *LWT: Food Science and Technology* 64:405–411.

Lauro, M.R., De Simone, F., Sansone, F., Iannelli, P., and Aquino, R.P. 2007. Preparations and release characteristics of naringin and naringenin gastro-resistant microparticles by spray-drying. *Journal of Drug Delivery Science and Technology* 17:119–124.

Lee, J., Durst, R.W., and Wrolstad, R.E. 2005. Determination of total monomeric anthocyanin pigment content of fruit juices, beverages, natural colorants, and wines by the pH differential method: Collaborative study. *Journal of AOAC International* 88:1269–1278.

Leite-Legatti, A.V., Batista, Â.G., Vicente Dragano, N.R. et al. 2012. Jaboticaba peel: Antioxidant compounds, antiproliferative and antimutagenic activities. *Food Research International* 49:596–603.

Lohachoompol, V.M., Mulholland, G., and Srzednicki, C.J. 2008. Determination of anthocyanins in various cultivars of highbush and rabbiteye blueberries. *Food Chemistry* 111:249–254.

Mahdavi, S.A., Jafari, S.M., Ghorbani, M., and Assadpoor, E. 2014. Spray-drying microencapsulation of anthocyanins by natural biopolymers: A review. *Dry Technology* 32:509–18.

Manach, C., Scalbert, A., Morand, C., Rémésy, C., and Jiménez, L. 2004. Polyphenols: Food sources and bioavailability. *American Journal of Clinical Nutrition* 79:727–747.

Manhita, A.C., Teixeira, D.M., and da Costa, C.T. 2006. Application of sample disruption methods in the extraction of anthocyanins from solid or semi-solid vegetable samples. *Journal of Chromatography A* 1129:14–20.

Masters, K. 1972. *Spray Drying: An Introduction to Principles, Operational Practice and Applications*. London, UK: Leonard Hill Books.

Meksi, N., Haddar, W., Hammami, S., and Mhenni, M.F. 2012. Olive mill wastewater: A potential source of natural dyes for textile dyeing. *Industrial Crops and Products* 40:103–109.

Montes, M. and Wilkomirsky, T. 1987. *Medicina tradicional chilena*. Concepción, Chile: Editorial de la Universidad de Concepción.

Mourtzinos, I., Salta, F., Yannakopoulou, K., Chiou, A., and Karathanos, V.T. 2007. Encapsulation of olive leaf extract in β-cyclodextrin. *Journal of Agricultural and Food Chemistry* 55:8088–8094.

Mylonaki, S., Kiassos, E., Makris, D.P., and Kefalas, P. 2008. Optimisation of the extraction of olive (*Olea europaea*) leaf phenolics using water/ethanol-based solvent systems and response surface methodology. *Analytical and Bioanalytical Chemistry* 392:977–985.

Nasini, L., Gigliotti, G., Balduccini, M.A., Federici, E., Cenci, G., and Proietti, P. 2012. Effect of solid olive-mill waste amendment on soil fertility and olive (*Olea europaea* L.) tree activity. *Agriculture, Ecosystems and Environment* 164:292–297.

Nayak, C.A. and Rastogi, N.K. 2010a. Effect of selected additives on microencapsulation of anthocyanin by spray drying. *Dry Technology* 28:1396–1404.

Nayak, C.A., Srinivas, P., and Rastogi, N.K. 2010b. Characterisation of anthocyanins from *Garcinia indica* Choisy. *Food Chemistry* 118:719–724.

Nazzaro, F., Orlando, P., Fratianni, F., and Coppola, R. 2012. Microencapsulation in food science and biotechnology. *Current Opinion in Biotechnology* 23:182–186.

Oidtmann, J., Schantz, M., Mäder, K. et al. 2012. Preparation and comparative release characteristics of three anthocyanin encapsulation systems. *Journal of Agricultural and Food Chemistry* 60:844–51.

Osorio, C., Acevedo, B., Hillebrand, S., Carriazo, J., Winterhalter, P., and Morales, A.L. 2010. Microencapsulation by spray-drying of anthocyanin pigments from corozo (*Bactris guineensis*) fruit. *Journal of Agricultural and Food Chemistry* 58:6977–6985.

Osorio, C., Hurtado, N., Dawid, C., Hofmann, T., Heredia-Mira, F.J., and Morales, A.L. 2012. Chemical characterisation of anthocyanins in tamarillo (*Solanum betaceum* Cav.) and Andes berry (*Rubus glaucus* Benth.) fruits. *Food Chemistry* 132(4), 1915–1921.

Palma, M., García, P., Márquez-Ruiz, G., Vergara, C., and Robert, P. 2014. Release kinetics of flavonoids in methyl linoleate from microparticles designed with inulin and channelizing agent. *Food Research International* 64:99–105.

Park, M., Cho, H., Jung, H., Lee, H., and Hwang, K.T. 2014. Antioxidant and anti-inflammatory activities of tannin fraction of the extract from black raspberry seeds compared to grape seeds. *Journal of Food Biochemistry* 38:259–270.

Pasrija, D., Ezhilarasi, P.N., Indrani, D., and Anandharamakrishnan, C. 2015. Microencapsulation of green tea polyphenols and its effect on incorporated bread quality. *LWT: Food Science and Technology* 64:289–296.

Porter, L.J., Hrstich, L.N., and Chan, B.G. 1986. The conversion of procyanidins and prodelphinidins to cyanidin and delphinidin. *Phytochemistry* 25:223–230.

Radunic, M., Jukic Špika, M., Goreta Ban, S., Gadze, J., Díaz-Pérez, J.C., and MacLean, D. 2015. Physical and chemical properties of pomegranate fruit accessions from Croatia. *Food Chemistry* 177:53–60.

Robert, P. and Fredes, C. 2015. The encapsulation of anthocyanins from berry-type fruits: Trends in foods. *Molecules* 20:5875–5888.

Robert, P., García, P., and Palma, M. 2013. Microencapsulación en la industria alimentaria. *Indualimentos* 82:58–62.

Robert, P., García, P., Reyes, N., Chávez, J., and Santos, J. 2012. Acetylated starch and inulin as encapsulating agents of gallic acid and their release behaviour in a hydrophilic system. *Food Chemistry* 134:1–8.

Robert, P., Gorena, T., Romero, N., Sepúlveda, E., Chávez, J., and Sáenz, C. 2010. Encapsulation of polyphenols and anthocyanins from pomegranate (*Punica granatum*) by spray drying. *International Journal of Food Science and Technology* 45:1386–1394.

Rocha-Guzmán, N.E., Gallegos-Infante, J.A., González-Laredo, R.F., Harte, F., Medina-Torres, L., Ochoa-Martínez, L.A., and Soto-García, M. 2010. Effect of high-pressure homogenization on the physical and antioxidant properties of *Quercus resinosa* infusions encapsulated by spray-drying. *Journal of Food Science* 75:N57–N61.

Santos, D.T., Albarelli, J.Q., Beppu, M.M., and Meireles, M.A. 2013. Stabilization of anthocyanin extract from jabuticaba skins by encapsulation using supercritical CO_2 as solvent. *Food Research International* 50:617–624.

Silva, P.I., Stringheta, P.C., Teófilo, R.F., and de Oliveira, I.R.N. 2013. Parameter optimization for spray-drying microencapsulation of jaboticaba (*Myrciaria jaboticaba*) peel extracts using simultaneous analysis of responses. *Journal of Food Engineering* 117:538–544.

Singleton, V.L. and Rossi, J.R.J.A. 1965. Colorimetry of total phenolics with phosphomolybdic-phosphotungstic acid reagents. *American Journal of Enology and Viticulture* 16:144–158.

Speisky, H., Peña, A., Gómez, M., Fredes, C., Hurtado, M., Gotteland, M., and Brunser, O. 2008. Antioxidants in Chilean berries. *Acta Horticulture* 777:485–492.

Sun-Waterhouse, D., Wadhwa, S.S., and Waterhouse, G.I.N. 2013. Spray-drying microencapsulation of polyphenol bioactives: A comparative study using different natural fibre polymers as encapsulants. *Food and Bioprocess Technology* 6:2376–2388.

Swain, T. and Hillis, W.E. 1959. The phenolic constituents of *Prunus domestica*. I. The quantitative analysis of phenolic constituents. *Journal of the Science of Food and Agriculture* 10:63–68.

Teixeira, A.S., Navarro, A.S., Molina-García, A.D., Martino, M., and Deladino, L. 2015. Corn starch systems as carriers for yerba mate (*Ilex paraguariensis*) antioxidants: Effect of mineral addition. *Food and Bioproducts Processing* 94:39–49.

Tonon, R.V., Brabet, C., and Hubinger, M.D. 2010. Anthocyanin stability and antioxidant activity of spray-dried açai (*Euterpe oleracea* Mart.) juice produced with different carrier agents. *Food Research International* 43:907–914.

Veberic, R., Slatnar, A., Bizjak, J., Stampar, F., and Mikulic-Petkovsek, M. 2015. Anthocyanin composition of different wild and cultivated berry species. *LWT: Food Science and Technology* 60:509–517.

Villacrez, J.L., Carriazo, J.G., and Osorio, C. 2014. Microencapsulation of Andes Berry (*Rubus glaucus* Benth.) aqueous extract by spray drying. *Food and Bioprocess Technology* 7:1445–1456.

Voos, P., Faas, M., Spasojevic, M., and Sikkema, J. 2010. Encapsulation for preservation of functionality and targeted delivery of bioactive food components. *International Dairy Journal* 20:293–302.

Vuong, Q.V., Golding, J.B., Stathopoulos, C.E., Nguyen, M.H., and Roach, P.D. 2011. Optimizing conditions for the extraction of catechins from green tea using hot water. *Journal of Separation Science* 34:3099–3106.

Wang, S.Y., Camp, M.J., and Ehlenfeldt, M.K. 2012a. Antioxidant capacity and α-glucosidase inhibitory activity in peel and flesh of blueberry (*Vaccinium* spp.) cultivars. *Food Chemistry* 132:1759–1768.

Wang, Z., Li, Y., Chen, L., Xin, X., and Yuan, Q. 2013. A study of controlled uptake and release of anthocyanins by oxidized starch microgels. *Journal of Agricultural and Food Chemistry* 61:5880–5887.

Xia, E.Q., Deng, G.F., Guo, Y.J., and Li, H.B. 2010. Biological activities of polyphenols from grapes. *International Journal of Molecular Sciences* 11:622–646.

Yatsu, F.K.J., Borghetti, G.S., and Bassani, V.L. 2011. Technological characterization and stability of *Ilex paraguariensis* St. Hil. Aquifoliaceae (maté) spray-dried powder. *Journal of Medicinal Food* 14:413–419.

Zhang, B., Mao, G., Zheng, D., Zhao, T., Zou, Y., Qu, H., Lia, F., Zhu, B., Yang, L., and Wu, X. 2014. Separation, identification, antioxidant, and anti-tumor activities of *Hibiscus sabdariffa* L. extracts. *Separation Science and Technology* 49:1379–1388.

Zhang, L., Mou, D., and Du, Y. 2007. Procyanidins: Extraction and micro-encapsulation. *Journal of the Science of Food and Agriculture* 87:2192–2197.

Zheng, L., Ding, Z., Zhang, M., and Sun, J. 2011. Microencapsulation of bayberry polyphenols by ethyl cellulose: Preparation and characterization. *Journal of Food Engineering* 104:89–95.

10 Inclusion of Polyphenols into Food Matrices

Verónica Dueik and Pedro Bouchon

CONTENTS

10.1 INTRODUCTION

Polyphenols form a very complex group of molecules present in most plants and are the most abundant antioxidants in the diet. Their total dietary intake could be as high as 1 g/day, which is much higher than that of all other classes of phytochemicals and known dietary antioxidants (Scalbert et al., 2005). Their main dietary sources are fruits and plant-derived beverages such as fruit juices, tea, coffee, and red wine. Vegetables, cereals, chocolate, and dry legumes also contribute to the total polyphenol intake. Phenolic acids comprise the dominant group of polyphenols (75% of total intake) followed by proanthocyanidins (14%) and anthocyanidins and other flavonoids (10%) (Ovaskainen et al., 2008).

Based on current studies, a consensus has been achieved to sustain the hypothesis that the specific intake of foods and beverages containing relatively high concentrations of polyphenols may play a meaningful role in reducing cardiovascular disease (CVD) risk through an improvement in vascular function and a modulation of inflammation. Furthermore, long-term consumption of diets rich in plant polyphenols offers some protection against the development of some types of cancer, diabetes, osteoporosis, and neurodegenerative diseases (Pandey and Rizvi, 2009). This protection relates to the antioxidant power of phenolics due to their redox properties and abundance of OH functional groups, thereby allowing them to act as reducing agents, hydrogen donors, and singlet oxygen quenchers. Phenolic compounds are also known for their strong metal-chelating capabilities (Piljac-Zegarac et al., 2013). Thus, polyphenols and other food phenolics are a subject of increasing scientific interest because of their beneficial effects on human health.

The association between increased consumption of fruits and vegetables and protection against chronic and degenerative diseases has driven interest in phytochemical extracts for applications in novel functional foods and beverages. As a result, polyphenols are being incorporated into many new functional food products (Crowe, 2013). Processed foods, particularly fabricated snacks, are excellent vehicles to develop these food products. Particularly, the increasing interest in the health benefits of tea has led to the inclusion of tea extracts in dietary supplements and functional foods enriched with tea polyphenols (Higdon and Frei, 2003). However, polyphenols are unstable and may be degraded when exposed to high temperatures, oxygen, light, and enzymes, among other factors. These conditions should be taken into account when processing polyphenol-rich food matrices. Regarding snacks processing, the most commonly used unit operations are baking, frying, and extrusion. Frying has been widely used for potatoes and formulated matrices processing (Bouchon et al., 2003; Dueik et al., 2014; Dueik and Bouchon, 2015; Gazmuri and Bouchon, 2009). A novel frying technology has been recently introduced for processing sensitive fruits and vegetables (Dueik and Bouchon, 2011a), which preserves beneficial compounds present in the raw materials and adds them during processing. Losses in polyphenol of up to 65% have been reported when frying, at 170°C, formulated gluten–starch matrices enriched with olive leaf polyphenols (Dueik and Bouchon, 2015). The authors reported that when lowering the frying temperature by 10°C, polyphenol losses were reduced to 57%, and when vacuum frying (106°C, 91.4 kPa), losses were minimized to 29%.

The main ingredient for the fabrication of formulated snacks is wheat flour, which is composed of starch and gluten. Wheat flour, with the addition of water and mixing energy, forms a cohesive and viscoelastic dough, thanks to starch gelatinization and gluten proteins, which confer elasticity and extensibility to the dough (Shewry et al., 2002). During normal thermal processing, starch granules are heated in excess water and they start to undergo irreversible transformations that initiate the gelatinization process. Gelatinization involves the disruption of the crystalline structure of starch, allowing amorphous regions to become more accessible to water and swell. This process is particularly important for conferring textural properties to formulated snacks. The inclusion of polyphenols in such matrices may have an impact on some of the properties of snacks, as they interact with starch, facilitating their hydration, while their interaction with proteins may modify their functionality.

Unfortunately, the amount of bioaccessible and bioavailable food polyphenols may differ quantitatively and qualitatively from that of polyphenols included in food databases. Bioaccessible polyphenols are those released from the food matrix by the action of digestive enzymes (small intestine). Potentially bioavailable polyphenols are accessible in the gut (large intestine) due to the action of the bacterial microflora. Food microstructure plays a critical role in the bioaccessibility and bioavailability of bioactive compounds, as their release from the food matrix is controlled by possible interactions with other food components, the presence of suppressors or cofactors, and the formation of stable compounds that are metabolized slowly. The concept of a "food matrix" points to the fact that nutrients are contained in a large continuous medium that may be of cellular origin (in fruits and vegetables) or a microstructure produced by processing, where they may interact at different length scales with the components and structures of the medium (Parada and Aguilera, 2007). There is ample evidence that the physical state of the food matrix plays a key role in the release, mass transfer, accessibility, and biochemical stability of many food components (Aguilera, 2005). Antioxidants are often located in cellular compartments when present in matrices structured by nature or within assemblies produced during processing. In either case, they need to be released during digestion so that they can be absorbed in the gut (Parada and Aguilera, 2007).

10.2 FOOD PRODUCT DESIGN PRINCIPLES

The creation of new food (micro) structures that respond to new consumer demands is the target of many food product developers. In fact, food manufacturing is facing new challenges, such as providing products that fulfill new demands of very well-informed consumers (Aguilera, 2006). Consumers not only ask for products that contribute to their wellness and health but also require specific textures, flavors, and colors, among other parameters, and certainly a consistent food. Consumers expect minimal variations in food products from batch to batch. These demands are redirecting the focus of the food industry and food research ("fork to farm" approach), imposing a challenge to the food industry to develop new products or modify existing ones to meet their needs. This *product-driven process engineering era* requires building controlled right structures, and therefore understanding the functionality of the structural components formed prior to or during processing. In light of this approach, product formulation appears as a good option to develop new products with controlled attributes. Formulated products (also known as fabricated products) are gaining importance in the snack industry as a good alternative for the use of raw natural materials. Formulated products offer the advantages of reproducibility, uniformity, and lack of defects (Gebhardt, 1996), in contrast to raw vegetables, whose heterogeneity can cause major variations in final products (Baumann and Escher, 1995).

The development of formulated products requires specific knowledge about ingredients' functionality, assemblage, effects of processing conditions, and microstructural aspects to design the right structures (Gazmuri and Bouchon, 2009). Most food components that contribute to food microstructure are below the 100 μm range. Some of these microstructural elements include gas bubbles, oil droplets, hydrated starch granules, debonded cells, fibers, gel and crystal networks, particulate solids, crystals of many types, and assemblies of colloidal nature, among others (Aguilera, 2006).

An important amount of formulated products is wheat flour based. The main components or "building blocks" of wheat flour are wheat gluten and starch; the dough is formed by the addition of water and mechanical input. Wheat popularity is largely determined by the ability of wheat flour to be processed into different foods, which is mainly given by the unique properties of wheat gluten proteins (Anjum et al., 2007). Wheat gluten's unique viscoelastic properties improve dough strength, mixing tolerance, and handling properties. Its film-forming ability provides gas retention and controlled expansion for improved volume, uniformity, and texture, whereas its thermosetting properties contribute to structural rigidity and its water absorption capacity improves baked product yield, softness, and shelf life (Day et al., 2006). Starch occurs as highly organized structures, known as starch granules. Starch has unique thermal properties and functionality, which have permitted its wide use in food products and industrial applications (Ratnayake and Jackson, 2005). In the presence of adequate temperature and water, its crystalline structure is disrupted, allowing the amorphous regions to become more accessible to absorb water and swell, a process known as starch gelatinization (Ovalle et al., 2013). Water is a critical constituent of wheat flour dough that should be present in the right amount since the different flour components compete for it. In addition to wheat flour–based dough, potato flakes are also a major ingredient in several snack matrices. Applications extend to extruded products and also to laminated ones. In addition to potato flakes, potato granules may also be used. They can be employed to give a higher resistance to the final product during processing, which might be of interest in new processing such as low pressure–temperature processing.

Several changes take place during snack manufacturing, some of them occurring at the micrometric level and affecting the final microstructure of the food. Consequently, the design of new snack products requires understanding the properties of starch, since starchy foods represent an important product category. This includes the understanding of factors that may affect its gelatinization temperature as well as its microstructural changes during processing and storage, to better understand the relationship between macro- (finished product) and microproperties. In order to understand the link between formulation and food properties after processing, the microstructural approach is vital. In order to get a reliable characterization of the food structure, the exclusion of artifacts during sample preparation is of paramount importance. This can be achieved by using modern techniques such as confocal laser-scanning microscopy, variable pressure scanning electron microscopy, and laser-scanning microscopy, among others. Recently, Ovalle et al. (2013) used an online *in situ* vacuum hot-stage microscopy to study changes in starch microstructural components during heating under vacuum and atmospheric conditions. The procedure allowed for following structural changes in real time, using a noninvasive, well-controlled tool to associate microstructural changes to specific processing conditions.

Major snack manufacturing technologies are baking, extrusion, and frying. All of these normally include food exposure to high temperatures (above 100°C) and the presence of oxygen, processing conditions that may induce the degradation of nutritional compounds, oxidation of beneficial compounds, or even worse, formation of toxic compounds (Dueik and Bouchon, 2011a; Kubow, 1992). Among the toxic compounds, acrylamide is proven to be carcinogenic in animals and a probable human carcinogen, produced when heating starchy foods to temperatures higher than 120°C in a reaction known as the Maillard reaction. This reaction also produces brown compounds, affecting the overall quality of formulated products, as well as the generation of toxic compounds. Researchers have reported several strategies for reducing acrylamide formation during food processing, particularly in potato chips and fries. These strategies include asparaginase addition, showing reductions of up to 30% (Pedreschi et al., 2008); blanching, which has been reported to reduce acrylamide considerably in potato chips (Pedreschi et al., 2004); and vacuum frying, which is a novel frying technology able to reduce acrylamide content in potato chips by 94% when compared with their atmospheric counterpart (Granda et al., 2004). The addition of polyphenols has been reported to reduce acrylamide formation in several formulated products. Zhang and Zhang (2007) added tea polyphenol extracts from bamboo leaves and green tea to fried bread sticks. They found that the addition of 1 g of bamboo leaves extract or green tea extract to 1 kg of flour could reduce acrylamide formation during frying up to 82.9% and 72.5%, respectively, with no sensory changes when compared with the traditional fried product. Using a similar approach, Cheng et al. (2009) studied the inhibitory effect of apple extracts in acrylamide formation in model systems. The authors observed that the proanthocyanidin-rich fraction of apple extracts played a key role in mediating the inhibitory activity, which was finally corroborated in fried potato chips. Consequently, the addition of polyphenols to food matrices not only represents a health benefit for the consumer but may also provide technological benefits and reduce the formation of toxic compounds.

10.3 STABILITY OF POLYPHENOLS DURING FOOD PROCESSING

In recent years, increasing attention has been paid to the role of diet in human health. In addition to the nutritional and sensory properties of foods, their protective role has been recently recognized (Nicoli et al., 1999). Polyphenols are abundant micronutrients in our diet and alleged to play several roles in the prevention of degenerative diseases, and thus, they have been associated with the health benefits derived from consuming high levels of fruits, vegetables, and wine, mainly as antioxidants (Balasundram et al., 2006).

However, the health-promoting capacity of fruits and vegetables strictly depends on their processing history. This aspect has been generally scarcely considered in present nutritional and epidemiological studies. Processing affects content, activity, bioaccessibility, and bioavailability of bioactive compounds (Nicoli et al., 1999). For example, simple peeling of fruits and vegetables can eliminate a significant portion of polyphenols because these substances are often present in higher concentrations in the skin than in the pulp (Manach et al., 2005). Moreover, foods have natural enzymes that can degrade natural polyphenols (e.g., polyphenoloxidase, peroxidase, glycosidases, and esterases). These enzymes are released when plant cells are broken. Glycosidase and esterase enzymes catalyze polyphenol degradation during hydrolysis reactions. Oxidation reactions caused by polyphenoloxidase and peroxidase enzymes tend to produce black pigments, thereby having negative effects on the visual aspect of fresh vegetables and fruits (Shoji, 2007). Even though these enzymes can be effectively deactivated during food processing at high temperatures, nonenzymatic pathways can also cause polyphenol degradation.

Snack categories that are showing the greatest growth rates are those that offer a wide range of alternatives and coincide with convenience, sensory, and health trends. In relation to health, salty snack products that are low in fat, calories, carbohydrates, and sodium and rich in fiber and vitamins are becoming of great interest, as they are organic and offer some health-promoting benefit (Dueik and Bouchon, 2011a). However, it is of paramount importance to predict the effect that processing conditions during snack production may have on the health-promoting components; these components may be those that are added to foods or are found naturally in foods such as polyphenols. The most important conditions that affect these compounds during food processing are temperature and the presence of oxygen. We will refer briefly to the effect of different snack manufacturing technologies in the stability of polyphenols.

10.3.1 Baking

Baking is a complex process that brings about a series of physical, chemical, and biochemical changes in a product such as volume expansion, water evaporation, formation of a porous structure, denaturation of protein, gelatinization of starch, crust formation, and a browning reaction (Sablani et al., 1998).

Rodriguez-Mateos et al. (2013) studied the degradation of polyphenols in blueberries added to muffins; they found that anthocyanin levels decreased 42% after baking. Sun-Waterhouse et al. (2011) prepared breads in the absence or presence of a high methoxyl (HM) apple pectin and apple polyphenol extracts and subjected them to texture measurements (crumb firmness) and chemical analyses. The results showed that adding apple polyphenol extract and pectin to bread modified bread firmness, specific volume, recovery of added polyphenols, and amount of free thiol groups of gluten proteins. The recovery of added polyphenols was the lowest for the apple pectin bread, which had the highest bread firmness. High-performance liquid chromatography–mass spectrometry (HPLC-MS) polyphenol profiling revealed that baking caused oxidation of some added polyphenols such as quercetin and myricetin. The added pectin and polyphenols, and possibly their oxidation products, affected the interactions between water and bread components (such as gluten proteins and starch) during dough development and bread baking, thereby causing differences in bread microstructure and textural properties, which ultimately influenced the extractability of added polyphenols.

The addition of fruit processing by-products for improving the antioxidant capacity of foods is a new trend to reduce waste and the severe loss of valuable substances, for example, dietary fiber and polyphenols, among others (Altan et al., 2009). Apple skins are by-products of apple processing and a rich source of dietary fiber and phenolics. When adding apple skin powder to a muffin model, the retention of polyphenols was 55.9% for chlorogenic acid, 24.8% for anthocyanins, and 55%–65%

for flavonols. When a green tea extract was incorporated into the bread dough, the recovery of catechins from freshly baked bread was about 84% (Wang and Zhou, 2004). These catechin losses could be due to the combined effect of oxidation, isomerization, and degradation that occur during the various stages of bread making. In addition, the interaction of polyphenols with wheat proteins and other components of the dough via hydrogen bonding can also cause polyphenol losses (Wang and Zhou, 2004); these interactions will be discussed further in this chapter.

10.3.2 Frying

Atmospheric deep-fat frying can be defined as a process for the cooking of foods by immersing them in an edible fluid (fat) at a temperature above the boiling point of water under atmospheric conditions. This complex unit operation involves simultaneous heat and mass transfer, resulting in counterflows of water vapor (bubbles) and oil at the surface of the piece (Bouchon et al., 2003). Deep-fat frying may induce several changes in the foodstuff and the oil itself. Degradation of beneficial compounds depends on specific parameters during deep-frying, such as temperature, oxygen, light, moisture content, and obviously length of exposure (Lešková et al., 2006). Among them, high temperature and the presence of oxygen are the most detrimental factors, severely affecting beneficial compounds such as polyphenols of fried foods.

Anthocyanins are one of the most abundant flavonoid constituents of fruits and vegetables, such as black carrots and blue potatoes. The conjugated bonds in their structure are the basis for their red, blue, and purple colors. Anthocyanin pigments easily degrade in the presence of oxygen and high temperatures. This process results in the formation of polyphenolic degradation products, whose antioxidant activity is still under investigation (Patras et al., 2010). In the presence of oxygen, anthocyanins can lose their attractive color due to the development of colorless degradation compounds, but they can also react with chlorogenoquinone (oxidation product of chlorogenic acid) to form brown condensation products, affecting the quality of the final product (Patras et al., 2010). Da Silva and Moreira (2008) found that anthocyanin retention after frying blue potato slices was just 26%.

In addition, the degradation of oil polyphenols during frying has been extensively studied. Particularly, olive oil has been reported to exhibit higher oxidative stability, compared to that of other vegetable oils, which is attributed to the high oleic acid (monounsaturated), low polyunsaturated fatty acid content, and high level of natural phenolic components with antioxidant activity. The main components of the phenolic fraction are hydroxytyrosol and its derivatives. Gomez-Alonso et al. (2003) reported that the concentration of hydroxytyrosol in virgin olive oil decreased rapidly when the oil was repeatedly used for preparing French fries in deep-fat frying operations. At the end of the first frying process (10 min at 180°C), the concentration of the dihydroxyphenol components was reduced to 50%–60% of the original value; after six frying operations, only about 10% of the initial components remained.

Due to the lack of consumers' trust toward synthetic antioxidants in foods, the inclusion of natural extracts to enhance the oxidative stability of fried products has been intensively studied. Polyphenols have been reported to reduce the formation of heterocyclic aromatic amines (HCAs) in fried meats (Balogh et al., 2000;

Wang et al., 1982). In animal models, mostly in rats, HCAs may cause cancer of the mammary gland, colon, pancreas, and prostate. HCAs are formed during frying due to the Maillard reaction occurring at temperatures between 120°C and 250°C. The scavenging effect of antioxidants on the cation radicals participating in the formation of HCAs has been verified by electron spin resonance spectroscopy (Kikugawa, 1999). Polyphenolic compounds in tea effectively reduced the concentration of HAs in model systems and when applied to the surface of meat before frying (Weisburger et al., 2002). Green tea and green coffee extracts have been added to fine yeast pastry fried products (donuts) for improving frying oil stability (Budryn et al., 2013) and reducing acrylamide formation, as explained earlier.

10.3.3 Extrusion

Extrusion cooking is one of the most important food processing technologies, which has been used for the production of breakfast cereals, ready-to-eat snack foods, and other textured foods. Extrusion cooking is a high-temperature, short-time process in which moistened, expansive, starchy, and proteinaceous food materials are plasticized and cooked in a tube by a combination of moisture, pressure, temperature, and mechanical shear, resulting in molecular transformation and chemical reactions (Singh et al., 2007). The majority of extruded food products are prepared from food grains with few studies focusing on the incorporation of fruits and vegetables to produce snacks enriched with bioactive compounds (Brennan et al., 2011).

Like other heat treatment food processes, extrusion cooking may have both beneficial and undesirable effects on the final product. The beneficial effects include destruction of antinutritional factors and gelatinization of starch. On the other hand, Maillard reactions between protein and reducing sugars may generate toxic compounds and decrease the nutritional value of foods. Moreover, heat-labile compounds may be degraded (Björck and Asp, 1983; Cheftel, 1986).

Several extrusion process variables can influence the composition of finished products. Critical process variables are temperature, screw speed, and initial moisture content. Several studies have shown that extrusion processing significantly reduces the polyphenol content. Chaovanalikit (1999) reported that blueberry anthocyanins are significantly reduced by extrusion in breakfast cereals containing cornmeal and sucrose. Repo-Carrasco-Valencia et al. (2009) reported that the extrusion process reduced the total phenolic compounds content of *Amaranthus* by 80.3% and 64.4% for Centenario and Oscar Blanco varieties, respectively. Altan et al. (2009) developed extruded barley–fruits and vegetables using by-products such as grape and tomato pomace, respectively, and evaluated the stability of polyphenols during the extrusion process. They found that extrusion cooking decreased the antioxidant capacity and total polyphenols in all the extruded products by 60%–68% and 46%–60%, respectively, compared with that of the unprocessed barley flour. Reductions in phenolic content have also been reported for extruded bean, sorghum, and oat cereals (Anton et al., 2009; Dlamini et al., 2007; Viscidi et al., 2004).

Alonso et al. (2000) suggested that processing conditions during extrusion could alter the molecular structure of phenolic compounds, reducing their antioxidant activity or decreasing their extractability due to a certain degree of polymerization.

10.3.4 Improving Polyphenol Stability through the Modification of Processing Conditions

Food processing produces losses of beneficial compounds mainly caused by high temperatures and high oxygen levels. Vacuum technology can protect heat-sensitive foods during processing by reducing the boiling point of the product's moisture and the environmental oxygen level. Among the different methods available, we find traditional applications such as vacuum drying and freeze drying, as well as some recent developments such as microwave vacuum drying (Vega-Mercado et al., 2001) and vacuum frying (Dueik and Bouchon, 2011a). In solid matrices, most of the studies on vacuum processing focus on agricultural products, such as potatoes that are naturally structured, but few deal with fabricated or formulated products. This is a challenge, since specific processing conditions are required to form a structure during processing.

In vacuum deep-fat frying, the process is carried out at pressures well below atmospheric conditions. Preferably, operating pressure should be lower than 7 kPa; hence, frying temperatures lower than 90°C can be reached (Garayo and Moreira, 2002). However, if frying temperatures are too low, the required structural changes may not take place (Dueik and Bouchon, 2011a). As we explained in the previous section, adequate moisture contents and temperatures are necessary to produce starch gelatinization. Even though frying temperatures below 60°C–70°C are high enough to attain gelatinization in the crust region, the loss of liquid water during immersion prevents it. In turn, even if the core region has enough liquid water, these low frying temperatures cannot overcome the boiling point of water, impairing starch gelatinization (Sobukola et al., 2013). The replacement of 5% of the native starch by pregelatinized starch improves the structure formation in formulated matrices (Dueik et al., 2014).

Dueik and Bouchon (2011b) studied the evolution of the antioxidant capacity of apples during vacuum and atmospheric frying. They observed that during atmospheric frying, the antioxidant capacity decreased in the early stages of atmospheric frying but increased significantly as the frying process proceeds. In turn, during vacuum frying, the antioxidant capacity remained constant. These authors argued that during atmospheric frying, new compounds with strong antioxidant capacity are formed. These compounds have been associated with nonenzymatic browning and the formation of brown melanoidins in model systems. Nonenzymatic browning reactions have been associated with the formation of potentially carcinogenic compounds such as acrylamide and furan. This increase in antioxidant capacity was not observed in vacuum-fried apples, indicating that low temperature processing does not induce browning reactions.

Several researchers have shown that vacuum frying retains high levels of polyphenols. Da Silva and Moreira (2008) observed that in blue potatoes, retention of anthocyanins was 60% higher in vacuum frying compared with atmospheric frying. Fang et al. (2011), when studying the evolution of phenolic compounds during vacuum frying of Chinese purple yam, obtained similar results. Vacuum frying retained 69% of total phenolic compounds and 63% of the anthocyanins. To compare vacuum and atmospheric frying adequately, Mariscal and Bouchon (2008) defined the "equivalent thermal driving forces" as the difference between the oil temperature and the boiling point of water at the working pressure. Recently, Dueik and Bouchon (2015) studied the preservation of polyphenols during vacuum and atmospheric frying in polyphenol-enriched formulated matrices. Polyphenols were extracted from olive leaves, freeze-dried, and added to the dry ingredients. The authors used equivalent thermal driving forces of 60°C and 70°C, giving rise to frying temperatures of 106°C and 116°C for vacuum frying (P = 9.8 kPa, water boiling point = 46°C) and 160°C and 170°C for atmospheric frying. The polyphenol retention in vacuum-fried products was 70.6% and 62.2% when using a driving force of 60°C and 70°C, respectively, while the atmospherically fried products retained just 42.5% and 34.9%.

10.4 POLYPHENOL INTERACTION WITH FOOD COMPONENTS

Significant research has been carried out on the interactions of polyphenols with compounds present in foods, such as carbohydrates, proteins, or lipids, because these food constituents may interact with polyphenols and reduce their activity and consequently their health benefits.

10.4.1 Carbohydrates

In food systems, carbohydrates can be broadly classified as either digestible or non-digestible. The former category can be further broken down to sugars, glucose oligomers (such as maltodextrins), and starch, while the latter category typically covers dietary fibers.

10.4.2 Digestible Carbohydrates

Polyphenols in foods can interact with digestible carbohydrates, producing changes in their properties and affecting their digestibility and absorption. Due to their interaction with amylopectin, tea polyphenols favor gelatinization by considerably decreasing the gelatinization temperature and enthalpy of starch. In addition, these polyphenols affect the retrogradation properties of starch (Wu et al., 2009; Xiao et al., 2011). Xiao et al. (2011) explained that the hydrophilic character of the OH groups in tea polyphenols might interact with side chains of amylopectin and bind to the amorphous region of starch granules to various degrees and thus change the coupling forces between the crystallites and the amorphous matrix. This results in easier hydration of starch granules, requiring less thermal energy for gelatinization. Dueik and Bouchon (2015) reported that polyphenol addition to a gluten–starch matrix reduced the oil content in vacuum-fried matrices, and they attributed these results to better structure formation due to facilitated starch gelatinization. As mentioned before, starch gelatinization during vacuum frying might be impaired because of the absence of proper conditions of water and temperature for starch gelatinization. Regarding starch digestibility, Thompson and Yoon (1984) studied wheat starch digestibility in the presence or absence of polyphenols at concentrations found in legumes using an *in vitro* dialysis system. Addition of tannic acid reduced the starch digestibility by 13%. Later, the same research group (Thompson et al., 1984) found a negative correlation between glycemic index and the total intake of polyphenols. They explained that polyphenols, especially the large polymeric type or condensed tannins, appear to be responsible in part for the reduced glycemic response of legumes compared with cereal products. The underlying mechanisms responsible for the negative relationship between the polyphenol intake and glycemic index is not clear. It could be related to direct starch–polyphenol interactions and inhibition of α-amylase (Tamir and Alumot, 1969).

Sugars participate in the stability of polyphenols, acting as inhibitors of the oxidation of polyphenols by trace metals. Peinado et al. (2010) studied the addition of sugars in a model system simulating sweet wine, which contained catechin and trace metals such as iron, thereby finding that a mixture of glucose and fructose clearly inhibited the iron-catalyzed oxidation of catechin. Moreover, there is some evidence suggesting that dietary polyphenols might provide a convenient dietary mechanism for regulating the rate of intestinal sugar absorption. Johnston et al. (2005) investigated the absorption of sugars using Caco-2 cells in the presence of polyphenols, thereby finding that they can inhibit glucose transport through the cells. They concluded that the consumption of polyphenols might provide a convenient dietary mechanism for regulating the rate of intestinal sugar absorption that is an important factor in the management of diabetes and in the long term offers some protection against developing the metabolic syndrome or type II diabetes.

10.4.3 Nondigestible Carbohydrates

These complex carbohydrates seem to interfere with the adequate assimilation of food antioxidants. Nondigestible carbohydrates interact with components of the cell wall, such as pectin, cellulose, or dietary fibers. Hydrogen bonds can be formed between the hydroxyl groups of polyphenols and the oxygen atoms of the glycosidic linkages of polysaccharides. In addition, covalent bonds can be formed between phenolic acids and polysaccharides. Complex porous structure and surface properties of cell walls can also be important (Jakobek, 2015). Padayachee et al. (2012a,b) observed that chlorogenic, caffeic, as well as ferulic acids and anthocyanins bound *in vitro* to cell wall models (cellulose and pectin–cellulose models) depleted phenolic acids and anthocyanins.

Dietary fiber limits the absorption of fruit and vegetable polyphenols, reducing their ability to prevent and treat some chronic human diseases. Some studies showed that the bioavailability decreases due to the associations of carbohydrates and polyphenols, which capture polyphenols into their structure. In addition, dietary fiber increases the viscosity of the luminal contents, reducing the antioxidants' absorption rates by physically trapping them within the fiber matrix in chyme. The chyme is a 2-phase system with a discontinuous particulate phase dispersed in a continuous liquid phase. Antioxidants trapped within the particles must first be released into the continuous solution phase before they can be absorbed through the gut wall (Palafox-Carlos et al., 2011).

During digestion, the initial phase includes the mechanical disruption of the food matrix. Here, the cells of the food components are broken down, releasing first those phenolic compounds contained in the vacuoles and those linked weakly to the cell wall. Polyphenols linked strongly to the cell wall, especially in skin cells, are released chemically and enzymatically (Tagliazucchi et al., 2009). Only a minor portion of the low-molecular-weight phenolic compounds (monomers or oligomers) is able to pass through the gut wall. The rest, and those polyphenols associated with dietary fiber, reach the colon where they become fermentable substrates (Manach et al., 2005). Nonabsorbable metabolites of polyphenol colonic fermentation and nonfermented phenolic compounds remain in the colonic lumen, where they may contribute to a healthy antioxidant environment by scavenging free radicals and counteracting the effects of dietary prooxidants; the remaining polyphenols are excreted in the feces (Palafox-Carlos et al., 2011).

10.4.4 Proteins

Polyphenol–protein interactions might influence the sensory properties of foods such as astringency sensation in the mouth, reduce proteins digestibility, change the functional properties of proteins, reduce digestive enzymes activity, and protect polyphenols in the gastrointestinal (GI) tract, thereby improving their bioaccessibility and bioavailability.

The interaction happens through both covalent and noncovalent interactions. Protein binding by flavonoids is associated with several sensory properties of foods and beverages such as color, flavor, and taste (bitterness and astringency). Flavonoids from tea, cocoa, grape seeds, and fruits, in general, reduce the activity of digestive enzymes, including α-amylase, pancreatic lipase, and phospholipase A (Bordenave et al., 2014; Gonçalves et al., 2011). Rohn et al. (2003) reported an inhibition of α-chymotrypsin by chlorogenic acid with a reduced affinity of the enzyme and substrate and reduced protein hydrolysis. Tea polyphenols interact with milk proteins, particularly casein. A recent *in vitro* study demonstrated that polyphenols from green tea bind to milk protein, generating a hydrophobic complex, thereby reducing the absorption of milk proteins (Jakobek, 2015). On the other hand, this interaction protects polyphenols from the severe digestion conditions. Lamothe et al. (2014) observed that green tea extracts significantly decreased proteolysis of casein in the gastric phase but had no effect in the intestinal phase. Transition from the gastric phase to the intestinal phase induced a 50% decrease in the antioxidant activity of the control (tea extract dispersed in water) due to the degradation of polyphenols. However, the presence of dairy matrices significantly improved polyphenol stability in the intestinal phase and increased the antioxidant activity by 29% with cheese and by 42% with milk, compared to the control. These results suggest that simultaneous consumption of green tea and dairy products helps to maintain the antioxidant activity of polyphenols during digestion.

Protein–polyphenol interactions may have various sensory effects. Polyphenols are responsible for the bitterness and astringency of beverages such as wine, coffee, and tea. Interactions between salivary proteins and tannins have been studied extensively in relation to astringency. However, the presence of preexisting polyphenol–protein complexes in foods and beverages is believed to directly affect subsequent interactions in the oral cavity and as such modulate astringency and the overall sensorial properties of foods. A specific example is the attenuation of tea astringency by addition of milk (de Freitas and Mateus, 2012).

Protein–polyphenol interactions can also change the functionality of the protein. Interactions can influence protein structures as polyphenols bind to the hydrophobic sites of proteins; therefore, the protein structure may change because only weak hydrophobic sites may remain on the surface of the protein (Yuksel et al., 2010), leading to a possible change in the protein folding and its functionality. Rawel et al. (2002) reported a change in the isoelectric point of soy glycinin, a major storage protein in soybeans, when adding polyphenols. This was a consequence of the decrease in the positive charge of the amino groups of glycinin, changing the protein solubility. Solubility is a prime requisite for a functional ingredient protein and is critically necessary for products such as beverages; other functional properties of proteins (emulsifying and foaming properties, gel forming, etc.) are also affected by solubility. Polyphenol–protein interactions have been associated with improving egg albumen foaming properties, degrading the textural properties of cheese and skim milk gels (Bordenave et al., 2014).

10.4.5 Oils

Polyphenols have been investigated as potential inhibitors of lipid absorption. Shishikura et al. (2006) studied the *in vitro* digestibility of lipids and the influence of polyphenols in this process. It was found that green and black tea polyphenols affected emulsions by increasing droplet size and decreasing specific surface area, which can be attributed to the interaction of polyphenols with the polar heads of the emulsion. An increased droplet size could cause a decrease in the lipase action and fat absorption (McDougall et al., 2009). Lamothe et al. (2014) observed that when adding green tea to milk, the kinetics of fatty acid release after digestion was reduced. *In vivo* studies suggest that triglyceride levels in the rat plasma can be reduced by polyphenols from tea (Uchiyama et al., 2011), green tea (Koo and Noh, 2007) and apple (Sugiyama et al., 2007). It has been proposed that polyphenols form hydrogen bonds with membranes, where phenolic hydroxyl groups act as bond donors and oxygen atoms as bond acceptors (Sirk et al., 2008). The interaction of polyphenols with phospholipids is one of the mechanisms responsible for their claimed health benefits. In addition, polyphenols create a positive antioxidant environment during lipid absorption in the gastrointestinal tract, reacting with harmful lipid peroxidation products (Jakobek, 2015). Kanner and Lapidot (2001) suggested that the propagation of lipid peroxidation products derived from the acidic conditions of the stomach could be inhibited by the presence of catechin or red wine polyphenols; they may also reduce β-carotene cooxidation under acidic conditions. The authors concluded that potentially harmful effects of oxidized fats could be reduced by including plant dietary antioxidants in the meal. *In vivo* studies performed by Gladine et al. (2007) suggest that plant extracts, rich in polyphenols, can inhibit plasma lipids peroxidation in rats fed with n-3 PUFA-supplemented diets.

Interaction with fats does not reduce significantly the accessibility of polyphenols for absorption; therefore, their bioavailability is unaffected by fats. Ortega et al. (2009) evaluated the effect of the food matrix composition, particularly fat content in the bioaccessibility and digestibility of cocoa polyphenols (procyanidins, phenolic acids, and flavones). The higher fat content in the cocoa beverage, as compared to cocoa powder, had a protective effect that favors the stability of polyphenols during their passage through the gastrointestinal tract, which was related to better micellarization. This protection allows polyphenols to act in the gastrointestinal tract by decreasing the propagation of peroxidation products and reducing fat absorption due to a decrease in the surface area of oil droplets, which hinders lipase action.

Microencapsulation of polyphenols creates a physical barrier that reduces their interactions with other food components in formulated matrices and improves their bioavailability.

10.4.6 Minerals

Polyphenols are well-known inhibitors of the absorption of some minerals. Evidence abounds in the literature indicating that polyphenolic compounds inhibit iron absorption (Brune et al., 1989, 1991). The degree of iron absorption inhibition can be related to the amount and type of phenolic compounds, type of iron compound (for nonheme iron), and pH. Polyphenols contain two aromatic rings as a functional group with two or more hydroxyl groups. Molecules with an aromatic ring with two hydroxyl groups (catechol group) or three hydroxyl groups (galloyl group) positioned at adjacent carbon atoms have *in vitro* iron-binding capacities (Brune et al., 1989; Khokhar and Owusu Apenten, 2003; Perron and Brumaghim, 2009; Zijp et al., 2000). Brune et al. (1989) showed that the inhibition of iron absorption by phenolic compounds in foods is mainly attributable to the galloyl group; the catechol group also interfered with iron absorption, although to a lesser extent. Brune et al. (1991) observed that these complexes (Fe-galloyl and Fe-catechol) had different color and absorbance maxima; thus, they can be determined separately by readings at two wavelengths.

Avoiding polyphenol/iron interaction is a major challenge. Dueik et al. (2015) studied tea fortification with iron for overcoming iron deficiency in developing countries. Tea is consumed primarily in South and East Asia among all socioeconomic strata, as it is a relatively low-cost beverage and seems like a logical substrate for iron fortification. However, its fortification with iron presents technical challenges as tea polyphenols form a blue complex with iron that makes both unavailable for absorption. The authors developed a methodology for the iron–tea polyphenol complex determination, observing that complexation was pH dependent and that soluble iron compounds were more susceptible to complex formation. Furthermore, they developed a formula for the fortification of tea with highly bioavailable iron and no sensory changes (patent pending).

10.5 POLYPHENOL BIOAVAILABILITY IN FOOD MATRICES

The *in vivo* effects of antioxidants depend not only on their concentrations in the food matrix, but also on their bioaccessibility and bioavailability after ingestion. Bioaccessibility corresponds to the amount of an ingested nutrient that is available for absorption in the gut after digestion; this means that it has been released from the food matrix during the digestion process. On the other hand, bioavailability refers to the fraction of the bioaccessible ingested nutrient or active compound that is available for use in normal physiologic functions and for storage. Thus, bioaccessibility and bioavailability are not equivalent concepts (Parada and Aguilera, 2007).

The methods to determine bioavailability can be classified into *in vivo* (humans/animals) and *in vitro* (simulated experiments performed in a laboratory) tests. Bioaccessibility can be determined by using just an *in vitro* digestion process. *In vivo* methods provide direct data of bioavailability; however, cost, ethical restrictions, and the lengthy approval process of protocols when humans and animals are used in biological or clinical research severely limit the use of this type of method. Another drawback of *in vivo* tests relates to variations in the physiologic state of participating individuals and possible interactions of the target nutrient with other components in the diet. Thus, as an alternative to *in vivo* tests, *in vitro* methods are extensively used in the food industry since they are rapid, safe, cost-efficient, and do not raise ethical or legal issues (Li et al., 2014; Parada and Aguilera, 2007). However, they might not represent real conditions.

This section will describe each method for the determination of polyphenol bioavailability in food matrices and an analysis of the obtained data.

10.5.1 *In Vitro* Bioavailability of Polyphenols

In vitro methods simulate either the digestion process alone (for bioaccessibility) or the digestion and absorption processes (for bioavailability); the measured response is the concentration of the nutrient in the extract obtained after digestion (bioaccessibility) or absorption (bioavailability).

An understanding of the basic physicochemical and physiologic processes that occur as a delivery system passing through the human gastrointestinal tract is required to develop effective *in vitro* models that accurately simulate digestion (Li et al., 2014). In a typical *in vitro* test, the digestion process is simulated under controlled conditions using a variety of enzymes, for example, lingual lipase, amylase, protease, pepsin, pancreatin, and so on; the final transport and absorption process is commonly assessed using differentiated cell monolayers obtained from human intestinal epithelium (Caco-2 cell cultures) (Gangloff et al., 1996). Caco-2 cells are grown as monolayers on porous membranes in bicameral chambers; they have been used to study the transport of several micronutrients from the apical (lumenal) chamber to the basolateral (serosal) chamber. By performing two-way (apical to basolateral [A–B] and basolateral to apical [B–A]) permeation experiments, it is possible to determine the existence of an active efflux/uptake or a passive diffusion of bioactive compounds into the intestinal mucosa (Yu and Huang, 2011). In addition, mechanical models of the whole digestive system are under development, for example, by Dr. Paul Singh's laboratory at UC Davis. Figure 10.1 shows the simulation of the digestion and Caco-2 cells' transport and absorption processes in the lab. The amount of a given nutrient present in the extract (before the Caco-2 cell step) is the bioavailable part (Glahn et al., 1998), while the amount of the given nutrient in the liquid in the basolateral compartment (after the Caco-2 cell step) is the bioaccessible part.

Several protocols following the scheme of Figure 10.1 have been adapted to assess the bioavailability of specific nutrients *in vitro*. A major problem when using an *in vitro* method is the great number of factors that are not taken into account, such as the relation among physical changes of food, transit time, digestion rate, and the

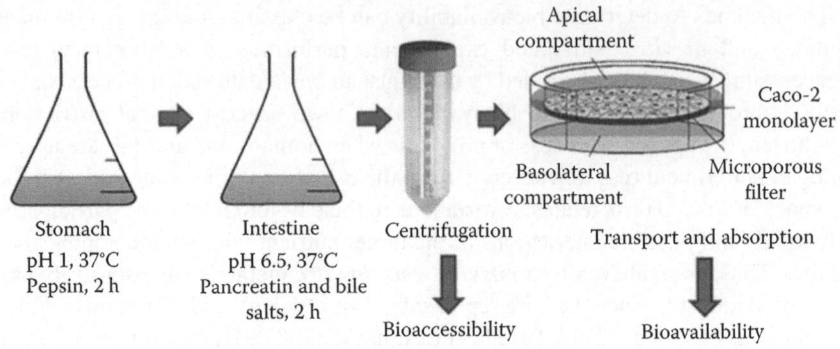

FIGURE 10.1 Schematics of the simulation of the digestion process and Caco-2 cell transport and absorption processes.

effect of intestinal bacteria on transformation of nutrients into metabolites that are the absorbable active form (Hur et al., 2011). To avoid erroneous conclusions, strict control of experimental conditions and correct analysis and interpretation of results are critical.

Several *in vitro* digestion models have been applied to different food matrices to measure the release of polyphenols during digestion. Gastrointestinal digestion and intestinal absorption depend on the polyphenol structure and the composition of the food matrix. Molecular weight, glycosylation, and esterification are the most studied structural parameters. Relatively low amounts of polyphenols are absorbed. Isoflavones and gallic acid are the most absorbed, followed by catechins, flavanones, and quercetin glucosides. Proanthocyanidins, galloylated tea catechins, and anthocyanins are the least absorbed (Manach et al., 2005).

In vitro solubility is the most used method to measure the bioaccessibility of polyphenols and has been applied in many foods such as pomegranate juices (Pérez-Vicente et al., 2002), cocoa beverages (Ortega et al., 2009), and raspberries (McDougall et al., 2005). Studies on Caco-2 cell bioavailability are also common, since these cells are able to metabolize some polyphenols and carry out normal metabolic processes such as glucuronidation, sulfation, and methylation; polyphenols undergo these processes in the small intestine and in the liver. Yi et al. (2006) added anthocyanins from blueberries to Caco-2 cells grown on Transwell membranes and suggested that anthocyanins can be degraded and demethylated during absorption and transport by the cells.

Dueik and Bouchon (2015) showed that olive leaf polyphenols added into a vacuum-fried or atmospherically fried gluten–starch matrix are highly bioaccessible (more than 90% for all samples). Hence, digestive enzymes can make them available for absorption. Bioavailability was low, however, due to the small intestinal permeability of oleuropein across the intestine (12.4%; see Figure 10.2). The authors explained that polyphenols in our diet are not necessarily those having the best bioavailability profile, either because they are poorly absorbed or highly metabolized into other compounds.

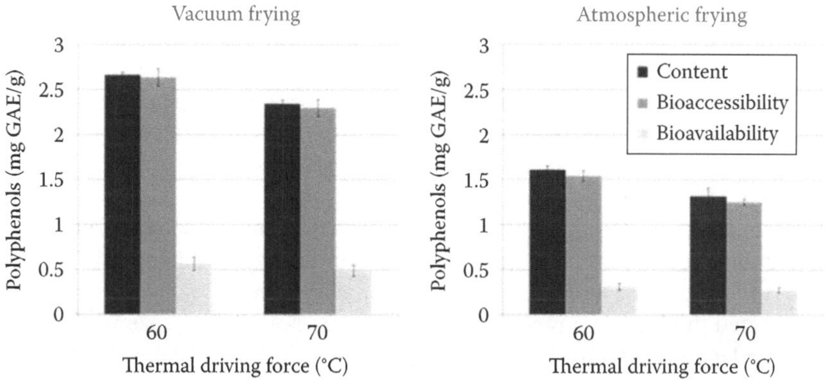

FIGURE 10.2 Polyphenol bioaccessibility and bioavailability in vacuum-fried and atmospherically fried gluten–starch matrices. (From Dueik, V. and Bouchon, P., Bioavailability of polyphenols in vacuum and atmospherically fried starchy snacks, *Delivery of Functionality in Complex Food Systems*, Paris, France, 2015.)

10.5.2 *In Vivo* Bioavailability of Polyphenols

In vivo methods have been used for a large variety of nutrients and foods. Usually a response is measured after consumption of a pure nutrient (natural or synthetic) by living beings, either humans (most common) or animals, and compared to an equivalent nutrient dose found in a food source (Parada and Aguilera, 2007). Many researchers have investigated the kinetics and extent of polyphenol absorption by measuring plasma concentrations and urinary excretion among adults after the ingestion of a single dose of polyphenol, provided as a pure compound, plant extract, or whole food or beverage (Manach et al., 2005).

The few bioavailability studies on humans show that the quantities of polyphenols found intact in urine vary from one phenolic compound to another. For instance, values lower than 0.1% have been reported for most anthocyanins, 1%–5% for quercetin and flavanols, and 30%–50% for isoflavones and gallic acid (red wine, tea, and various fruits) (Parada and Aguilera, 2007).

Edgecombe et al. (2000) developed a novel *in situ* intestinal perfusion technique for studying the absorption of oleuropein, a major polyphenol compound in olive oil. They determined that oleuropein could be absorbed, albeit poorly (12.4%), from isolated perfused rat intestine.

Cerdá et al. (2004) evaluated, in healthy humans, the bioavailability and metabolism of pomegranate juice ellagitannins. Neither of the polyphenols present in the juice were detected in both plasma and urine. However, three ellagitannin-derived metabolites produced by the colonic microflora were detected, which did not show significant antioxidant activity compared to polyphenols from pomegranate juice. The authors concluded that the potential systemic biological effects of pomegranate juice ingestion should be attributed to the colonic microflora metabolites rather than to the polyphenols present in the juice.

10.6 CONCLUDING REMARKS

Polyphenol consumption is associated with health benefits; it may play a meaningful role in reducing cardiovascular diseases and may offer some protection against the development of cancer, diabetes, osteoporosis, and neurodegenerative diseases. Due to these healthy characteristics, there is increasing interest in the development of functional foods.

Unfortunately, polyphenols are highly unstable during processing due to their susceptibility to high temperatures and presence of oxygen, among others. Novel technologies providing mild operating conditions, including vacuum frying, have demonstrated the ability to preserve polyphenols and other phytochemicals in food matrices adequately. Moreover, polyphenols may interact with other food components, such as dietary fibers and minerals, which may reduce their bioavailability. Microencapsulated polyphenol extracts have shown higher stability and bioactivity, reducing the detrimental effects of processing and interactions with other food components. Bioavailability of polyphenols can be measured *in vivo* and *in vitro*, and when using either method, the bioavailability seems to be low.

REFERENCES

Aguilera, J.M. 2005. Why food microstructure? *Journal of Food Engineering* 67(1–2):3–11.

Aguilera, J.M. 2006. Seligman lecture 2005 food product engineering: Building the right structures. *Journal of the Science of Food and Agriculture* 86(8):1147–1155.

Alonso, R., Grant, G., Dewey, P., and Marzo, F. 2000. Nutritional assessment in vitro and in vivo of raw and extruded peas (*Pisum sativum* L.). *Journal of Agricultural and Food Chemistry* 48:2286–2290.

Altan, A., McCarthy, K.L., and Maskan, M. 2009. Effect of extrusion process on antioxidant activity, total phenolics and β-glucan content of extrudates developed from barley-fruit and vegetable by-products. *International Journal of Food Science and Technology* 44(6):1263–1271.

Anjum, F.M., Khan, M.R., and Din, A. 2007. Wheat gluten: High molecular weight glutenin subunits-structure, genetics and relation to dough elasticity. *Journal of Food Science* 72(3):R56–R61.

Anton, A.A., Fulcher, R.G., and Arntfield, S.D. 2009. Physical and nutritional impact of fortification of corn starch-based extruded snacks with common bean (*Phaseolus vulgaris* L.) flour: Effects of bean addition and extrusion cooking. *Food Chemistry* 113(4):989–996.

Balasundram, N., Sundram, K., and Samman, S. 2006. Phenolic compounds in plants and agri-industrial by-products: Antioxidant activity, occurrence, and potential uses. *Food Chemistry* 99:191–203.

Balogh, Z., Gray, J.I., Gomaa, E.A., and Booren, A.M. 2000. Formation and inhibition of heterocyclic aromatic amines in fried ground beef patties. *Food and Chemical Toxicology* 38(5):395–401.

Baumann, B. and Escher, F. 1995. Mass and heat transfer during deep-fat frying of potato slices. I. Rate of drying and oil uptake. *Lebensmittel-Wissenschaft und Technologie* 28:395–403.

Björck, I. and Asp, N.G. 1983. The effects of extrusion cooking on nutritional value—A literature review. *Journal of Food Engineering* 2(4):281–308.

Bordenave, N., Hamaker, B.R., and Ferruzzi, M.G. 2014. Nature and consequences of non-covalent interactions between flavonoids and macronutrients in foods. *Food and Function* 5(1):18–34.

Bouchon, P., Aguilera, J.M., and Pyle, D.L. 2003. Structure oil-absorption relationships during deep-fat frying. *Journal of Food Science* 68:2711–2716.

Brennan, C., Brennan, M., Derbyshire, E., and Tiwari, B.K. 2011. Effects of extrusion on the polyphenols, vitamins and antioxidant activity of foods. *Trends in Food Science and Technology* 22(10):570–575.

Brune, M., Hallberg, L., and Skanberg, A.N.N.B. 1991. Determination of iron-binding phenolic groups in foods. *Journal of Food Science* 56:128–131.

Brune, M., Rossander, L., and Hallberg, L. 1989. Iron absorption and phenolic compounds: Importance of different phenolic structures. *European Journal of Clinical Nutrition* 43:547.

Budryn, G., ŻyŻelewicz, D., Nebesny, E., Oracz, J., and Krysiak, W. 2013. Influence of addition of green tea and green coffee extracts on the properties of fine yeast pastry fried products. *Food Research International* 50(1):149–160.

Cerdá, B., Espín, J.C., Parra, S., Martínez, P., and Tomás-Barberán, F.A. 2004. The potent in vitro antioxidant ellagitannins from pomegranate juice are metabolised into bioavailable but poor antioxidant hydroxy–6H–dibenzopyran–6–one derivatives by the colonic microflora of healthy humans. *European Journal of Nutrition* 43(4):205–220.

Chaovanalikit, A. 1999. Anthocyanin stability during extrusion cooking. MS thesis. University of Maine, Orono, ME.

Cheftel, J.C. 1986. Nutritional effects of extrusion-cooking. *Food Chemistry* 20(4): 263–283.

Cheng, K.W., Shi, J.J., Ou, S.Y., Wang, M., and Jiang, Y. 2009. Effects of fruit extracts on the formation of acrylamide in model reactions and fried potato crisps. *Journal of Agricultural and Food Chemistry* 58(1):309–312.

Crowe, K.M. 2013. Designing functional foods with bioactive polyphenols: Highlighting lessons learned from original plant matrices. *Journal of Human Nutrition and Food Science* 1(3):1018.

Da Silva, P. and Moreira, R. 2008. Vacuum frying of high-quality fruit and vegetable-based snacks. *Lebensmittel-Wissenschaft und Technologie* 41:1758–1767.

Day, L., Augustin, M.A., Batey, I.L., and Wrigley, C.W. 2006. Wheat-gluten uses and industry needs. *Trends in Food Science and Technology* 17(2):82–90.

de Freitas, V. and Mateus, N. 2012. Protein/polyphenol interactions: Past and present contributions. Mechanisms of astringency perception. *Current Organic Chemistry* 16(6):724–746.

Dlamini, N.R., Taylor, J.R.N., and Rooney, L.W. 2007. The effect of sorghum type and processing on the antioxidant properties of African sorghum-based foods. *Food Chemistry* 105(4):1412–1419.

Dueik, V. and Bouchon, P. 2011a. Development of healthy low-fat snacks: Understanding the mechanisms of quality changes during atmospheric and vacuum frying. *Food Reviews International* 27:408–432.

Dueik, V. and Bouchon, P. 2011b. Vacuum frying as a route to produce novel snacks with desired quality attributes according to new health trends. *Journal of Food Science* 76:E188–E195.

Dueik, V. and Bouchon, P. 2015. Bioavailability of polyphenols in vacuum and atmospherically fried starchy snacks. *Delivery of Functionality in Complex Food Systems*, Paris, France.

Dueik, V., Chen, B.K., and Diosady, L. 2015. Effective technology for iron fortification of tea. 6th International Symposium on Delivery of Functionality in Complex food Systems, 14-17 July 2015, Paris, France.

Dueik, V., Sobukola, O., and Bouchon, P. 2014. Development of low-fat gluten and starch fried matrices with high fiber content. *Lebensmittel-Wissenschaft und-Technologie* 59:6–11.

Edgecombe, S.C., Stretch, G.L., and Hayball, P.J. 2000. Oleuropein, an antioxidant polyphenol from olive oil, is poorly absorbed from isolated perfused rat intestine. *Journal of Nutrition* 130(12):2996–3002.

Fang, Z., Wu, D., Yü, D., Ye, X., Liu, D., and Chen, J. 2011. Phenolic compounds in Chinese purple yam and changes during vacuum frying. *Food Chemistry* 128(4):943–948.

Gangloff, M.B., Glahn, R.P., Miller, D.D., and Van Campen, D.R. 1996. Assessment of iron availability using combined in vitro digestion and Caco-2 cell culture. *Nutrition Research* 16:479487.

Garayo, J. and Moreira, R. 2002. Vacuum frying of potato chips. *Journal of Food Engineering* 55:181–191.

Gazmuri, A.M. and Bouchon, P. 2009. Analysis of wheat gluten and starch matrices during deep-fat frying. *Food Chemistry* 115:999–1005.

Gebhardt, B. 1996. Oils and fats in snack foods. In *Bailey's Industrial Oil and Fat Products*, ed. Y.I. Hui, 5th ed., Vol. 3. New York: Wiley.

Gladine, C., Morand, C., Rock, E., Bauchart, D., and Durand, D. 2007. Plant extracts rich in polyphenols (PERP) are efficient antioxidants to prevent lipoperoxidation in plasma lipids from animals fed n − 3 PUFA supplemented diets. *Animal Feed Science and Technology* 136(3):281–296.

Glahn, R.P., Lee, O.A., Yeung, A., Goldman, M.I., and Miller, D.D. 1998. Caco-2 cell ferritin formation predicts nonradiolabeled food iron availability in an in vitro digestion/Caco-2 cell culture model. *Journal of Nutrition* 128:1555–1561.

Gómez-Alonso, S., Fregapane, G., Salvador, M.D., and Gordon, M.H. 2003. Changes in phenolic composition and antioxidant activity of virgin olive oil during frying. *Journal of Agricultural and Food Chemistry* 51(3):667–672.

Gonçalves, R., Mateus, N., and de Freitas, V. 2011. Inhibition of α-amylase activity by condensed tannins. *Food Chemistry* 125:665.

Granda, C., Moreira, R.G., and Tichy, S.E. 2004. Reduction of acrylamide formation in potato chips by low-temperature vacuum frying. *Journal of Food Science* 69(8):E405–E411.

Higdon, J.V. and Frei, B. 2003. Tea catechins and polyphenols: Health effects, metabolism, and antioxidant functions. *Critical Reviews in Food Science and Nutrition* 43:89–143.

Hur, S.J., Lim, B.O., Decker, E.A., and McClements, D.J. 2011. In vitro human digestion models for food applications. *Food Chemistry* 125:1–12.

Jakobek, L. 2015. Interactions of polyphenols with carbohydrates, lipids and proteins. *Food Chemistry* 175:556–567.

Johnston, K., Sharp, P., Clifford, M., and Morgan, L. 2005. Dietary polyphenols decrease glucose uptake by human intestinal Caco-2 cells. *FEBS Letters* 579(7):1653–1657.

Kanner, J. and Lapidot, T. 2001. The stomach as a bioreactor: Dietary lipid peroxidation in the gastric fluid and the effects of plant-derived antioxidants. *Free Radical Biology and Medicine* 31(11):1388–1395.

Kikugawa, K. 1999. Involvement of free radicals in the formation of heterocyclic amines and prevention by antioxidants. *Cancer Letters* 143(2):123–126.

Khokhar, S. and Owusu Apenten, R.K. 2003. Iron binding characteristics of phenolic compounds: Some tentative structure–activity relations. *Food Chemistry* 81:133–140.

Koo, S.I. and Noh, S.K. 2007. Green tea as inhibitor of the intestinal absorption of lipids: Potential mechanism for its lipid-lowering effect. *Journal of Nutritional Biochemistry* 18(3):179–183.

Kubow, D. 1992. Routes of formation and toxic consequences of lipid oxidation products in foods. *Free Radical Biology and Medicine* 12:63–81.

Lamothe, S., Azimy, N., Bazinet, L., Couillard, C., and Britten, M. 2014. Interaction of green tea polyphenols with dairy matrices in a simulated gastrointestinal environment. *Food and Function* 5(10):2621–2631.

Lešková, E., Kubíková, J., Kováčiková, E., Košická, M., Porubská, J., and Holčíková, K. 2006. Vitamin losses: Retention during heat treatment and continual changes expressed by mathematical models. *Journal of Food Composition and Analysis* 19:252–276.

Li, O., Dueik, V., and Diosady, L. 2014. Microencapsulation of vitamins, minerals, and nutraceuticals for food applications. In *The Art and Science of Microencapsulation: An Application Handbook for Food Industry*, eds. A. Gaonkar, N. Vasisht, A. Khare, and R. Sobel, pp. 501–522. Academic Press, Elsevier Science, San Diego, CA.

Manach, C., Williamson, G., Morand, C., Scalbert, A., and Remesy, C. 2005. Bioavailability and bioefficacy of polyphenols in humans. I. Review of 97 bioavailability studies. *American Journal of Clinical Nutrition* 81:230S–42S.

Mariscal, M. and Bouchon, P. 2008. Comparison between atmospheric and vacuum frying of apple slices. *Food Chemistry* 107:1561–1569.

McDougall, G.J., Dobson, P., Smith, P., Blake, A., and Stewart, D. 2005. Assessing potential bioavailability of raspberry anthocyanins using an in vitro digestion system. *Journal of Agricultural and Food Chemistry* 53:5896–5904.

McDougall, G.J., Kulkarni, N.N., and Stewart, D. 2009. Berry polyphenols inhibit pancreatic lipase activity in vitro. *Food Chemistry* 115(1):193–199.

Nicoli, M.C., Anese, M., and Parpinel, M. 1999. Influence of processing on the antioxidant properties of fruits and vegetables. *Trends in Food Science and Technology* 10:94–100.

Ortega, N., Reguant, J., Romero, M.P., Macià, A., and Motilva, M.J. 2009. Effect of fat content on the digestibility and bioaccessibility of cocoa polyphenol by an in vitro digestion model. *Journal of Agricultural and Food Chemistry* 57(13):5743–5749.

Ovalle, N., Cortés, P., and Bouchon, P. 2013. Understanding microstructural changes of starch during atmospheric and vacuum heating in water and oil through online in situ vacuum hot-stage microscopy. *Innovative Food Science and Emerging Technologies* 17:135–143.

Ovaskainen, M.L., Törrönen, R., Koponen, J.M., Sinkko, H., Hellström, J., Reinivuo, H., and Mattila, P. 2008. Dietary intake and major food sources of polyphenols in Finnish adults. *Journal of Nutrition* 138(3):562–566.

Padayachee, A., Netzel, G., Netzel, M. et al. 2012a. Binding of polyphenols to plant cell wall analogues—Part 1: Anthocyanins. *Food Chemistry* 134:155–161.

Padayachee, A., Netzel, G., Netzel, M. et al. 2012b. Binding of polyphenols to plant cell wall analogues—Part 2: Phenolic acids. *Food Chemistry* 135:2287–2292.

Palafox-Carlos, H., Ayala-Zavala, J.F., and González-Aguilar, G.A. 2011. The role of dietary fiber in the bioaccessibility and bioavailability of fruit and vegetable antioxidants. *Journal of Food Science* 76(1):R6–R15.

Pandey, K.B. and Rizvi, S.I. 2009. Plant polyphenols as dietary antioxidants in human health and disease. *Oxidative Medicine and Cellular Longevity* 2(5):270–278.

Parada, J. and Aguilera, J.M. 2007. Food microstructure affects the bioavailability of several nutrients. *Journal of Food Science* 72:R21–R32.

Patras, A., Brunton, N., O'Donnell, C., and Tiwari, B.K. 2010. Effect of thermal processing on anthocyanin stability in foods; mechanism and kinetics of degradation. *Trends in Food Science and Technology* 21:3–11.

Pedreschi, F., Kaack, K., and Granby, K. 2004. Reduction of acrylamide formation in potato slices during frying. *LWT: Food Science and Technology* 37(6):679–685.

Pedreschi, F., Kaack, K., and Granby, K. 2008. The effect of asparaginase on acrylamide formation in French fries. *Food Chemistry* 109(2):386–392.

Peinado, J., De Lerma, N.L., and Peinado, R.A. 2010. Synergistic antioxidant interaction between sugars and phenolics from a sweet wine. *European Food Research and Technology* 231(3):363–370.

Pérez-Vicente, A., Gil-Izquierdo, A., and García-Viguera, C. 2002. In vitro gastrointestinal digestion study of pomegranate juice phenolic compounds, anthocyanins, and vitamin C. *Journal of Agricultural and Food Chemistry* 50:2308–2312.

Perron, N.R. and Brumaghim, J.L. 2009. A review of the antioxidant mechanisms of polyphenol compounds related to iron binding. *Cell Biochemistry and Biophysics* 53:75–100.

Piljac-Zegarac, J., Samec, D., and Piljac, A. 2013. Herbal teas: A focus on antioxidant properties. In Preedy, V. R. (Ed.) *Tea in Health and Diseases Prevention*, pp. 129–140. Academic Press, Elsevier Science, San Diego, CA.

Ratnayake, W. and Jackson, D. 2005. Starch gelatinization. In *Advances in Food and Nutrition Research*, ed. S.L. Taylor, Vol. 55, pp. 221–268. Amsterdam, the Netherlands: Academic Press.

Rawel, H.M., Czajka, D., Rohn, S., and Kroll, J. 2002. Interactions of different phenolic acids and flavonoids with soy proteins. *International Journal of Biological Macromolecules* 30(3):137–150.

Repo-Carrasco-Valencia, R., Peña, J., Kallio, H., and Salminen, S. 2009. Dietary fiber and other functional components in two varieties of crude and extruded kiwicha (*Amaranthus caudatus*). *Journal of Cereal Science* 49(2):219–224.

Rodriguez-Mateos, A., Cifuentes-Gomez, T., George, T.W., and Spencer, J.P. 2013. Impact of cooking, proving, and baking on the (poly) phenol content of wild blueberry. *Journal of Agricultural and Food Chemistry* 62(18):3979–3986.

Rohn, S., Rawel, H.M., Wollenberger, U., and Kroll, J. 2003. Enzyme activity of α-chymotrypsin after derivatization with phenolic compounds. *Food Nahrung* 47:325–329.

Sablani, S.S., Marcotte, M., Baik, O.D., and Castaigne, F. 1998. Modeling of simultaneous heat and water transport in the baking process. *LWT: Food Science and Technology* 31(3):201–209.

Scalbert, A., Johnson, I.T., and Saltmarsh, M. 2005. Polyphenols: Antioxidants and beyond. *American Journal of Clinical Nutrition* 81(1):215S–217S.

Shewry, P.R., Halford, N.G., Belton, P.S., and Tatham, A.S. 2002. The structure and properties of gluten: An elastic protein from wheat grain. *Philosophical Transactions of the Royal Society of London. Series B: Biological Sciences* 357:133–142.

Shishikura, Y., Khokhar, S., and Murray, B.S. 2006. Effects of tea polyphenols on emulsification of olive oil in a small intestine model system. *Journal of Agricultural and Food Chemistry* 54(5):1906–1913.

Shoji, T. 2007. Polyphenols as natural food pigments: Changes during food processing. *American Journal of Food Technology* 2:570–581.

Singh, S., Gamlath, S., and Wakeling, L. 2007. Nutritional aspects of food extrusion: A review. *International Journal of Food Science and Technology* 42(8):916–929.

Sirk, T.W., Brown, E.F., Sum, A.K., and Friedman, M. 2008. Molecular dynamics study on the biophysical interactions of seven green tea catechins with lipid bilayers of cell membranes. *Journal of Agricultural and Food Chemistry* 56(17):7750–7758.

Sobukola, O.P., Dueik, V., Muñoz, L., and Bouchon, P. 2013. Comparison of vacuum and atmospheric deep-fat frying of wheat starch and gluten based snacks. *Food Science and Biotechnology* 22(S):177–182.

Sugiyama, H., Akazome, Y., Shoji, T., Yamaguchi, A., Yasue, M., Kanda, T., and Ohtake, Y. 2007. Oligomeric procyanidins in apple polyphenol are main active components for inhibition of pancreatic lipase and triglyceride absorption. *Journal of Agricultural and Food Chemistry* 55(11):4604–4609.

Sun-Waterhouse, D., Sivam, A.S., Cooney, J., Zhou, J., Perera, C.O., and Waterhouse, G.I.N. 2011. Effects of added fruit polyphenols and pectin on the properties of finished breads revealed by HPLC/LC-MS and size-exclusion HPLC. *Food Research International* 44(9):3047–3056.

Tagliazucchi, D., Verzelloni, E., Bertolini, D., and Conte, A. 2009. In vitro bio-accessibility and antioxidant activity of grape polyphenols. *Food Chemistry* 120(2):599–606.

Tamir, M. and Alumot, E. 1969. Inhibition of digestive enzymes by condensed tannins from green and ripe carobs. *Journal of the Science of Food and Agriculture* 20(4):199–202.

Thompson, L.U. and Yoon, J.H. 1984. Starch digestibility as affected by polyphenols and phytic acid. *Journal of Food Science* 49(4):1228–1229.

Thompson, L.U., Yoon, J.H., Jenkins, D.J., Wolever, T.M., and Jenkins, A.L. 1984. Relationship between polyphenol intake and blood glucose response of normal and diabetic individuals. *American Journal of Clinical Nutrition* 39(5):745–751.

Uchiyama, S., Taniguchi, Y., Saka, A., Yoshida, A., and Yajima, H. 2011. Prevention of diet-induced obesity by dietary black tea polyphenols extract in vitro and in vivo. *Nutrition* 27(3):287–292.

Vega-Mercado, H.; Góngora-Nieto, M., and Barbosa-Cánovas, G. 2001. Advances in dehydration of foods. *Journal of Food Engineering* 49:271–289.

Viscidi, K.A., Dougherty, M.P., Briggs, J., and Camire, M.E. 2004. Complex phenolic compounds reduce lipid oxidation in extruded oat cereals. *Lebensmittel-Wissenschaft und-Technologie—Food Science and Technology* 37:789–796.

Wang, R. and Zhou, W. 2004. Stability of tea catechins in the bread making process. *Journal of Agricultural and Food Chemistry* 52:8224–8229.

Wang, Y.Y., Vuolo, L.L., Spingarn, N.E., and Weisburger, J.H. 1982. Formation of mutagens in cooked foods V. The mutagen reducing effect of soy protein concentrates and antioxidants during frying of beef. *Cancer Letters* 16(2):179–189.

Weisburger, J.H., Veliath, E., Larios, E., Pittman, B., Zang, E., and Hara, Y. 2002. Tea polyphenols inhibit the formation of mutagens during the cooking of meat. *Mutation Research/Genetic Toxicology and Environmental Mutagenesis* 516(1):19–22.

Wu, Y., Chen, Z., Li, X., and Li, M. 2009. Effect of tea polyphenols on the retrogradation of rice starch. *Food Research International* 42(2):221–225.

Xiao, H., Lin, Q., Liu, G.Q., Wu, Y., Tian, W., Wu, W., and Fu, X. 2011. Effect of green tea polyphenols on the gelatinization and retrogradation of rice starches with different amylose contents. *Journal of Medicinal Plants Research* 5(17):4298–4303.

Yi, W., Akoh, C.C., Fischer, J., and Krewer, G. 2006. Absorption of anthocyanins from blueberry extracts by Caco-2 human intestinal cell monolayers. *Journal of Agricultural and Food Chemistry* 54:5651–5658.

Yu, H. and Huang, Q. 2011. Investigation of the absorption mechanism of solubilized curcumin using Caco-2 cell monolayers. *Journal of Agricultural and Food Chemistry* 59:9120–9126.

Yuksel, Z., Avci, E., and Erdem, Y.K. 2010. Characterization of binding interactions between green tea flavanoids and milk proteins. *Food Chemistry* 121(2):450–456.

Zhang, Y. and Zhang, Y. 2007. Study on reduction of acrylamide in fried bread sticks by addition of antioxidant of bamboo leaves and extract of green tea. *Asia Pacific Journal of Clinical Nutrition* 16(1):131–136.

Zijp, I.M., Korver, O., and Tijburg, L.B.M. 2000. Effect of tea and other dietary factors on iron absorption. *Critical Reviews in Food Science and Nutrition* 40:371–398.

Index

D

E

F

G

N

O

P

T

U

V

W

Y